T0192424

Hidden Markov Models for Time Series

An Introduction Using R

Second Edition

MONOGRAPHS ON STATISTICS AND APPLIED PROBABILITY

General Editors

F. Bunea, V. Isham, N. Keiding, T. Louis, R. L. Smith, and H. Tong

1. Stochastic Population Models in Ecology and Epidemiology *M.S. Barlett* (1960)
2. Queues *D.R. Cox and W.L. Smith* (1961)
3. Monte Carlo Methods *J.M. Hammersley and D.C. Handscomb* (1964)
4. The Statistical Analysis of Series of Events *D.R. Cox and P.A.W. Lewis* (1966)
5. Population Genetics *W.J. Ewens* (1969)
6. Probability, Statistics and Time *M.S. Barlett* (1975)
7. Statistical Inference *S.D. Silvey* (1975)
8. The Analysis of Contingency Tables *B.S. Everitt* (1977)
9. Multivariate Analysis in Behavioural Research *A.E. Maxwell* (1977)
10. Stochastic Abundance Models *S. Engen* (1978)
11. Some Basic Theory for Statistical Inference *E.J.G. Pitman* (1979)
12. Point Processes *D.R. Cox and V. Isham* (1980)
13. Identification of Outliers *D.M. Hawkins* (1980)
14. Optimal Design *S.D. Silvey* (1980)
15. Finite Mixture Distributions *B.S. Everitt and D.J. Hand* (1981)
16. Classification *A.D. Gordon* (1981)
17. Distribution-Free Statistical Methods, 2nd edition *J.S. Maritz* (1995)
18. Residuals and Influence in Regression *R.D. Cook and S. Weisberg* (1982)
19. Applications of Queueing Theory, 2nd edition *G.F. Newell* (1982)
20. Risk Theory, 3rd edition *R.E. Beard, T. Pentikäinen and E. Pesonen* (1984)
21. Analysis of Survival Data *D.R. Cox and D. Oakes* (1984)
22. An Introduction to Latent Variable Models *B.S. Everitt* (1984)
23. Bandit Problems *D.A. Berry and B. Fristedt* (1985)
24. Stochastic Modelling and Control *M.H.A. Davis and R. Vinter* (1985)
25. The Statistical Analysis of Composition Data *J. Aitchison* (1986)
26. Density Estimation for Statistics and Data Analysis *B.W. Silverman* (1986)
27. Regression Analysis with Applications *G.B. Wetherill* (1986)
28. Sequential Methods in Statistics, 3rd edition *G.B. Wetherill and K.D. Glazebrook* (1986)
29. Tensor Methods in Statistics *P. McCullagh* (1987)
30. Transformation and Weighting in Regression *R.J. Carroll and D. Ruppert* (1988)
31. Asymptotic Techniques for Use in Statistics *O.E. Bandorff-Nielsen and D.R. Cox* (1989)
32. Analysis of Binary Data, 2nd edition *D.R. Cox and E.J. Snell* (1989)
33. Analysis of Infectious Disease Data *N.G. Becker* (1989)
34. Design and Analysis of Cross-Over Trials *B. Jones and M.G. Kenward* (1989)
35. Empirical Bayes Methods, 2nd edition *J.S. Maritz and T. Lwin* (1989)
36. Symmetric Multivariate and Related Distributions *K.T. Fang, S. Kotz and K.W. Ng* (1990)
37. Generalized Linear Models, 2nd edition *P. McCullagh and J.A. Nelder* (1989)
38. Cyclic and Computer Generated Designs, 2nd edition *J.A. John and E.R. Williams* (1995)
39. Analog Estimation Methods in Econometrics *C.F. Manski* (1988)
40. Subset Selection in Regression *A.J. Miller* (1990)
41. Analysis of Repeated Measures *M.J. Crowder and D.J. Hand* (1990)
42. Statistical Reasoning with Imprecise Probabilities *P. Walley* (1991)
43. Generalized Additive Models *T.J. Hastie and R.J. Tibshirani* (1990)
44. Inspection Errors for Attributes in Quality Control *N.L. Johnson, S. Kotz and X. Wu* (1991)
45. The Analysis of Contingency Tables, 2nd edition *B.S. Everitt* (1992)
46. The Analysis of Quantal Response Data *B.J.T. Morgan* (1992)
47. Longitudinal Data with Serial Correlation—A State-Space Approach *R.H. Jones* (1993)

48. Differential Geometry and Statistics *M.K. Murray and J.W. Rice* (1993)

49. Markov Models and Optimization *M.H.A. Davis* (1993)

50. Networks and Chaos—Statistical and Probabilistic Aspects
O.E. Barndorff-Nielsen, J.L. Jensen and W.S. Kendall (1993)

51. Number-Theoretic Methods in Statistics *K.-T. Fang and Y. Wang* (1994)

52. Inference and Asymptotics *O.E. Barndorff-Nielsen and D.R. Cox* (1994)

53. Practical Risk Theory for Actuaries *C.D. Daykin, T. Pentikäinen and M. Pesonen* (1994)

54. Biplots *J.C. Gower and D.J. Hand* (1996)

55. Predictive Inference—An Introduction *S. Geisser* (1993)

56. Model-Free Curve Estimation *M.E. Tarter and M.D. Lock* (1993)

57. An Introduction to the Bootstrap *B. Efron and R.J. Tibshirani* (1993)

58. Nonparametric Regression and Generalized Linear Models *P.J. Green and B.W. Silverman* (1994)

59. Multidimensional Scaling *T.F. Cox and M.A.A. Cox* (1994)

60. Kernel Smoothing *M.P. Wand and M.C. Jones* (1995)

61. Statistics for Long Memory Processes *J. Beran* (1995)

62. Nonlinear Models for Repeated Measurement Data *M. Davidian and D.M. Giltinan* (1995)

63. Measurement Error in Nonlinear Models *R.J. Carroll, D. Rupert and L.A. Stefanski* (1995)

64. Analyzing and Modeling Rank Data *J.J. Marden* (1995)

65. Time Series Models—In Econometrics, Finance and Other Fields
D.R. Cox, D.V. Hinkley and O.E. Barndorff-Nielsen (1996)

66. Local Polynomial Modeling and its Applications *J. Fan and I. Gijbels* (1996)

67. Multivariate Dependencies—Models, Analysis and Interpretation *D.R. Cox and N. Wermuth* (1996)

68. Statistical Inference—Based on the Likelihood *A. Azzalini* (1996)

69. Bayes and Empirical Bayes Methods for Data Analysis *B.P. Carlin and T.A Louis* (1996)

70. Hidden Markov and Other Models for Discrete-Valued Time Series *I.L. MacDonald and W. Zucchini* (1997)

71. Statistical Evidence—A Likelihood Paradigm *R. Royall* (1997)

72. Analysis of Incomplete Multivariate Data *J.L. Schafer* (1997)

73. Multivariate Models and Dependence Concepts *H. Joe* (1997)

74. Theory of Sample Surveys *M.E. Thompson* (1997)

75. Retrial Queues *G. Falin and J.G.C. Templeton* (1997)

76. Theory of Dispersion Models *B. Jørgensen* (1997)

77. Mixed Poisson Processes *J. Grandell* (1997)

78. Variance Components Estimation—Mixed Models, Methodologies and Applications *P.S.R.S. Rao* (1997)

79. Bayesian Methods for Finite Population Sampling *G. Meeden and M. Ghosh* (1997)

80. Stochastic Geometry—Likelihood and computation
O.E. Barndorff-Nielsen, W.S. Kendall and M.N.M. van Lieshout (1998)

81. Computer-Assisted Analysis of Mixtures and Applications—Meta-Analysis, Disease Mapping and Others
D. Böhning (1999)

82. Classification, 2nd edition *A.D. Gordon* (1999)

83. Semimartingales and their Statistical Inference *B.L.S. Prakasa Rao* (1999)

84. Statistical Aspects of BSE and vCJD—Models for Epidemics *C.A. Donnelly and N.M. Ferguson* (1999)

85. Set-Indexed Martingales *G. Ivanoff and E. Merzbach* (2000)

86. The Theory of the Design of Experiments *D.R. Cox and N. Reid* (2000)

87. Complex Stochastic Systems *O.E. Barndorff-Nielsen, D.R. Cox and C. Klüppelberg* (2001)

88. Multidimensional Scaling, 2nd edition *T.F. Cox and M.A.A. Cox* (2001)

89. Algebraic Statistics—Computational Commutative Algebra in Statistics
G. Pistone, E. Riccomagno and H.P. Wynn (2001)

90. Analysis of Time Series Structure—SSA and Related Techniques
N. Golyandina, V. Nekrutkin and A.A. Zhigljavsky (2001)

91. Subjective Probability Models for Lifetimes *Fabio Spizzichino* (2001)

92. Empirical Likelihood *Art B. Owen* (2001)

93. Statistics in the 21st Century *Adrian E. Raftery, Martin A. Tanner, and Martin T. Wells* (2001)

94. Accelerated Life Models: Modeling and Statistical Analysis
Vilijandas Bagdonavicius and Mikhail Nikulin (2001)

95. Subset Selection in Regression, Second Edition *Alan Miller* (2002)

96. Topics in Modelling of Clustered Data *Marc Aerts, Helena Geys, Geert Molenberghs, and Louise M. Ryan* (2002)

97. Components of Variance *D.R. Cox and P.J. Solomon* (2002)

98. Design and Analysis of Cross-Over Trials, 2nd Edition *Byron Jones and Michael G. Kenward* (2003)

99. Extreme Values in Finance, Telecommunications, and the Environment
    Bärbel Finkenstädt and Holger Rootzén (2003)

100. Statistical Inference and Simulation for Spatial Point Processes
     Jesper Møller and Rasmus Plenge Waagepetersen (2004)

101. Hierarchical Modeling and Analysis for Spatial Data
     Sudipto Banerjee, Bradley P. Carlin, and Alan E. Gelfand (2004)

102. Diagnostic Checks in Time Series *Wai Keung Li* (2004)

103. Stereology for Statisticians *Adrian Baddeley and Eva B. Vedel Jensen* (2004)

104. Gaussian Markov Random Fields: Theory and Applications *Håvard Rue and Leonhard Held* (2005)

105. Measurement Error in Nonlinear Models: A Modern Perspective, Second Edition
     Raymond J. Carroll, David Ruppert, Leonard A. Stefanski, and Ciprian M. Crainiceanu (2006)

106. Generalized Linear Models with Random Effects: Unified Analysis via H-likelihood
     Youngjo Lee, John A. Nelder, and Yudi Pawitan (2006)

107. Statistical Methods for Spatio-Temporal Systems
     Bärbel Finkenstädt, Leonhard Held, and Valerie Isham (2007)

108. Nonlinear Time Series: Semiparametric and Nonparametric Methods *Jiti Gao* (2007)

109. Missing Data in Longitudinal Studies: Strategies for Bayesian Modeling and Sensitivity Analysis
     Michael J. Daniels and Joseph W. Hogan (2008)

110. Hidden Markov Models for Time Series: An Introduction Using R
     Walter Zucchini and Iain L. MacDonald (2009)

111. ROC Curves for Continuous Data *Wojtek J. Krzanowski and David J. Hand* (2009)

112. Antedependence Models for Longitudinal Data *Dale L. Zimmerman and Vicente A. Núñez-Antón* (2009)

113. Mixed Effects Models for Complex Data *Lang Wu* (2010)

114. Intoduction to Time Series Modeling *Genshiro Kitagawa* (2010)

115. Expansions and Asymptotics for Statistics *Christopher G. Small* (2010)

116. Statistical Inference: An Integrated Bayesian/Likelihood Approach *Murray Aitkin* (2010)

117. Circular and Linear Regression: Fitting Circles and Lines by Least Squares *Nikolai Chernov* (2010)

118. Simultaneous Inference in Regression *Wei Liu* (2010)

119. Robust Nonparametric Statistical Methods, Second Edition
     Thomas P. Hettmansperger and Joseph W. McKean (2011)

120. Statistical Inference: The Minimum Distance Approach
     Ayanendranath Basu, Hiroyuki Shioya, and Chanseok Park (2011)

121. Smoothing Splines: Methods and Applications *Yuedong Wang* (2011)

122. Extreme Value Methods with Applications to Finance *Serguei Y. Novak* (2012)

123. Dynamic Prediction in Clinical Survival Analysis *Hans C. van Houwelingen and Hein Putter* (2012)

124. Statistical Methods for Stochastic Differential Equations
     Mathieu Kessler, Alexander Lindner, and Michael Sørensen (2012)

125. Maximum Likelihood Estimation for Sample Surveys
     R. L. Chambers, D. G. Steel, Suojin Wang, and A. H. Welsh (2012)

126. Mean Field Simulation for Monte Carlo Integration *Pierre Del Moral* (2013)

127. Analysis of Variance for Functional Data *Jin-Ting Zhang* (2013)

128. Statistical Analysis of Spatial and Spatio-Temporal Point Patterns, Third Edition *Peter J. Diggle* (2013)

129. Constrained Principal Component Analysis and Related Techniques *Yoshio Takane* (2014)

130. Randomised Response-Adaptive Designs in Clinical Trials *Anthony C. Atkinson and Atanu Biswas* (2014)

131. Theory of Factorial Design: Single- and Multi-Stratum Experiments *Ching-Shui Cheng* (2014)

132. Quasi-Least Squares Regression *Justine Shults and Joseph M. Hilbe* (2014)

133. Data Analysis and Approximate Models: Model Choice, Location-Scale, Analysis of Variance, Nonparametric
     Regression and Image Analysis *Laurie Davies* (2014)

134. Dependence Modeling with Copulas *Harry Joe* (2014)

135. Hierarchical Modeling and Analysis for Spatial Data, Second Edition *Sudipto Banerjee, Bradley P. Carlin,
     and Alan E. Gelfand* (2014)

136. Sequential Analysis: Hypothesis Testing and Changepoint Detection *Alexander Tartakovsky, Igor Nikiforov, and Michèle Basseville* (2015)

137. Robust Cluster Analysis and Variable Selection *Gunter Ritter* (2015)

138. Design and Analysis of Cross-Over Trials, Third Edition *Byron Jones and Michael G. Kenward* (2015)

139. Introduction to High-Dimensional Statistics *Christophe Giraud* (2015)

140. Pareto Distributions: Second Edition *Barry C. Arnold* (2015)

141. Bayesian Inference for Partially Identified Models: Exploring the Limits of Limited Data *Paul Gustafson* (2015)

142. Models for Dependent Time Series *Granville Tunnicliffe Wilson, Marco Reale, John Haywood* (2015)

143. Statistical Learning with Sparsity: The Lasso and Generalizations *Trevor Hastie, Robert Tibshirani, and Martin Wainwright* (2015)

144. Measuring Statistical Evidence Using Relative Belief *Michael Evans* (2015)

145. Stochastic Analysis for Gaussian Random Processes and Fields: With Applications *Vidyadhar S. Mandrekar and Leszek Gawarecki* (2015)

146. Semialgebraic Statistics and Latent Tree Models *Piotr Zwiernik* (2015)

147. Inferential Models: Reasoning with Uncertainty *Ryan Martin and Chuanhai Liu* (2016)

148. Perfect Simulation *Mark L. Huber* (2016)

149. State-Space Methods for Time Series Analysis: Theory, Applications and Software *Jose Casals, Alfredo Garcia-Hiernaux, Miguel Jerez, Sonia Sotoca, and A. Alexandre Trindade* (2016)

150. Hidden Markov Models for Time Series: An Introduction Using R, Second Edition *Walter Zucchini, Iain L. MacDonald, and Roland Langrock* (2016)

Monographs on Statistics and Applied Probability 150

Hidden Markov Models for Time Series

An Introduction Using R

Second Edition

Walter Zucchini
University of Göttingen, Germany

Iain L. MacDonald
University of Cape Town, South Africa

Roland Langrock
Bielefeld University, Germany

CRC Press
Taylor & Francis Group
Boca Raton London New York

CRC Press is an imprint of the
Taylor & Francis Group, an **informa** business
A CHAPMAN & HALL BOOK

CRC Press
Taylor & Francis Group
6000 Broken Sound Parkway NW, Suite 300
Boca Raton, FL 33487-2742

First issued in paperback 2021

© 2016 by Walter Zucchini, Iain L. MacDonald, Roland Langrock
CRC Press is an imprint of Taylor & Francis Group, an Informa business

No claim to original U.S. Government works

ISBN-13: 978-1-4822-5383-2 (hbk)
ISBN-13: 978-1-03-217949-0 (pbk)
DOI: 10.1201/b20790

Library of Congress Cataloging-in-Publication Data

Names: Zucchini, W. | MacDonald, Iain L. | Langrock, Roland, 1983-
Title: Hidden markov models for time series : an introduction using R.
Description: Second edition / Walter Zucchini, Iain L. MacDonald, and Roland
Langrock. | Boca Raton : Taylor & Francis, 2016. | Series: Monographs on
statistics and applied probability ; 150 | "A CRC title." | Includes
bibliographical references and index.
Identifiers: LCCN 2016008476 | ISBN 9781482253832 (alk. paper)
Subjects: LCSH: Time-series analysis. | Markov processes. | R (Computer
program language)
Classification: LCC QA280 .Z83 2016 | DDC 519.2/33--dc23
LC record available at https://lccn.loc.gov/2016008476

Visit the Taylor & Francis Web site at
http://www.taylorandfrancis.com

and the CRC Press Web site at
http://www.crcpress.com

Für Hanne und Werner,
mit herzlichem Dank für Eure Unterstützung
bei der Suche nach den versteckten Ketten.

Für Andrea, Johann und Hendrik,
dass Ihr Euch freut.

Contents

Preface   xxi

Preface to first edition   xxiii

Notation and abbreviations   xxvii

I   Model structure, properties and methods   **1**

1   Preliminaries: mixtures and Markov chains   **3**
  1.1 Introduction   3
  1.2 Independent mixture models   6
    1.2.1 Definition and properties   6
    1.2.2 Parameter estimation   9
    1.2.3 Unbounded likelihood in mixtures   11
    1.2.4 Examples of fitted mixture models 12
  1.3 Markov chains   14
    1.3.1 Definitions and example   14
    1.3.2 Stationary distributions   17
    1.3.3 Autocorrelation function   18
    1.3.4 Estimating transition probabilities 19
    1.3.5 Higher-order Markov chains   20
  Exercises   23

2   Hidden Markov models: definition and properties   **29**
  2.1 A simple hidden Markov model   29
  2.2 The basics   30
    2.2.1 Definition and notation   30
    2.2.2 Marginal distributions   32
    2.2.3 Moments   33
  2.3 The likelihood   34
    2.3.1 The likelihood of a two-state Bernoulli–HMM   35
    2.3.2 The likelihood in general   36
    2.3.3 HMMs are not Markov processes   39
    2.3.4 The likelihood when data are missing 40

2.3.5	The likelihood when observations are interval-censored	41
Exercises		41

3 Estimation by direct maximization of the likelihood **47**
3.1	Introduction	47
3.2	Scaling the likelihood computation	48
3.3	Maximization of the likelihood subject to constraints	50
3.3.1	Reparametrization to avoid constraints	50
3.3.2	Embedding in a continuous-time Markov chain	52
3.4	Other problems	53
3.4.1	Multiple maxima in the likelihood	53
3.4.2	Starting values for the iterations	53
3.4.3	Unbounded likelihood	53
3.5	Example: earthquakes	54
3.6	Standard errors and confidence intervals	56
3.6.1	Standard errors via the Hessian	56
3.6.2	Bootstrap standard errors and confidence intervals	58
3.7	Example: the parametric bootstrap applied to the three-state model for the earthquakes data	59
Exercises		60

4 Estimation by the EM algorithm **65**
4.1	Forward and backward probabilities	65
4.1.1	Forward probabilities	66
4.1.2	Backward probabilities	67
4.1.3	Properties of forward and backward probabilities	68
4.2	The EM algorithm	69
4.2.1	EM in general	70
4.2.2	EM for HMMs	70
4.2.3	M step for Poisson- and normal-HMMs	72
4.2.4	Starting from a specified state	73
4.2.5	EM for the case in which the Markov chain is stationary	73
4.3	Examples of EM applied to Poisson-HMMs	74
4.3.1	Earthquakes	74
4.3.2	Foetal movement counts	76
4.4	Discussion	77
Exercises		78

5 Forecasting, decoding and state prediction **81**
5.1	Introduction	81

	5.2	Conditional distributions	82
	5.3	Forecast distributions	83
	5.4	Decoding	85
		5.4.1 State probabilities and local decoding	86
		5.4.2 Global decoding	88
	5.5	State prediction	92
	5.6	HMMs for classification	93
	Exercises	94
6	**Model selection and checking**	**97**
	6.1	Model selection by AIC and BIC	97
	6.2	Model checking with pseudo-residuals	101
		6.2.1 Introducing pseudo-residuals	101
		6.2.2 Ordinary pseudo-residuals	105
		6.2.3 Forecast pseudo-residuals	105
	6.3	Examples	106
		6.3.1 Ordinary pseudo-residuals for the earthquakes	106
		6.3.2 Dependent ordinary pseudo-residuals	108
	6.4	Discussion	109
	Exercises	109
7	**Bayesian inference for Poisson–hidden Markov models**	**111**
	7.1	Applying the Gibbs sampler to Poisson–HMMs	111
		7.1.1 Introduction and outline	111
		7.1.2 Generating sample paths of the Markov chain	113
		7.1.3 Decomposing the observed counts into regime contributions	114
		7.1.4 Updating the parameters	114
	7.2	Bayesian estimation of the number of states	114
		7.2.1 Use of the integrated likelihood	115
		7.2.2 Model selection by parallel sampling	116
	7.3	Example: earthquakes	116
	7.4	Discussion	119
	Exercises	120
8	**R packages**	**123**
	8.1	The package `depmixS4`	123
		8.1.1 Model formulation and estimation	123
		8.1.2 Decoding	124
	8.2	The package `HiddenMarkov`	124
		8.2.1 Model formulation and estimation	124
		8.2.2 Decoding	126
		8.2.3 Residuals	126
	8.3	The package `msm`	126

8.3.1   Model formulation and estimation   126
8.3.2   Decoding   128
8.4   The package R2OpenBUGS   128
8.5   Discussion   129

II Extensions   131

9 HMMs with general state-dependent distribution   133
9.1   Introduction   133
9.2   General univariate state-dependent distribution 133
9.2.1   HMMs for unbounded counts   133
9.2.2   HMMs for binary data   134
9.2.3   HMMs for bounded counts   134
9.2.4   HMMs for continuous-valued series   135
9.2.5   HMMs for proportions   135
9.2.6   HMMs for circular-valued series   136
9.3   Multinomial and categorical HMMs   136
9.3.1   Multinomial–HMM   136
9.3.2   HMMs for categorical data   137
9.3.3   HMMs for compositional data   138
9.4   General multivariate state-dependent distribution 138
9.4.1   Longitudinal conditional independence   138
9.4.2   Contemporaneous conditional independence 140
9.4.3   Further remarks on multivariate HMMs   141
Exercises   142

10 Covariates and other extra dependencies   145
10.1   Introduction   145
10.2   HMMs with covariates   145
10.2.1   Covariates in the state-dependent distributions 146
10.2.2   Covariates in the transition probabilities 147
10.3   HMMs based on a second-order Markov chain   148
10.4   HMMs with other additional dependencies   150
Exercises   152

11 Continuous-valued state processes   155
11.1   Introduction   155
11.2   Models with continuous-valued state process 156
11.2.1   Numerical integration of the likelihood 157
11.2.2   Evaluation of the approximate likelihood via
         forward recursion   158
11.2.3   Parameter estimation and related issues 160
11.3   Fitting an SSM to the earthquake data   160
11.4   Discussion   162

12 Hidden semi-Markov models and their representation as HMMs   **165**
    12.1   Introduction   165
    12.2   Semi-Markov processes, hidden semi-Markov models and approximating HMMs   165
    12.3   Examples of HSMMs represented as HMMs   167
           12.3.1   A simple two-state Poisson–HSMM   167
           12.3.2   Example of HSMM with three states   169
           12.3.3   A two-state HSMM with general dwell-time distribution in one state   171
    12.4   General HSMM   173
    12.5   **R** code   176
    12.6   Some examples of dwell-time distributions   178
           12.6.1   Geometric distribution   178
           12.6.2   Shifted Poisson distribution   178
           12.6.3   Shifted negative binomial distribution   179
           12.6.4   Shifted binomial distribution   180
           12.6.5   A distribution with unstructured start and geometric tail   180
    12.7   Fitting HSMMs via the HMM representation   181
    12.8   Example: earthquakes   182
    12.9   Discussion   184
    Exercises   184

13 HMMs for longitudinal data   **187**
    13.1   Introduction   187
    13.2   Models that assume some parameters to be constant across component series   189
    13.3   Models with random effects   190
           13.3.1   HMMs with continuous-valued random effects   191
           13.3.2   HMMs with discrete-valued random effects   193
    13.4   Discussion   195
    Exercises   196

III   Applications   **197**

14 Introduction to applications   **199**

15 Epileptic seizures   **201**
    15.1   Introduction   201
    15.2   Models fitted   201
    15.3   Model checking by pseudo-residuals   204
    Exercises   206

16 Daily rainfall occurrence   **207**
  16.1 Introduction   207
  16.2 Models fitted   207

17 Eruptions of the Old Faithful geyser   **213**
  17.1 Introduction   213
  17.2 The data   213
  17.3 The binary time series of short and long eruptions   214
      17.3.1 Markov chain models   214
      17.3.2 Hidden Markov models   216
      17.3.3 Comparison of models   219
      17.3.4 Forecast distributions   219
  17.4 Univariate normal–HMMs for durations and waiting
       times   220
  17.5 Bivariate normal–HMM for durations and waiting times 223
  Exercises   224

18 HMMs for animal movement   **227**
  18.1 Introduction   227
  18.2 Directional data   228
      18.2.1 Directional means   228
      18.2.2 The von Mises distribution   228
  18.3 HMMs for movement data   229
      18.3.1 Movement data   229
      18.3.2 HMMs as multi-state random walks   230
  18.4 A basic HMM for *Drosophila* movement   232
  18.5 HMMs and HSMMs for bison movement   235
  18.6 Mixed HMMs for woodpecker movement   238
  Exercises   242

19 Wind direction at Koeberg   **245**
  19.1 Introduction   245
  19.2 Wind direction classified into 16 categories   245
      19.2.1 Three HMMs for hourly averages of wind
              direction   245
      19.2.2 Model comparisons and other possible models 248
  19.3 Wind direction as a circular variable   251
      19.3.1 Daily at hour 24: von Mises–HMMs   251
      19.3.2 Modelling hourly change of direction   253
      19.3.3 Transition probabilities varying with lagged
              speed   253
      19.3.4 Concentration parameter varying with lagged
              speed   254
  Exercises   257

20 Models for financial series   **259**

20.1   Financial series I: A multivariate normal–HMM for returns on four shares   259

20.2   Financial series II: Discrete state-space stochastic volatility models   262

    20.2.1   Stochastic volatility models without leverage   263

    20.2.2   Application: FTSE 100 returns   265

    20.2.3   Stochastic volatility models with leverage   265

    20.2.4   Application: TOPIX returns   268

    20.2.5   Non-standard stochastic volatility models   270

    20.2.6   A model with a mixture AR(1) volatility process   271

    20.2.7   Application: S&P 500 returns   272

Exercises   273

21 Births at Edendale Hospital   **275**

21.1   Introduction   275

21.2   Models for the proportion Caesarean   275

21.3   Models for the total number of deliveries   282

21.4   Conclusion   285

22 Homicides and suicides in Cape Town, 1986–1991   **287**

22.1   Introduction   287

22.2   Firearm homicides as a proportion of all homicides, suicides and legal intervention homicides   287

22.3   The number of firearm homicides   289

22.4   Firearm homicides as a proportion of all homicides, and firearm suicides as a proportion of all suicides   291

22.5   Proportion in each of the five categories   295

23 A model for animal behaviour which incorporates feedback   **297**

23.1   Introduction   297

23.2   The model   298

23.3   Likelihood evaluation   300

    23.3.1   The likelihood as a multiple sum   301

    23.3.2   Recursive evaluation   301

23.4   Parameter estimation by maximum likelihood   302

23.5   Model checking   302

23.6   Inferring the underlying state   303

23.7   Models for a heterogeneous group of subjects   304

    23.7.1   Models assuming some parameters to be constant across subjects   304

    23.7.2   Mixed models   305

	23.7.3	Inclusion of covariates	306
23.8		Other modifications or extensions	306
	23.8.1	Increasing the number of states	306
	23.8.2	Changing the nature of the state-dependent distribution	306
23.9		Application to caterpillar feeding behaviour	307
	23.9.1	Data description and preliminary analysis	307
	23.9.2	Parameter estimates and model checking	307
	23.9.3	Runlength distributions	311
	23.9.4	Joint models for seven subjects	313
23.10		Discussion	314

24 Estimating the survival rates of Soay sheep from mark–recapture–recovery data — **317**

24.1	Introduction	317
24.2	MRR data without use of covariates	318
24.3	MRR data involving individual-specific time-varying continuous-valued covariates	321
24.4	Application to Soay sheep data	324
24.5	Conclusion	328

A Examples of R code — **331**

A.1		The functions	331
	A.1.1	Transforming natural parameters to working	332
	A.1.2	Transforming working parameters to natural	332
	A.1.3	Computing minus the log-likelihood from the working parameters	332
	A.1.4	Computing the MLEs, given starting values for the natural parameters	333
	A.1.5	Generating a sample	333
	A.1.6	Global decoding by the Viterbi algorithm	334
	A.1.7	Computing log(forward probabilities)	334
	A.1.8	Computing log(backward probabilities)	334
	A.1.9	Conditional probabilities	335
	A.1.10	Pseudo-residuals	336
	A.1.11	State probabilities	336
	A.1.12	State prediction	336
	A.1.13	Local decoding	337
	A.1.14	Forecast probabilities	337
A.2		Examples of code using the above functions	338
	A.2.1	Fitting Poisson–HMMs to the earthquakes series	338
	A.2.2	Forecast probabilities	339

B Some proofs   **341**
    B.1   A factorization needed for the forward probabilities   341
    B.2   Two results needed for the backward probabilities   342
    B.3   Conditional independence of $\mathbf{X}_1^t$ and $\mathbf{X}_{t+1}^T$ 343

References   **345**

Author index   **359**

Subject index   **365**

Preface

The biggest change we have made for this edition is that the single, rather terse, chapter on extensions (Chapter 8) has been replaced by Part II, which consists of five chapters. The main additions here are entirely new chapters on models with continuous-valued state process, hidden Markov representations of hidden semi-Markov models, and hidden Markov models for longitudinal data.

Since the first edition appeared in 2009 there has been much interest in applying hidden Markov models in ecology and the environment, and this is reflected in most of the new applications in Part III. There are new chapters on the modelling of rainfall and animal movement, and on modelling of the survival of a wild population of sheep. The latter application demonstrates the usefulness of hidden Markov models in the analysis of mark–recapture–recovery data of a somewhat complex nature. The discussion of hidden Markov approximations to the stochastic volatility models of finance has also been expanded to include a brief account of some recent work of the authors.

One change we have made in response to readers' suggestions is to include snippets of code in the text of several chapters, close to the formulae being coded, so that readers can easily follow the rationale of our code. But we also include in an appendix the code which will perform basic analyses in respect of a Poisson–hidden Markov model. It is intended that this code, with examples of its use, will be available on the website of the book, currently `www.hmms-for-time-series.de`. **R** packages are available that can perform many of the analyses reported in this book, for instance `hiddenMarkov`, `msm` and `depmixS4`. In the new Chapter 8 we provide simple examples of the use of such packages, and comparisons with our results, plus an example of the use of `R2OpenBUGS`.

To the list of people who helped us produce the first edition, or commented on drafts thereof, we now wish to add the following: Thomas Rooney (who checked passages and wrote code), Victoria Goodall, Brendon Lapham, and Nia and Patrick Reen (all of whom participated enthusiastically in the HMMs group at the University of Cape Town), and Aman Parker (who set up the server on which we kept our repository of files for the book). We thank the James M. Kilts Center, University of Chicago Booth School of Business, for making available the sales data

given on p. 25. We thank Ann E. McKellar and Dylan C. Kesler for providing the woodpecker data (see p. 238), and Tim Coulson for providing
the Soay sheep data (see p. 317). We thank the reviewers of the first
edition for their many helpful suggestions, not all of which we have been
able to implement. We thank Andrea Langrock for helping us to finalize
the cover design. We also thank Rob Calver of Chapman & Hall/CRC
Press, who for years has been supportive of and enthusiastic about our
books, and Richard Leigh, who did a superb job of copy-editing. It has
been a pleasure for us to work with them. Like many others, we are
much indebted to the developers and maintainers of the software **R**,
LaTeX, and Subversion; we thank them especially.

We are most grateful to all of the above for their interest, suggestions
or assistance.

Göttingen
February 2016

Preface to first edition

In the eleven years since the publication of our book *Hidden Markov and Other Models for Discrete-valued Time Series* it has become apparent that most of the 'other models', though undoubtedly of theoretical interest, have led to few published applications. This is in marked contrast to hidden Markov models, which are of course applicable to more than just *discrete-valued* time series. These observations have led us to write a book with different objectives.

Firstly, our emphasis is no longer principally on discrete-valued series. We have therefore removed Part One of the original text, which covered the 'other models' for such series. Our focus here is exclusively on hidden Markov models, but applied to a wide range of types of time series: continuous-valued, circular, multivariate, for instance, in addition to the types of data we previously considered, namely binary data, bounded and unbounded counts and categorical observations.

Secondly, we have attempted to make the models more accessible by illustrating how the computing environment **R** can be used to carry out the computations, e.g., for parameter estimation, model selection, model checking, decoding and forecasting. In our previous book we used proprietary software to perform numerical optimization, subject to linear constraints on the variables, for parameter estimation. We now show how one can use standard **R** functions instead. The **R** code that we used to carry out the computations for some of the applications is given, and can be applied directly in similar applications. We do not, however, supply a ready-to-use package; packages that cover 'standard' cases already exist. Rather, it is our intention to show the reader how to go about constructing and fitting application-specific variations of the standard models, variations that may not be covered in the currently available software. The programming exercises are intended to encourage readers to develop expertise in this respect.

The book is intended to illustrate the wonderful plasticity of hidden Markov models as general-purpose models for time series. We hope that readers will find it easy to devise for themselves 'customized' models that will be useful in summarizing and interpreting their data. To this end we offer a range of applications and types of data – Part Two is entirely devoted to applications. Some of the applications appeared in the original text, but these have been extended or refined.

Our intended readership is applied statisticians, students of statistics, and researchers in fields in which time series arise that are not amenable to analysis by the standard time series models such as Gaussian ARMA models. Such fields include animal behaviour, epidemiology, finance, hydrology and sociology. We have tried to write for readers who wish to acquire a general understanding of the models and their uses, and who wish to apply them. Researchers primarily interested in developing the theory of hidden Markov models are likely to be disappointed by the lack of generality of our treatment, and by the dearth of material on specific issues such as identifiability, hypothesis testing, properties of estimators and reversible jump Markov chain Monte Carlo methods. Such readers would find it more profitable to refer to alternative sources, such as Cappé, Moulines and Rydén (2005) or Ephraim and Merhav (2002). Our strategy has been to present most of the ideas by using a single running example and a simple model, the Poisson–hidden Markov model. In Chapter 8, and in Part Two of the book, we illustrate how this basic model can be progressively and variously extended and generalized.

We assume only a modest level of knowledge of probability and statistics: the reader is assumed to be familiar with the basic probability distributions such as the Poisson, normal and binomial, and with the concepts of dependence, correlation and likelihood. While we would not go as far as Lindsey (2004, p. ix) and state that 'Familiarity with classical introductory statistics courses based on point estimation, hypothesis testing, confidence intervals [...] will be a definite handicap', we hope that extensive knowledge of such matters will not prove necessary. No prior knowledge of Markov chains is assumed, although our coverage is brief enough that readers may wish to supplement our treatment by reading the relevant parts of a book such as Grimmett and Stirzaker (2001). We have also included exercises of a theoretical nature in many of the chapters, both to fill in the details and to illustrate some of the concepts introduced in the text. All the datasets analysed in this book can be accessed at the following address: http://134.76.173.220/hmm-with-r/data .

This book contains some material which has not previously been published, either by ourselves or (to the best of our knowledge) by others. If we have anywhere failed to make appropriate acknowledgement of the work of others, or misquoted their work in any way, we would be grateful if the reader would draw it to our attention. The applications described in Chapters 14, 15 and 16 contain material which first appeared in (respectively) the *South African Statistical Journal*, the *International Journal of Epidemiology* and *Biometrics*. We are grateful to the editors of these journals for allowing us to reuse such material.

We wish to thank the following researchers for giving us access to their data, and in some cases spending much time discussing it with us: David Bowie, Graeme Fick, Linda Haines, Len Lerer, Frikkie Potgieter, David Raubenheimer and Max Suster.

We are especially indebted to Andreas Schlegel and Jan Bulla for their important inputs, particularly in the early stages of the project; to Christian Gläser, Oleg Nenadić and Daniel Adler, for contributing their computing expertise; and to Antony Unwin and Ellis Pender for their constructive comments on and criticisms of different aspects of our work. The second author wishes to thank the Institute for Statistics and Econometrics of Georg-August-Universität, Göttingen, for welcoming him on many visits and placing facilities at his disposal. Finally, we are most grateful to our colleague and friend of many years, Linda Haines, whose criticism has been invaluable in improving this book.

Göttingen
November 2008

Note added in second edition: The internet address and chapter numbers shown above do not apply to the second edition, nor does 'Part Two'. All the material relevant to the first edition can now be found at `www.hmms-for-time-series.de/first/index_v1.html`.

Notation and abbreviations

Since the underlying mathematical ideas are the important quantities, no notation should be adhered to slavishly. It is all a question of who is master.

Bellman (1960, p. 82)

[M]any writers have acted as though they believe that the success of the Box–Jenkins models is largely due to the use of the acronyms.

Granger (1982)

Notation

Although notation is defined as it is introduced, it may also be helpful to list here the most common meanings of symbols, and the pages on which they are introduced. Matrices and vectors are denoted by bold type. Transposition of matrices and vectors is indicated by the prime symbol: $'$. All vectors are row vectors unless indicated otherwise.

Symbol	Meaning	Page
$A_n(\kappa)$	$I_n(\kappa)/I_0(\kappa)$	243
$\mathbf{B}_t$	$\boldsymbol{\Gamma}\mathbf{P}(x_t)$	82
C_t	state occupied by Markov chain at time t	14
$\mathbf{C}^{(t)}$	$(C_1, C_2, \ldots, C_t)$	14
$\mathbf{e}_i$	$(0, \ldots, 0, 1, 0, \ldots, 0)$, with the 1 in the ith position	93
$\{g_t\}$	parameter process of a stochastic volatility model	263
I_n	modified Bessel function of the first kind of order n	242
l	log-likelihood	19
L or L_T	likelihood	19, 34
$\log$	logarithm to the base e	
m	number of states in a Markov chain,	16
	or number of components in a mixture	7
$\mathbb{N}$	the set of all positive integers	
N_t	nutrient level	298
$N(\cdot; \mu, \sigma^2)$	distribution function of general normal distribution	264
$n(\cdot; \mu, \sigma^2)$	density of general normal distribution	264
p_i	probability mass or density function in state i	31
$\mathbf{P}(x)$	diagonal matrix with ith diagonal element $p_i(x)$	32
$\mathbb{R}$	the set of all real numbers	

T	length of a time series	34
$\mathbf{U}$	square matrix with all elements equal to 1	18
$\mathbf{u}(t)$	vector $(\Pr(C_t = 1), \ldots, \Pr(C_t = m))$	16
$u_i(t)$	$\Pr(C_t = i)$, i.e. ith element of $\mathbf{u}(t)$	32
w_t	$\boldsymbol{\alpha}_t \mathbf{1}' = \sum_i \alpha_t(i)$	48
X_t	observation at time t, or just tth observation	30
$\mathbf{X}^{(t)}$	$(X_1, X_2, \ldots, X_t)$	30
$\mathbf{X}^{(-t)}$	$(X_1, \ldots, X_{t-1}, X_{t+1}, \ldots, X_T)$	82
$\mathbf{X}_a^b$	$(X_a, X_{a+1}, \ldots, X_b)$	67
$\boldsymbol{\alpha}_t$	(row) vector of forward probabilities	38
$\alpha_t(i)$	forward probability, i.e. $\Pr(\mathbf{X}^{(t)} = \mathbf{x}^{(t)}, C_t = i)$	65
$\boldsymbol{\beta}_t$	(row) vector of backward probabilities	66
$\beta_t(i)$	backward probability, i.e. $\Pr(\mathbf{X}_{t+1}^T = \mathbf{x}_{t+1}^T \mid C_t = i)$	66
$\boldsymbol{\Gamma}$	transition probability matrix of Markov chain	15
γ_{ij}	(i, j) element of $\boldsymbol{\Gamma}$; probability of transition from state i to state j in a Markov chain	15
$\boldsymbol{\delta}$	stationary or initial distribution of Markov chain, or vector of mixing probabilities	17 7
$\boldsymbol{\phi}_t$	vector of forward probabilities, normalized to have sum equal to 1, i.e. $\boldsymbol{\alpha}_t / w_t$	48
Φ	distribution function of standard normal distribution	
$\boldsymbol{\Omega}$	t.p.m. driving state switches in an HSMM	165
$\mathbf{1}$	(row) vector of ones	18

Abbreviations

ACF	autocorrelation function
AIC	Akaike's information criterion
BIC	Bayesian information criterion
CDLL	complete-data log-likelihood
c.o.d.	change of direction
c.v.	coefficient of variation
HM(M)	hidden Markov (model)
HSMM	hidden semi-Markov model
MC	Markov chain
MCMC	Markov chain Monte Carlo
ML	maximum likelihood
MLE	maximum likelihood estimator or estimate
p.d.f.	probability density function
p.m.f.	probability mass function
qq-plot	quantile–quantile plot
SV	stochastic volatility
t.p.m.	transition probability matrix

## PART I

Model structure, properties and methods

## CHAPTER 1

Preliminaries: mixtures and Markov chains

1.1 Introduction

Hidden Markov models (HMMs) are models in which the distribution that generates an observation depends on the state of an underlying and unobserved Markov process. They provide flexible general-purpose models for univariate and multivariate time series, especially for discrete-valued series, including categorical series and series of counts.

The purposes of this chapter are to provide a brief and informal introduction to HMMs, and to their many potential uses, and then to discuss two topics that will be fundamental in understanding the structure of such models. In Section 1.2 we give an account of (finite) mixture distributions, because the marginal distribution of a hidden Markov model is a mixture distribution. Then, in Section 1.3, we introduce Markov chains, which provide the underlying 'parameter process' of a hidden Markov model.

Consider, as an example, the series of annual counts of major earthquakes (i.e. magnitude 7 and above) for the years 1900–2006, both inclusive, displayed in Table 1.1 and Figure 1.1.* For this series, the application of standard models such as autoregressive moving-average (ARMA) models would be inappropriate, because such models are based on the normal distribution. Instead, the usual model for unbounded counts is the Poisson distribution, but, as will be demonstrated later, the series displays considerable overdispersion relative to the Poisson distribution, and strong positive serial dependence. A model consisting of independent Poisson random variables would therefore for two reasons also be inappropriate. An examination of Figure 1.1 suggests that there may be some periods with a low rate of earthquakes, and some with a relatively high rate. HMMs, which allow the probability distribution of each observation to depend on the unobserved (or 'hidden') state of a Markov chain, can accommodate both overdispersion and serial dependence. We

---

* These data were downloaded from http://neic.usgs.gov/neis/eqlists on 25 July 2007. Note, however, that the US Geological Survey undertook a systematic review, and there may be minor differences between the information now available and the data we present here.

Table 1.1 *Number of major earthquakes (magnitude 7 or greater) in the world, 1900–2006; to be read across rows.*

13	14	8	10	16	26	32	27	18	32	36	24	22	23	22	18	25	21	21	14
8	11	14	23	18	17	19	20	22	19	13	26	13	14	22	24	21	22	26	21
23	24	27	41	31	27	35	26	28	36	39	21	17	22	17	19	15	34	10	15
22	18	15	20	15	22	19	16	30	27	29	23	20	16	21	21	25	16	18	15
18	14	10	15	8	15	6	11	8	7	18	16	13	12	13	20	15	16	12	18
15	16	13	15	16	11	11													

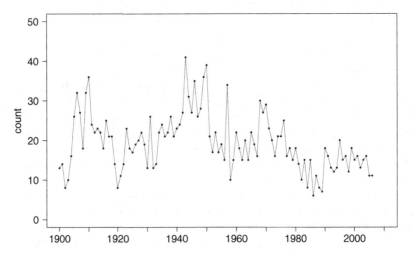

Figure 1.1 *Number of major earthquakes (magnitude 7 or greater) in the world, 1900–2006.*

shall use this series of earthquake counts as a running example in Part I of the book, in order to illustrate the fitting of a Poisson–HMM and many other aspects of that model.

HMMs have been used for at least three decades in signal-processing applications, especially in the context of automatic speech recognition, but interest in their theory and application has expanded to other fields, for example:

- all kinds of recognition – face, gesture, handwriting, signature;
- bioinformatics – biological sequence analysis;
- environment – rainfall, earthquakes, wind direction;

- finance – series of daily returns;
- biophysics – ion channel modelling;
- ecology – animal behaviour.

Attractive features of HMMs include their simplicity, their general mathematical tractability, and specifically the fact that the likelihood is relatively straightforward to compute. The main aim of this book is to illustrate how HMMs can be used as general-purpose models for time series.

Following this preliminary chapter, the book introduces what we shall call the **basic HMM**: basic in the sense that it is univariate, is based on a homogeneous Markov chain, and has neither trend nor seasonal variation. The observations may be either discrete- or continuous-valued, but we initially ignore information that may be available on covariates. We focus on the following issues:

- parameter estimation (Chapters 3 and 4);
- point and interval forecasting (Chapter 5);
- decoding, i.e. estimating the sequence of hidden states (Chapter 5);
- model selection, model checking and outlier detection (Chapter 6).

In Chapter 7 we give one example of the Bayesian approach to inference. In Chapter 8 we give examples of how several **R** packages can be used to fit basic HMMs to data and to decode.

In Part II we discuss the many possible extensions of the basic HMM to a wider range of models. These include HMMs for series with trend and seasonal variation, methods to include covariate information from other time series, multivariate models of various types, HMM approximations to hidden semi-Markov models and to models with continuous-valued state process, and HMMs for longitudinal data.

Part III of the book offers fairly detailed applications of HMMs to time series arising in a variety of subject areas. These are intended to illustrate the theory covered in Parts I and II, and also to demonstrate the versatility of HMMs. Indeed, so great is the variety of HMMs that it is hard to imagine this diversity being exhaustively covered by any single software package. In some applications the model needs to accommodate some special features of the time series, which makes it necessary to write one's own code. We have found the computing environment **R** (Ihaka and Gentleman, 1996; R Core Team, 2015) to be particularly convenient for this purpose.

Many of the chapters contain exercises, some theoretical and some practical. Because one always learns more about models by applying them in practice, and because some aspects of the theory of HMMs are covered only in these exercises, we regard these as an important part of

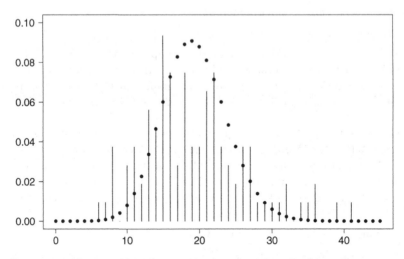

Figure 1.2 *Major earthquakes, 1900–2006: bar plot of relative frequencies of counts, and fitted Poisson distribution.*

the book. As regards the practical exercises, our strategy has been to give examples of **R** functions for some important but simple cases, and to encourage readers to learn to write their own code, initially just by modifying the functions given in Appendix A.

1.2 Independent mixture models

1.2.1 Definition and properties

Consider again the series of earthquake counts displayed in Figure 1.1. A standard model for unbounded counts is the Poisson distribution, with its probability function $p(x) = e^{-\lambda}\lambda^x/x!$ and the property that the variance equals the mean. However, for the earthquakes series the sample variance, $s^2 \approx 52$, is much larger than the sample mean, $\bar{x} \approx 19$, which indicates strong overdispersion relative to the Poisson distribution. The lack of fit is confirmed by Figure 1.2, which displays the fitted Poisson distribution and a bar plot of the relative frequencies of the counts.

One method of dealing with overdispersed observations with a bimodal or (more generally) multimodal distribution is to use a mixture model. Mixture models are designed to accommodate unobserved heterogeneity in the population; that is, the population may consist of unobserved groups, each having a distinct distribution for the observed variable.

Consider, for example, the distribution of the number, X, of packets of cigarettes bought by the customers of a supermarket. The customers can be divided into groups, for example, non-smokers, occasional smokers, and regular smokers. Now even if the number of packets bought by customers within each group were Poisson-distributed, the distribution of X would not be Poisson; it would be overdispersed relative to the Poisson, and maybe even multimodal.

Analogously, suppose that each count in the earthquakes series is generated by one of two Poisson distributions, with means λ_1 and λ_2, where the choice of mean is determined by some other random mechanism which we call the **parameter process**. Suppose also that λ_1 is selected with probability δ_1 and λ_2 with probability $\delta_2 = 1 - \delta_1$. We shall see later in this chapter that the variance of the resulting distribution exceeds the mean by $\delta_1 \delta_2 (\lambda_1 - \lambda_2)^2$. If the parameter process is a series of independent random variables, the counts are also independent, hence the term 'independent mixture'.

In general, an independent mixture distribution consists of a finite number, say m, of component distributions and a 'mixing distribution' which selects from these components. The component distributions may be either discrete or continuous. In the case of two components, the mixture distribution depends on two probability or density functions:

component	1	2
probability or density function	$p_1(x)$	$p_2(x)$.

To specify the component, one needs a discrete random variable C which performs the mixing:

$$C = \begin{cases} 1 & \text{with probability } \delta_1 \\ 2 & \text{with probability } \delta_2 = 1 - \delta_1. \end{cases}$$

The structure of that process for the case of two continuous component distributions is illustrated in Figure 1.3. In that example one can think of C as the outcome of tossing a coin with probability 0.75 of 'heads': if the outcome is 'heads', then $C = 1$ and an observation is drawn from p_1; if it is 'tails', then $C = 2$ and an observation is drawn from p_2. We suppose that we do not know the value C, that is, which of p_1 or p_2 was active when the observation was generated.

The extension to m components is straightforward. Let $\delta_1, \ldots, \delta_m$ denote the probabilities assigned to the different components, and let $p_1, \ldots, p_m$ denote their probability or density functions. Let X denote the random variable which has the mixture distribution. In the discrete

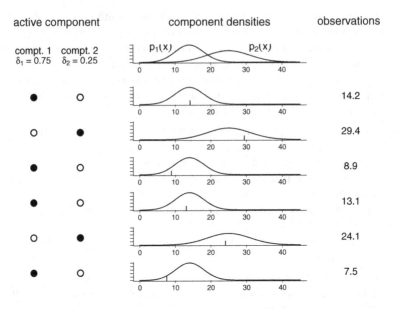

Figure 1.3 *Process structure of a two-component mixture distribution. From top to bottom, the states are* $1, 2, 1, 1, 2, 1$. *The corresponding component distributions are shown in the middle. The observations are generated from the active component density.*

case the probability function of X is given by

$$p(x) \quad = \quad \sum_{i=1}^{m} \Pr(X = x \mid C = i) \Pr(C = i)$$

$$= \quad \sum_{i=1}^{m} \delta_i p_i(x).$$

The continuous case is analogous. The expectation of the mixture can be given in terms of the expectations of the component distributions. Letting Y_i denote the random variable with probability function p_i, we have

$$\mathrm{E}(X) = \sum_{i=1}^{m} \Pr(C = i) \, \mathrm{E}(X \mid C = i) = \sum_{i=1}^{m} \delta_i \, \mathrm{E}(Y_i).$$

The same result holds for a mixture of continuous distributions.

More generally, for a mixture the kth moment about the origin is simply a linear combination of the kth moments of its components Y_i:

$$E(X^k) = \sum_{i=1}^{m} \delta_i\, E(Y_i^k), \quad k = 1, 2, \ldots.$$

Note that the analogous result does not hold for central moments. In particular, the variance of X is not a linear combination of the variances of its components Y_i. Exercise 1 asks the reader to prove that, in the two-component case, the variance of the mixture is given by

$$\mathrm{Var}(X) = \delta_1 \mathrm{Var}(Y_1) + \delta_2 \mathrm{Var}(Y_2) + \delta_1 \delta_2 \big(E(Y_1) - E(Y_2) \big)^2.$$

1.2.2 Parameter estimation

The estimation of the parameters of a mixture distribution is often performed by maximum likelihood (ML). The likelihood of a mixture model with m components is given, for both discrete and continuous cases, by

$$L(\boldsymbol{\theta}_1, \ldots, \boldsymbol{\theta}_m, \delta_1, \ldots, \delta_m \mid x_1, \ldots, x_n) = \prod_{j=1}^{n} \sum_{i=1}^{m} \delta_i p_i(x_j, \boldsymbol{\theta}_i). \quad (1.1)$$

Here $\boldsymbol{\theta}_1, \ldots, \boldsymbol{\theta}_m$ are the parameter vectors of the component distributions, $\delta_1, \ldots, \delta_m$ are the mixing parameters, totalling 1, and $x_1, \ldots, x_n$ are the n observations. Thus, in the case of component distributions each specified by one parameter, $2m - 1$ independent parameters have to be estimated. Except perhaps in special cases, analytic maximization of such a likelihood is not possible, but it is in general straightforward to evaluate it fast; see Exercise 3. Numerical maximization will be illustrated here by considering the case of a mixture of Poisson distributions.

Suppose that $m = 2$ and the two components are Poisson-distributed with means λ_1 and λ_2. Let δ_1 and δ_2 be the mixing parameters (with $\delta_1 + \delta_2 = 1$). The mixture distribution p is then given by

$$p(x) = \delta_1 \frac{\lambda_1^x e^{-\lambda_1}}{x!} + \delta_2 \frac{\lambda_2^x e^{-\lambda_2}}{x!}.$$

Since $\delta_2 = 1 - \delta_1$, there are only three parameters to be estimated: λ_1, λ_2 and δ_1. The likelihood is

$$L(\lambda_1, \lambda_2, \delta_1 \mid x_1, \ldots, x_n) = \prod_{i=1}^{n} \left(\delta_1 \frac{\lambda_1^{x_i} e^{-\lambda_1}}{x_i!} + (1 - \delta_1) \frac{\lambda_2^{x_i} e^{-\lambda_2}}{x_i!} \right).$$

The analytic maximization of L with respect to λ_1, λ_2 and δ_1 would be awkward, as L is the product of n factors, each of which is a sum. First taking the logarithm and then differentiating does not greatly simplify matters either. Therefore parameter estimation is more conveniently carried out by direct numerical maximization of the likelihood (or its logarithm), although the EM algorithm is a commonly used alternative; see, for example, McLachlan and Peel (2000) or Frühwirth-Schnatter (2006). (We shall in Chapter 4 discuss the EM algorithm more fully in the context of the estimation of HMMs.) A useful **R** package for estimation in mixture models is `flexmix` (Leisch, 2004). However, it is straightforward to write one's own **R** code to evaluate, and then maximize, mixture likelihoods in simple cases.

This log-likelihood can then be maximized by using (for example) the **R** function `nlm`. However, the parameters δ and λ are constrained by $\sum_{i=1}^{m} \delta_i = 1$ and (for $i = 1, \ldots, m$) $\delta_i > 0$ and $\lambda_i > 0$. It is therefore necessary to reparametrize if one wishes to use an unconstrained optimizer such as `nlm`. One possibility is to maximize the likelihood with respect to the $2m - 1$ unconstrained 'working parameters'

$$\eta_i = \log \lambda_i \quad (i = 1, \ldots, m)$$

and

$$\tau_i = \log \left(\frac{\delta_i}{1 - \sum_{j=2}^{m} \delta_j} \right) \quad (i = 2, \ldots, m).$$

One recovers the original 'natural parameters' via

$$\lambda_i = e^{\eta_i} \quad (i = 1, \ldots, m),$$

$$\delta_i = \frac{e^{\tau_i}}{1 + \sum_{j=2}^{m} e^{\tau_j}} \quad (i = 2, \ldots, m),$$

and $\delta_1 = 1 - \sum_{j=2}^{m} \delta_i$. The following code implements the above ideas in order to fit a mixture of four Poisson distributions to the earthquake counts. The results are given for $m = 1, 2, 3, 4$ in Table 1.2.

```
# Function to compute -log(likelihood)
mllk <- function(wpar,x){ zzz <- w2n(wpar)
      -sum(log(outer(x,zzz$lambda,dpois)%*%zzz$delta)) }

# Function to transform natural to working parameters
n2w  <- function(lambda,delta)log(c(lambda,delta[-1]/(1-sum(delta[-1]))))

# Function to transform working to natural parameters
w2n  <- function(wpar){m <- (length(wpar)+1)/2
      lambda <- exp(wpar[1:m])
      delta  <- exp(c(0,wpar[(m+1):(2*m-1)]))
return(list(lambda=lambda,delta=delta/sum(delta))) }

# Read data, specify starting values, minimize -log(likelihood),
# and transform to natural parameters
x         <- read.table("earthquakes.txt")[,2] # Set your own path.
wpar      <- n2w(c(10,20,25,30),c(1,1,1,1)/4)
w2n(nlm(mllk,wpar,x)$estimate)
```

Notice how, in this code, the use of the function `outer` makes it possible to evaluate a Poisson mixture log-likelihood in a single compact expression rather than a loop. But if the distributions being mixed were distributions with more than one parameter (e.g. normal), a slightly different approach would be needed.

1.2.3 Unbounded likelihood in mixtures

There is one aspect of mixtures of continuous distributions that differs from the discrete case and is worth highlighting. It is this: it can happen that, in the vicinity of certain parameter combinations, the likelihood is unbounded. For instance, in the case of a mixture of normal distributions, the likelihood becomes arbitrarily large if one sets a component mean equal to one of the observations and allows the corresponding variance to tend to zero. The problem has been extensively discussed in the literature on mixture models, and there are those who would say that, if the likelihood is thus unbounded, the ML estimates simply 'do not exist'; see, for instance, Scholz (2006, p. 4630).

The source of the problem, however, is just the use of densities rather than probabilities in the likelihood; it would not arise if one were to replace each density value in a likelihood by the probability of the interval corresponding to the recorded value. (For example, an observation recorded as '12.4' is associated with the interval $[12.35, 12.45)$.) In the context of independent mixtures one replaces the expression

$$\prod_{j=1}^{n}\sum_{i=1}^{m}\delta_i p_i(x_j, \boldsymbol{\theta}_i)$$

for the likelihood (see equation (1.1)) by the **discrete likelihood**

$$L = \prod_{j=1}^{n}\sum_{i=1}^{m}\delta_i \int_{a_j}^{b_j} p_i(x, \boldsymbol{\theta}_i)\,\mathrm{d}x, \qquad (1.2)$$

where the interval (a_j, b_j) consists of those values which, if observed, would be recorded as x_j. This simply amounts to acknowledging explicitly the interval nature of all supposedly continuous observations. More generally, the discrete likelihood of observations on a set of random variables $X_1, X_2, \ldots, X_n$ is a probability of the form $\Pr(a_t < X_t < b_t$, for all $t)$. We use the term **continuous likelihood** for the joint density evaluated at the observations.

Another way of avoiding the problem is to impose a lower bound on the variances and search for the best local maximum subject to that bound. It can happen, though, that one is fortunate enough to avoid the likelihood 'spikes' when searching for a local maximum; in this respect

Table 1.2 *Poisson independent mixture models fitted to the earthquakes series. The number of components is m, the mixing probabilities are denoted by δ_i, and the component means by λ_i. The maximized likelihood is L.*

Model	i	δ_i	λ_i	$-\log L$	Mean	Variance
$m = 1$	1	1.000	19.364	391.9189	19.364	19.364
$m = 2$	1	0.676	15.777	360.3690	19.364	46.182
	2	0.324	26.840			
$m = 3$	1	0.278	12.736	356.8489	19.364	51.170
	2	0.593	19.785			
	3	0.130	31.629			
$m = 4$	1	0.093	10.584	356.7337	19.364	51.638
	2	0.354	15.528			
	3	0.437	20.969			
	4	0.116	32.079			
observations					19.364	51.573

good starting values can help. The phenomenon of unbounded likelihood does not arise for discrete-valued observations because the likelihood is in that case a probability and thereby bounded by 0 and 1.

For a thorough account of the unbounded likelihood 'problem', see Liu, Wu and Meeker (2015). Liu *et al.* use the terms 'density-approximation likelihood' and 'correct likelihood' for what we call the continuous likelihood and discrete likelihood, respectively.

1.2.4 Examples of fitted mixture models

Mixtures of Poisson distributions

If one uses `nlm` to fit a mixture of m Poisson distributions ($m = 1, 2, 3, 4$) to the earthquakes data, one obtains the results displayed in Table 1.2. Notice that there is a very clear improvement in likelihood resulting from the addition of a second component, and very little improvement from addition of a fourth – apparently insufficient to justify the additional two parameters. Section 6.1 will discuss the model selection problem in more detail. Figure 1.4 presents a histogram of the observed counts and the

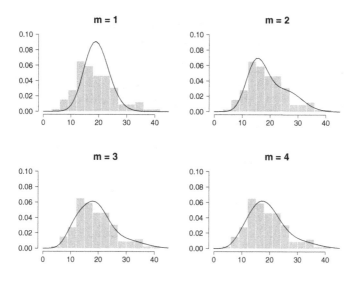

Figure 1.4 *Earthquakes data: histogram of counts, compared to mixtures of one, two, three and four Poisson distributions.*

four models fitted. It is clear that the mixtures fit the observations much better than does a single Poisson distribution, and visually the three- and four-state models seem adequate. The better fit of the mixtures is also evident from the variances of the four models as presented in Table 1.2. In computing the means and variances of the models we have used $E(X) = \sum_i \delta_i \lambda_i$ and $\mathrm{Var}(X) = E(X^2) - (E(X))^2$, with $E(X^2) = \sum_i \delta_i (\lambda_i + \lambda_i^2)$. For comparison we also used the **R** package `flexmix` to fit the same four models. The results corresponded closely except in the case of the four-component model, where the highest likelihood value that we found by `flexmix` was 356.7759 and the component means differed somewhat.

Note, however, that the above discussion ignores the possibility of serial dependence in the earthquakes data, a point we shall take up in Chapter 2.

A mixture of normal distributions

As a very simple example of the fitting of an independent mixture of normal distributions, consider the data presented in Table 8.1 of Hastie, Tibshirani and Friedman (2009, p. 273); see our Table 1.3. Hastie *et al.* use the EM algorithm to fit a mixture model with two normal components.

Table 1.3 *Data of Hastie* et al. *(2009), plus two mixture models. The first model was fitted by direct numerical maximization in* **R***, the second is the model fitted by EM by Hastie* et al.

| −0.39 | 0.12 | 0.94 | 1.67 | 1.76 | 2.44 | 3.72 | 4.28 | 4.92 | 5.53 |
| 0.06 | 0.48 | 1.01 | 1.68 | 1.80 | 3.25 | 4.12 | 4.60 | 5.28 | 6.22 |

i	δ_i	μ_i	σ_i^2	$-\log L$
1	0.4454	4.656	0.8188	38.9134
2	0.5546	1.083	0.8114	
1	0.454	4.62	0.87	
2	0.546	1.06	0.77	

Our two-component model, fitted by direct numerical maximization of the log-likelihood in **R**, has log-likelihood -38.9134, and is also displayed in Table 1.3. (Here we used the continuous likelihood, i.e. the joint density of the observations, not the discrete likelihood.) The parameter estimates are close to those given by Hastie *et al.*, but not identical.

1.3 Markov chains

We now introduce Markov chains, a second building-block of hidden Markov models. Our treatment is restricted to those few aspects of discrete-time Markov chains that we need. Thus, although we shall make passing reference to properties such as irreducibility and aperiodicity, we shall not dwell on such technical issues. For a general account of the topic, see Grimmett and Stirzaker (2001, Chapter 6), or Feller's classic text (Feller, 1968).

1.3.1 Definitions and example

A sequence of discrete random variables $\{C_t : t \in \mathbb{N}\}$ is said to be a (discrete-time) **Markov chain** (MC) if, for all $t \in \mathbb{N}$, it satisfies the **Markov property**

$$\Pr(C_{t+1} \mid C_t, \ldots, C_1) = \Pr(C_{t+1} \mid C_t).$$

That is, conditioning on the 'history' of the process up to time t is equivalent to conditioning only on the most recent value C_t. For compactness we define $\mathbf{C}^{(t)}$ as the history $(C_1, C_2, \ldots, C_t)$, in which case the Markov

property can be written as

$$\Pr(C_{t+1} \mid \mathbf{C}^{(t)}) = \Pr(C_{t+1} \mid C_t).$$

The Markov property can be regarded as a first relaxation of the assumption of independence. The random variables $\{C_t\}$ are dependent in a specific way that is mathematically convenient, as displayed in the following directed graph in which the past and the future are dependent only through the present.

Important quantities associated with a Markov chain are the conditional probabilities called **transition probabilities**:

$$\Pr(C_{s+t} = j \mid C_s = i).$$

If these probabilities do not depend on s, the Markov chain is called **homogeneous**, otherwise non-homogeneous. Unless there is an explicit indication to the contrary, we shall assume that the Markov chain under discussion is homogeneous, in which case the transition probabilities will be denoted by

$$\gamma_{ij}(t) = \Pr(C_{s+t} = j \mid C_s = i).$$

Notice that the notation $\gamma_{ij}(t)$ does not involve s. The matrix $\mathbf{\Gamma}(t)$ is defined as the matrix with (i, j) element $\gamma_{ij}(t)$.

An important property of all finite state-space homogeneous Markov chains is that they satisfy the **Chapman–Kolmogorov equations**:

$$\mathbf{\Gamma}(t + u) = \mathbf{\Gamma}(t)\,\mathbf{\Gamma}(u).$$

The proof requires only the definition of conditional probability and the application of the Markov property: this is Exercise 10. The Chapman–Kolmogorov equations imply that, for all $t \in \mathbb{N}$,

$$\mathbf{\Gamma}(t) = \mathbf{\Gamma}(1)^t;$$

that is, the matrix of t-step transition probabilities is the tth power of $\mathbf{\Gamma}(1)$, the matrix of one-step transition probabilities. The matrix $\mathbf{\Gamma}(1)$, which will be abbreviated as $\mathbf{\Gamma}$, is a square matrix of probabilities with row sums equal to 1:

$$\mathbf{\Gamma} = \begin{pmatrix} \gamma_{11} & \cdots & \gamma_{1m} \\ \vdots & \ddots & \vdots \\ \gamma_{m1} & \cdots & \gamma_{mm} \end{pmatrix},$$

where (throughout this text) m denotes the number of states of the Markov chain. The statement that the row sums are equal to 1 can be written as $\mathbf{\Gamma 1'} = \mathbf{1'}$; that is, the column vector $\mathbf{1'}$ is a right eigenvector of $\mathbf{\Gamma}$ and corresponds to eigenvalue 1. We shall refer to $\mathbf{\Gamma}$ as the (one-step) **transition probability matrix** (t.p.m.). Many authors use instead the term 'transition matrix'; we avoid that term because of possible confusion with a matrix of transition counts, or a matrix of transition intensities.

The **unconditional probabilities** $\Pr(C_t = j)$ of a Markov chain being in a given state at a given time t are often of interest. We denote these by the row vector

$$\mathbf{u}(t) = (\Pr(C_t = 1), \dots, \Pr(C_t = m)), \quad t \in \mathbb{N}.$$

We refer to $\mathbf{u}(1)$ as the **initial distribution** of the Markov chain. To deduce the distribution at time $t + 1$ from that at t we postmultiply by the transition probability matrix $\mathbf{\Gamma}$:

$$\mathbf{u}(t + 1) = \mathbf{u}(t)\mathbf{\Gamma}. \tag{1.3}$$

The proof of this statement is left as an exercise.

Example. Imagine that the sequence of rainy and sunny days is such that each day's weather depends only on the previous day's, and the transition probabilities are given by the following table.

	day $t + 1$	
day t	rainy	sunny
rainy	0.9	0.1
sunny	0.6	0.4

That is, if today is rainy, the probability that tomorrow will be rainy is 0.9; if today is sunny, that probability is 0.6. The weather is then a two-state homogeneous Markov chain, with t.p.m. $\mathbf{\Gamma}$ given by

$$\mathbf{\Gamma} = \begin{pmatrix} 0.9 & 0.1 \\ 0.6 & 0.4 \end{pmatrix}.$$

Now suppose that today (time 1) is a sunny day. This means that the distribution of today's weather is

$$\mathbf{u}(1) = \big(\Pr(C_1 = 1), \Pr(C_1 = 2)\big) = (0, 1).$$

The distribution of the weather of tomorrow, the day after tomorrow, and so on, can be calculated by repeatedly postmultiplying $\mathbf{u}(1)$ by $\mathbf{\Gamma}$, the t.p.m.:

$$\begin{aligned}
\mathbf{u}(2) &= \big(\Pr(C_2 = 1), \Pr(C_2 = 2)\big) = \mathbf{u}(1)\mathbf{\Gamma} = (0.6, 0.4), \\
\mathbf{u}(3) &= \big(\Pr(C_3 = 1), \Pr(C_3 = 2)\big) = \mathbf{u}(2)\mathbf{\Gamma} = (0.78, 0.22), \text{ etc.}
\end{aligned}$$

1.3.2 Stationary distributions

A Markov chain with transition probability matrix $\mathbf{\Gamma}$ is said to have **stationary distribution** $\boldsymbol{\delta}$ (a row vector with non-negative elements) if $\boldsymbol{\delta}\mathbf{\Gamma} = \boldsymbol{\delta}$ and $\boldsymbol{\delta}\mathbf{1}' = 1$. The first of these requirements expresses the stationarity, the second is the requirement that $\boldsymbol{\delta}$ is indeed a probability distribution. For instance, the Markov chain with t.p.m. given by

$$\mathbf{\Gamma} = \begin{pmatrix} 1/3 & 1/3 & 1/3 \\ 2/3 & 0 & 1/3 \\ 1/2 & 1/2 & 0 \end{pmatrix}$$

has as stationary distribution $\boldsymbol{\delta} = \frac{1}{32}(15, 9, 8)$.

Since $\mathbf{u}(t+1) = \mathbf{u}(t)\mathbf{\Gamma}$, a Markov chain started from its stationary distribution will continue to have that distribution at all subsequent time points, and we shall refer to such a process as a **stationary Markov chain**. It is perhaps worth stating that this assumes more than merely homogeneity. Homogeneity alone would not be sufficient to render the Markov chain a stationary process, and we prefer to reserve the adjective 'stationary' for homogeneous Markov chains that have the additional property that the initial distribution $\mathbf{u}(1)$ is the stationary distribution and are therefore stationary processes. Not all authors use this terminology, however; see, for example, McLachlan and Peel (2000, p. 328), who use the word 'stationary' of a Markov chain where we would say 'homogeneous'.

An irreducible (homogeneous, discrete-time, finite state-space) Markov chain has a unique, strictly positive, stationary distribution. Note that although the technical assumption of irreducibility is needed for this conclusion, aperiodicity is not; see Grimmett and Stirzaker (2001, Lemma 6.3.5 on p. 225 and Theorem 6.4.3 on p. 227).

If, however, one does add the assumption of aperiodicity, it follows that a unique limiting distribution exists, and is precisely the stationary distribution; see Feller (1968, p. 394). Since we shall always assume aperiodicity and irreducibility, the terms 'limiting distribution' and 'stationary distribution' are for our purposes synonymous.

A general result that can conveniently be used to compute a stationary distribution (see Exercise 9(a)) is as follows. The vector $\boldsymbol{\delta}$ with non-negative elements is a stationary distribution of the Markov chain with t.p.m. $\boldsymbol{\Gamma}$ if and only if

$$\boldsymbol{\delta}(\mathbf{I}_m - \boldsymbol{\Gamma} + \mathbf{U}) = \mathbf{1},$$

where $\mathbf{1}$ is a row vector of ones, $\mathbf{I}_m$ is the $m \times m$ identity matrix, and $\mathbf{U}$ is the $m \times m$ matrix of ones. Alternatively, a stationary distribution can be found by deleting one of the equations in the system $\boldsymbol{\delta}\boldsymbol{\Gamma} = \boldsymbol{\delta}$ and replacing it by $\sum_i \delta_i = 1$.

1.3.3 Autocorrelation function

We shall have occasion, for example in Section 2.2.3 and in Exercise 4(f) in Chapter 2, to compare the autocorrelation function (ACF) of a hidden Markov model with that of its underlying Markov chain $\{C_t\}$, on the states $1, 2, \ldots, m$. We assume that these states are quantitative and not merely categorical. The ACF of $\{C_t\}$, assumed stationary and irreducible, may be obtained as follows.

Firstly, defining $\mathbf{v} = (1, 2, \ldots, m)$ and $\mathbf{V} = \mathrm{diag}(1, 2, \ldots, m)$, we have, for all non-negative integers k,

$$\mathrm{Cov}(C_t, C_{t+k}) = \boldsymbol{\delta}\mathbf{V}\boldsymbol{\Gamma}^k\mathbf{v}' - (\boldsymbol{\delta}\mathbf{v}')^2; \tag{1.4}$$

the proof is Exercise 11. Secondly, if $\boldsymbol{\Gamma}$ is diagonalizable, and its eigenvalues (other than 1) are denoted by $\omega_2, \omega_3, \ldots, \omega_m$, then $\boldsymbol{\Gamma}$ can be written as

$$\boldsymbol{\Gamma} = \mathbf{U}\boldsymbol{\Omega}\mathbf{U}^{-1},$$

where $\boldsymbol{\Omega}$ is $\mathrm{diag}(1, \omega_2, \omega_3, \ldots, \omega_m)$ and the columns of $\mathbf{U}$ are corresponding right eigenvectors of $\boldsymbol{\Gamma}$. We then have, for non-negative integers k,

$$
\begin{aligned}
\mathrm{Cov}(C_t, C_{t+k}) &= \boldsymbol{\delta}\mathbf{V}\mathbf{U}\boldsymbol{\Omega}^k\mathbf{U}^{-1}\mathbf{v}' - (\boldsymbol{\delta}\mathbf{v}')^2 \\
&= \mathbf{a}\boldsymbol{\Omega}^k\mathbf{b}' - a_1 b_1 \\
&= \sum_{i=2}^{m} a_i b_i \omega_i^k,
\end{aligned}
$$

where $\mathbf{a} = \boldsymbol{\delta}\mathbf{V}\mathbf{U}$ and $\mathbf{b}' = \mathbf{U}^{-1}\mathbf{v}'$. Hence $\mathrm{Var}(C_t) = \sum_{i=2}^{m} a_i b_i$ and, for non-negative integers k,

$$\rho(k) \equiv \mathrm{Corr}(C_t, C_{t+k}) = \sum_{i=2}^{m} a_i b_i \omega_i^k \Big/ \sum_{i=2}^{m} a_i b_i. \tag{1.5}$$

This is a weighted average of the kth powers of the eigenvalues ω_2, ω_3, $\ldots, \omega_m$, and somewhat similar to the ACF of a Gaussian autoregressive process of order $m-1$. Note that equation (1.5) implies in the case $m = 2$ that $\rho(k) = \rho(1)^k$ for all non-negative integers k, and that $\rho(1)$ is the eigenvalue other than 1 of $\boldsymbol{\Gamma}$.

1.3.4 Estimating transition probabilities

If we are given a realization of a Markov chain, and wish to estimate the transition probabilities, one approach – but not the only one – is to find the transition counts and estimate the transition probabilities as relative frequencies. For instance, if the MC has three states and the observed sequence is

2332111112 3132332122 3232332222 3132332212 3232132232
3132332223 3232331232 3232331222 3232132123 3132332121,

then the matrix of transition counts is

$$(f_{ij}) = \begin{pmatrix} 4 & 7 & 6 \\ 8 & 10 & 24 \\ 6 & 24 & 10 \end{pmatrix},$$

where f_{ij} denotes the number of transitions observed from state i to state j. Since the number of transitions from state 2 to state 3 is 24, and the total number of transitions from state 2 is 8+10+24, a relative frequency estimate of γ_{23} is 24/42. The t.p.m. Γ is therefore plausibly estimated by

$$\begin{pmatrix} 4/17 & 7/17 & 6/17 \\ 8/42 & 10/42 & 24/42 \\ 6/40 & 24/40 & 10/40 \end{pmatrix}.$$

We shall now show that this is in fact the conditional ML estimate of Γ, conditioned on the first observation.

Suppose, then, that we wish to estimate the $m^2 - m$ parameters γ_{ij} $(i \neq j)$ of an m-state Markov chain $\{C_t\}$ from a realization $c_1, c_2, \ldots, c_T$. The likelihood conditioned on the first observation is

$$L = \prod_{i=1}^{m} \prod_{j=1}^{m} \gamma_{ij}^{f_{ij}}.$$

The log-likelihood is

$$l = \sum_{i=1}^{m} \left(\sum_{j=1}^{m} f_{ij} \log \gamma_{ij} \right) = \sum_{i=1}^{m} l_i \text{ (say)},$$

and we can maximize l by maximizing each l_i separately. Substituting $1 - \sum_{k \neq i} \gamma_{ik}$ for γ_{ii}, differentiating l_i with respect to an off-diagonal transition probability γ_{ij}, and equating the derivative to zero yields

$$0 = \frac{-f_{ii}}{1 - \sum_{k \neq i} \gamma_{ik}} + \frac{f_{ij}}{\gamma_{ij}} = -\frac{f_{ii}}{\gamma_{ii}} + \frac{f_{ij}}{\gamma_{ij}}.$$

Hence, unless a denominator is zero in the above equation, $f_{ij}\gamma_{ii} = f_{ii}\gamma_{ij}$, and so $\gamma_{ii} \sum_{j=1}^{m} f_{ij} = f_{ii}$. This implies that, at a maximum of the likelihood,

$$\gamma_{ii} = f_{ii}\Big/\sum_{j=1}^{m} f_{ij} \quad \text{and} \quad \gamma_{ij} = f_{ij}\gamma_{ii}/f_{ii} = f_{ij}\Big/\sum_{j=1}^{m} f_{ij}.$$

(We could instead use Lagrange multipliers to express the constraints $\sum_{j=1}^{m}\gamma_{ij} = 1$ subject to which we seek to maximize the terms l_i and therefore the likelihood; see Exercise 12.)

The estimator $\widehat{\gamma}_{ij} = f_{ij}/\sum_{k=1}^{m} f_{ik}$ $(i,j = 1,\ldots,m)$ – which is just the empirical transition probability – is thereby seen to be a conditional ML estimator of γ_{ij}. This estimator of $\boldsymbol{\Gamma}$ satisfies the requirement that the row sums should equal 1.

The assumption of stationarity of the Markov chain was not used in the above derivation. If we wish to assume stationarity, we may use the unconditional likelihood. This is the conditional likelihood as above, multiplied by the stationary probability δ_{c_1}. The unconditional likelihood or its logarithm may then be maximized numerically, subject to non-negativity and row-sum constraints, in order to estimate the transition probabilities γ_{ij}. Bisgaard and Travis (1991) show in the case of a two-state Markov chain that, barring some extreme cases, the unconditional likelihood equations have a unique solution. For some non-trivial special cases of the two-state chain, they also derive explicit expressions for the unconditional maximum likelihood estimates (MLEs) of the transition probabilities. Since we use one such result later (in Section 17.3.1), we state it here.

Suppose the Markov chain $\{C_t\}$ takes the values 0 and 1, and that we wish to estimate the transition probabilities γ_{ij} from a sequence of observations in which there are f_{ij} transitions from state i to state j $(i,j = 0,1)$, and $f_{11} > 0$ but $f_{00} = 0$. So in the observations a zero is always followed by a one. Define $c = f_{10} + (1 - c_1)$ and $d = f_{11}$. Then the unconditional MLEs of the transition probabilities are given by

$$\widehat{\gamma}_{01} = 1 \quad \text{and} \quad \widehat{\gamma}_{10} = \frac{-(1+d) + \big((1+d)^2 + 4c(c+d-1)\big)^{\frac{1}{2}}}{2(c+d-1)}. \quad (1.6)$$

1.3.5 Higher-order Markov chains

This section is somewhat specialized, and the material is used only in Section 10.3 and parts of Sections 17.3.2 and 19.2.2. It will therefore not interrupt the continuity greatly if the reader should initially omit this section.

In cases where observations on a process with finite state space appear not to satisfy the Markov property, one possibility that suggests itself is to use a higher-order Markov chain, that is, a model $\{C_t\}$ satisfying the following generalization of the Markov property for some $l \geq 2$:

$$\Pr(C_t \mid C_{t-1}, C_{t-2}, \ldots) = \Pr(C_t \mid C_{t-1}, \ldots, C_{t-l}).$$

An account of such higher-order Markov chains may be found, for instance, in Lloyd (1980, Section 19.9). Although such a model is not in the usual sense a Markov chain (i.e. not a 'first-order' Markov chain), we can redefine the model in such a way as to produce an equivalent process which is. If we let $\mathbf{Y}_t = (C_{t-l+1}, C_{t-l+2}, \ldots, C_t)$, then $\{\mathbf{Y}_t\}$ is a first-order Markov chain on M^l, where M is the state space of $\{C_t\}$. Although some properties may be more awkward to establish, no essentially new theory is involved in analysing a higher-order Markov chain rather than a first-order one.

A *second-order* Markov chain, if stationary, is characterized by the transition probabilities

$$\gamma(i, j, k) = \Pr(C_t = k \mid C_{t-1} = j, C_{t-2} = i),$$

and has stationary bivariate distribution $u(j, k) = \Pr(C_{t-1} = j, C_t = k)$ satisfying

$$u(j, k) = \sum_{i=1}^{m} u(i, j) \gamma(i, j, k) \quad \text{and} \quad \sum_{j=1}^{m} \sum_{k=1}^{m} u(j, k) = 1.$$

For example, the most general stationary second-order Markov chain $\{C_t\}$ on the two states 1 and 2 is characterized by the following four transition probabilities:

$$
\begin{aligned}
a &= \Pr(C_t{=}2 \mid C_{t-1}{=}1, C_{t-2}{=}1), \\
b &= \Pr(C_t{=}1 \mid C_{t-1}{=}2, C_{t-2}{=}2), \\
c &= \Pr(C_t{=}1 \mid C_{t-1}{=}2, C_{t-2}{=}1), \\
d &= \Pr(C_t{=}2 \mid C_{t-1}{=}1, C_{t-2}{=}2).
\end{aligned}
$$

The process $\{\mathbf{Y}_t\} = \{(C_{t-1}, C_t)\}$ is then a first-order Markov chain, on the four states (1,1), (1,2), (2,1), (2,2), with transition probability matrix

$$
\begin{pmatrix}
1-a & a & 0 & 0 \\
0 & 0 & c & 1-c \\
1-d & d & 0 & 0 \\
0 & 0 & b & 1-b
\end{pmatrix}.
\tag{1.7}
$$

Notice the structural zeros appearing in this matrix. It is not possible, for instance, to make a transition directly from $(2, 1)$ to $(2, 2)$; hence the zero in row 3 and column 4 in the t.p.m. (1.7). The parameters a, b, c and d are bounded by 0 and 1 but are otherwise unconstrained. The stationary distribution of $\{\mathbf{Y}_t\}$ is proportional to the vector

$$\big(b(1-d), ab, ab, a(1-c)\big),$$

from which it follows that the matrix $(u(j,k))$ of stationary bivariate probabilities for $\{C_t\}$ is

$$\frac{1}{b(1-d)+2ab+a(1-c)} \begin{pmatrix} b(1-d) & ab \\ ab & a(1-c) \end{pmatrix}.$$

The use of a general higher-order Markov chain (instead of a first-order one) increases the number of parameters of the model; a general Markov chain of order l on m states has $m^l(m-1)$ independent transition probabilities. Pegram (1980) and Raftery (1985a,b) have therefore proposed certain classes of parsimonious models for higher-order chains. Pegram's models have $m+l-1$ parameters, and those of Raftery $m(m-1)+l-1$. For $m=2$ the models are equivalent, but for $m>2$ those of Raftery are more general and can represent a wider range of dependence patterns and autocorrelation structures. In both cases an increase of one in the order of the Markov chain requires only one additional parameter.

Raftery's models, which he terms 'mixture transition distribution' (MTD) models, are defined as follows. The process $\{C_t\}$ takes values in $M=\{1,2,\ldots,m\}$ and satisfies

$$\Pr(C_t{=}j_0 \mid C_{t-1}{=}j_1,\ldots,C_{t-l}{=}j_l) = \sum_{i=1}^{l}\lambda_i\, q(j_i,j_0), \qquad (1.8)$$

where $\sum_{i=1}^{l}\lambda_i = 1$, and $\mathbf{Q} = (q(j,k))$ is an $m \times m$ matrix with non-negative entries and row sums equal to one, such that the right-hand side of equation (1.8) is bounded by zero and one for all j_0, j_1, $\ldots$, $j_l \in M$. This last requirement, which generates m^{l+1} pairs of nonlinear constraints on the parameters, ensures that the conditional probabilities in equation (1.8) are indeed probabilities, and the condition on the row sums of $\mathbf{Q}$ ensures that the sum over j_0 of these conditional probabilities is one. Note that Raftery does not assume that the parameters λ_i are non-negative.

A variety of applications are presented by Raftery (1985a) and Raftery and Tavaré (1994). In several of the fitted models there are negative estimates of some of the coefficients λ_i. For further accounts of this class of models, see Haney (1993), Berchtold (2001), and Berchtold and Raftery (2002).

Azzalini and Bowman (1990) report the fitting of a second-order Markov chain model to the binary series they use to represent the lengths of successive eruptions of the Old Faithful geyser. Their analysis, and some alternative models, will be discussed in Chapter 17.

Exercises

1. (a) Let X be a random variable which is distributed as a (δ_1, δ_2)-mixture of two distributions with expectations μ_1, μ_2, and variances σ_1^2, σ_2^2, respectively, where $\delta_1 + \delta_2 = 1$.

   i. Show that $\mathrm{Var}(X) = \delta_1\sigma_1^2 + \delta_2\sigma_2^2 + \delta_1\delta_2(\mu_1 - \mu_2)^2$.

   ii. Show that a (non-trivial) mixture X of two Poisson distributions with distinct means is overdispersed, that is, $\mathrm{Var}(X) > \mathrm{E}(X)$.

   (b) Now suppose that X is a mixture of $m \geq 2$ distributions, with means μ_i and variances σ_i^2, for $i = 1, 2, \ldots, m$. The mixing distribution is $\boldsymbol{\delta}$.

   i. Show that

   $$\mathrm{Var}(X) = \sum_{i=1}^{m} \delta_i\sigma_i^2 + \sum_{i<j} \delta_i\delta_j(\mu_i - \mu_j)^2.$$

   Hint: use either $\mathrm{Var}(X) = \mathrm{E}(X^2) - (\mathrm{E}(X))^2$ or the conditional variance formula,

   $$\mathrm{Var}(X) = \mathrm{E}(\mathrm{Var}(X \mid C)) + \mathrm{Var}(\mathrm{E}(X \mid C)).$$

   ii. Describe the circumstances in which $\mathrm{Var}(X)$ equals the linear combination $\sum_{i=1}^{m} \delta_i\sigma_i^2$.

2. A zero-inflated Poisson distribution is sometimes used as a model for unbounded counts displaying an excessive number of zeros relative to the Poisson. Such a model is a mixture of two distributions: one is a Poisson and the other is identically zero.

   (a) Is it ever possible for such a model to display underdispersion relative to Poisson?

   (b) Now consider the zero-inflated binomial. Is it possible in such a model that the variance is less than the mean?

3. Brown and Buckley (2015, p. 308) consider a Poisson mixture likelihood of the form

   $$L = \prod_{i=1}^{n} \sum_{j=1}^{k} w_j f(x_i \mid \mu_j).$$

   (Here $f(\cdot \mid \mu)$ denotes a Poisson probability function with mean μ.) They write that 'Even for moderate values of n and k, this takes a long time to evaluate as there are k^n terms when the inner sums are expanded', and do not pursue maximum likelihood estimation.

   Explain why it is in fact possible to evaluate L or its logarithm in computations which are of order kn rather than k^n.

4. (a) Write an **R** function to minimize minus the log-likelihood of a normal mixture model with m components, using the nonlinear minimizer `nlm`.

   Hint: first write a function to transform the parameters (δ and the parameters of the m normal distributions) into unconstrained parameters. You will also need a function to reverse the transformation. (For the Poisson case, see the code on p. 10.)

   (b) Use your code to fit a mixture of two normals to the data appearing in Table 1.3, and compare your model with those displayed in that table.

5. Consider the following data, which appear in Lange (1995, 2002, 2004). (There they are quoted from Titterington, Smith and Makov (1985) and Hasselblad (1969), but the trail leads back via Schilling (1947) and Thorndike (1926) to Whitaker (1914), where in all eight similar data sets appear as Table XV on p. 67.)

   Here n_i denotes the number of days in 1910–1912 on which there appeared, in *The Times* of London, i death notices in respect of women aged 80 or over at death.

i	0	1	2	3	4	5	6	7	8	9
n_i	162	267	271	185	111	61	27	8	3	1

   (a) Use `nlm` or `optim` in **R** to fit a mixture of two Poisson distributions to these observations. (The parameter estimates reported by Lange (2002, p. 36; 2004, p. 151) are, in our notation: $\widehat{\delta}_1 = 0.3599$, $\widehat{\lambda}_1 = 1.2561$ and $\widehat{\lambda}_2 = 2.6634$.)

   (b) Fit also a single Poisson distribution to these data. Is a single Poisson distribution adequate as a model?

   (c) Fit a mixture of three Poisson distributions to these observations.

   (d) How many components do you think are necessary?

   (e) Repeat (a)–(d) for some of the other seven data sets of Whitaker.

6. Consider the series of weekly sales (in integer units) of a particular soap product in a supermarket, as shown in Table 1.4. The data were taken from a database[†] provided by the Kilts Center for Marketing, Graduate School of Business of the University of Chicago, at: http://gsbwww.uchicago.edu/kilts/research/db/dominicks. The product was 'Zest White Water 15 oz.', with code 3700031165, and the store number 67.

[†] That database is now at
http://research.chicagobooth.edu/kilts/marketing-databases/dominicks.

Table 1.4 *Weekly sales of the soap product; to be read across rows.*

1	6	9	18	14	8	8	1	6	7	3	3	1	3	4	12	8	10	8	2
17	15	7	12	22	10	4	7	5	0	2	5	3	4	4	7	5	6	1	3
4	5	3	7	3	0	4	5	3	3	4	4	4	4	4	3	5	5	5	7
4	0	4	3	2	6	3	8	9	6	3	4	3	3	3	3	2	1	4	5
5	2	7	5	2	3	1	3	4	6	8	8	5	7	2	4	2	7	4	15
15	12	21	20	13	9	8	0	13	9	8	0	6	2	0	3	2	4	4	6
3	2	5	5	3	2	1	1	3	1	2	6	2	7	3	2	4	1	5	6
8	14	5	3	6	5	11	4	5	9	9	7	9	8	3	4	8	6	3	5
6	3	1	7	4	9	2	6	6	4	6	6	13	7	4	8	6	4	4	4
9	2	9	2	2	2	13	13	4	5	1	4	6	5	4	2	3	10	6	15
5	9	9	7	4	4	2	4	2	3	8	15	0	0	3	4	3	4	7	5
7	6	0	6	4	14	5	1	6	5	5	4	9	4	14	2	2	1	5	2
6	4																		

Fit Poisson mixture models with one, two, three and four components. How many components do you think are necessary?

7. Consider a stationary two-state Markov chain with transition probability matrix given by

$$\boldsymbol{\Gamma} = \left(\begin{array}{cc} \gamma_{11} & \gamma_{12} \\ \gamma_{21} & \gamma_{22} \end{array} \right).$$

   (a) Show that the stationary distribution is

   $$(\delta_1, \delta_2) = \frac{1}{\gamma_{12} + \gamma_{21}} (\gamma_{21}, \gamma_{12}) \,.$$

   (b) Consider the case

   $$\boldsymbol{\Gamma} = \left(\begin{array}{cc} 0.9 & 0.1 \\ 0.2 & 0.8 \end{array} \right),$$

   and the following two sequences of observations that are assumed to be generated by the above Markov chain.

Sequence 1:	1	1	1	2	2	1
Sequence 2:	2	1	1	2	1	1

   Compute the probability of each of the sequences. Note that each sequence contains the same number of ones and twos. Why are these sequences not equally probable?

8. Consider a two-state Markov chain with transition probability matrix given by

$$\boldsymbol{\Gamma} = \left(\begin{array}{cc} \gamma_{11} & \gamma_{12} \\ \gamma_{21} & \gamma_{22} \end{array} \right).$$

Show that the k-step transition probability matrix, $\mathbf{\Gamma}^k$, is given by

$$\mathbf{\Gamma}^k = \begin{pmatrix} \delta_1 & \delta_2 \\ \delta_1 & \delta_2 \end{pmatrix} + w^k \begin{pmatrix} \delta_2 & -\delta_2 \\ -\delta_1 & \delta_1 \end{pmatrix},$$

where $w = 1 - \gamma_{12} - \gamma_{21}$ and δ_1 and δ_2 are as defined in Exercise 7. (Hint: one way of showing this is to diagonalize the transition probability matrix, but there is a quicker way.)

9. (a) This is one of several possible approaches to finding the stationary distribution of a Markov chain, plundered from Grimmett and Stirzaker (2001, Exercise 6.6.5).

   Suppose $\mathbf{\Gamma}$ is the transition probability matrix of a (discrete-time, homogeneous) Markov chain on m states, and that $\boldsymbol{\delta}$ is a nonnegative row vector with m components. Show that $\boldsymbol{\delta}$ is a stationary distribution of the Markov chain if and only if

   $$\boldsymbol{\delta}(\mathbf{I}_m - \mathbf{\Gamma} + \mathbf{U}) = \mathbf{1},$$

   where $\mathbf{1}$ is a row vector of ones, and $\mathbf{U}$ is an $m \times m$ matrix of ones.

   (b) Write an **R** function statdist(gamma) that computes the stationary distribution of the Markov chain with t.p.m. gamma.

   (c) Use your function to find stationary distributions corresponding to the following transition probability matrices. One of them should cause a problem!

   i. $\begin{pmatrix} 0.7 & 0.2 & 0.1 \\ 0 & 0.6 & 0.4 \\ 0.5 & 0 & 0.5 \end{pmatrix}$

   ii. $\begin{pmatrix} 0 & 1 & 0 \\ \frac{1}{3} & 0 & \frac{2}{3} \\ 0 & 1 & 0 \end{pmatrix}$

   iii. $\begin{pmatrix} 0 & 0.5 & 0 & 0.5 \\ 0.75 & 0 & 0.25 & 0 \\ 0 & 0.75 & 0 & 0.25 \\ 0.5 & 0 & 0.5 & 0 \end{pmatrix}$

   iv. $\begin{pmatrix} 0.25 & 0.25 & 0.25 & 0.25 \\ 0.25 & 0.25 & 0.5 & 0 \\ 0 & 0 & 0.25 & 0.75 \\ 0 & 0 & 0.5 & 0.5 \end{pmatrix}$

   v. $\begin{pmatrix} 1 & 0 & 0 & 0 \\ 0.5 & 0 & 0.5 & 0 \\ 0 & 0.75 & 0 & 0.25 \\ 0 & 0 & 0 & 1 \end{pmatrix}$

10. Prove the Chapman–Kolmogorov equations.

11. Prove equation (1.4).

12. Let the quantities a_i be non-negative, with $\sum_i a_i > 0$. Using a Lagrange multiplier, maximize $S = \sum_{i=1}^m a_i \log \delta_i$ over $\delta_i \geq 0$, subject to $\sum_i \delta_i = 1$. (Check the second- as well as the first-derivative condition.)

13. (This exercise is based on Example 2 of Bisgaard and Travis (1991).) Consider the following sequence of 21 observations, assumed to arise from a two-state (homogeneous) Markov chain:

$$11101\ 10111\ 10110\ 11111\ 1.$$

(a) Estimate the transition probability matrix by ML, conditional on the first observation.

(b) Estimate the t.p.m. by unconditional ML (assuming stationarity of the Markov chain).

(c) Use the **R** functions `contour` and `persp` to produce contour and perspective plots of the unconditional log-likelihood (as a function of the two off-diagonal transition probabilities).

14. Consider the following two transition probability matrices, neither of which is diagonalizable:

(a)

$$\Gamma = \begin{pmatrix} 1/3 & 1/3 & 1/3 \\ 2/3 & 0 & 1/3 \\ 1/2 & 1/2 & 0 \end{pmatrix} ;$$

(b)

$$\Gamma = \begin{pmatrix} 0.9 & 0.08 & 0 & 0.02 \\ 0 & 0.7 & 0.2 & 0.1 \\ 0 & 0 & 0.7 & 0.3 \\ 0 & 0 & 0 & 1 \end{pmatrix} .$$

In each case, write Γ in Jordan canonical form, and so find an explicit expression for the t-step transition probabilities ($t = 1, 2, \ldots$).

15. Consider the following (very) short DNA sequence, taken from Singh (2003, p. 358):

$$\text{AACGT CTCTA TCATG CCAGG ATCTG}$$

(a) Fit a homogeneous Markov chain to these data by:

   i. maximizing the likelihood conditioned on the first observation;
   ii. assuming stationarity and maximizing the unconditional likelihood of all 25 observations.

(b) Compare your estimates of the t.p.m. with each other and with the estimate displayed as Table 1 of Singh (2003, p. 360).

(c) Now repeat (a) for the following 50-nucleotide sequence, taken from Singh (2003, p. 367):

ATTAG GCACG CATTA TAATG GGCAC
CCGGA AATAA CCAGA GTTAC GGCCA.

16. Write an **R** function rMC(n,m,gamma,delta=NULL) that generates a series of length n from an m-state Markov chain with t.p.m. gamma. If the initial state distribution is given, then it should be used; otherwise the stationary distribution should be used as the initial distribution. (Use your function statdist from Exercise 9(b).)

## CHAPTER 2

Hidden Markov models: definition and properties

2.1 A simple hidden Markov model

Consider again the observed earthquake series displayed in Figure 1.1 on p. 4. The observations are unbounded counts, making the Poisson distribution a natural choice to describe them, but their distribution is clearly overdispersed relative to the Poisson. We saw in Chapter 1 that this feature can be accommodated by using a mixture of Poisson distributions with means $\lambda_1, \lambda_2, \ldots, \lambda_m$. The choice of mean is made by a second random process, the parameter process. The mean λ_i is selected with probability δ_i, where $i = 1, 2, \ldots, m$ and $\sum_{i=1}^{m} \delta_i = 1$.

An independent mixture model will not do for the earthquake series because – by definition – it does not allow for the serial dependence in the observations. The sample autocorrelation function (ACF), displayed in Figure 2.1, clearly indicates that the observations are serially dependent. One way of allowing for serial dependence in the observations is to relax the assumption that the parameter process is serially independent. A simple and mathematically convenient way to do so is to assume that it is a Markov chain. The resulting model for the observations is called a Poisson–hidden Markov model, a simple example of the class of

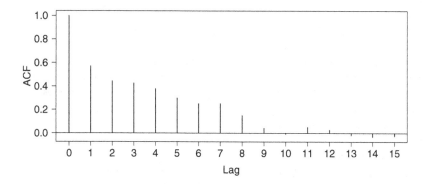

Figure 2.1 *Earthquakes series: sample autocorrelation function.*

models discussed in the rest of this book, namely hidden Markov models (HMMs).

We shall not give an account here of the (interesting) history of such models, but two valuable sources of information on HMMs that include accounts of the history are Ephraim and Merhav (2002) and Cappé *et al.* (2005).

2.2 The basics

2.2.1 Definition and notation

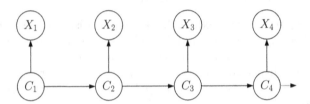

Figure 2.2 *Directed graph of basic HMM.*

A **hidden Markov model** $\{X_t : t \in \mathbb{N}\}$ is a particular kind of dependent mixture. With $\mathbf{X}^{(t)}$ and $\mathbf{C}^{(t)}$ representing the histories from time 1 to time t, one can summarize the simplest model of this kind by:

$$\Pr(C_t \mid \mathbf{C}^{(t-1)}) = \Pr(C_t \mid C_{t-1}), \quad t = 2, 3, \ldots \qquad (2.1)$$

$$\Pr(X_t \mid \mathbf{X}^{(t-1)}, \mathbf{C}^{(t)}) = \Pr(X_t \mid C_t), \quad t \in \mathbb{N}. \qquad (2.2)$$

The model consists of two parts: firstly, an unobserved 'parameter process' $\{C_t : t = 1, 2, \ldots \}$ satisfying the Markov property; and secondly, the 'state-dependent process' $\{X_t : t = 1, 2, \ldots \}$, in which the distribution of X_t depends only on the current state C_t and not on previous states or observations. This structure is represented by the directed graph in Figure 2.2.

If the Markov chain $\{C_t\}$ has m states, we call $\{X_t\}$ an m-state HMM. Although it is the usual terminology in speech-processing applications, the name 'hidden Markov model' is by no means the only one used for such models or similar ones. For instance, Ephraim and Merhav (2002) argue for 'hidden Markov process', Leroux and Puterman (1992) use 'Markov-dependent mixture', and others use 'Markov-switching model' (especially for models with extra dependencies at the level of the observations X_t), 'models subject to Markov regime', 'Markov mixture model', or 'latent Markov model'. Bartolucci, Farcomeni and Pennoni (2013) use the term 'latent Markov model' specifically for models for longitudinal data, as opposed to single time series.

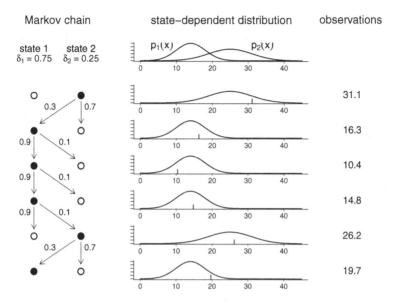

Figure 2.3 *Process generating the observations in a two-state HMM. The chain followed the path $2, 1, 1, 1, 2, 1$, as indicated on the left. The corresponding state-dependent distributions are shown in the middle. The observations are generated from the corresponding active distributions.*

The process generating the observations is demonstrated again in Figure 2.3, for state-dependent distributions p_1 and p_2, stationary distribution $\boldsymbol{\delta} = (0.75, 0.25)$, and t.p.m. $\boldsymbol{\Gamma} = \begin{pmatrix} 0.9 & 0.1 \\ 0.3 & 0.7 \end{pmatrix}$. In contrast to the case of an independent mixture, here the distribution of C_t, the state at time t, does depend on C_{t-1}. As is also true of independent mixtures, there is for each state a different distribution, discrete or continuous.

We now introduce some notation which will cover both discrete- and continuous-valued observations. In the case of discrete observations we define, for $i = 1, 2, \ldots, m$,

$$p_i(x) = \Pr(X_t = x \mid C_t = i).$$

That is, p_i is the probability mass function of X_t if the Markov chain is in state i at time t. The continuous case is treated similarly: there we define p_i to be the probability *density* function of X_t associated with state i. We refer to the m distributions p_i as the **state-dependent distributions** of the model. Many of our results are stated only for the discrete case, but, if probabilities are replaced by densities, apply also to the continuous case.

2.2.2 Marginal distributions

We shall often need the marginal distribution of X_t and also higher-order marginal distributions, such as that of (X_t, X_{t+k}). We shall derive the results for the case in which the Markov chain is homogeneous but not necessarily stationary, and then give them also for the special case in which the Markov chain is stationary. For convenience the derivation is given only for discrete state-dependent distributions; the continuous case can be derived analogously.

Univariate distributions

For discrete-valued observations X_t, defining $u_i(t) = \Pr(C_t = i)$ for $t = 1, \ldots, T$, we have

$$\Pr(X_t = x) = \sum_{i=1}^{m} \Pr(C_t = i)\Pr(X_t = x \mid C_t = i)$$

$$= \sum_{i=1}^{m} u_i(t)p_i(x).$$

This expression can conveniently be rewritten in matrix notation:

$$\Pr(X_t = x) = (u_1(t), \ldots, u_m(t)) \begin{pmatrix} p_1(x) & & 0 \\ & \ddots & \\ 0 & & p_m(x) \end{pmatrix} \begin{pmatrix} 1 \\ \vdots \\ 1 \end{pmatrix}$$

$$= \mathbf{u}(t)\mathbf{P}(x)\mathbf{1}',$$

where $\mathbf{P}(x)$ is defined as the diagonal matrix with ith diagonal element $p_i(x)$. It follows from equation (1.3) that $\mathbf{u}(t) = \mathbf{u}(1)\mathbf{\Gamma}^{t-1}$, and hence that

$$\Pr(X_t = x) = \mathbf{u}(1)\mathbf{\Gamma}^{t-1}\mathbf{P}(x)\mathbf{1}'. \tag{2.3}$$

Equation (2.3) holds if the Markov chain is merely homogeneous, and not necessarily stationary. If, as we shall often assume, the Markov chain is stationary, with stationary distribution $\boldsymbol{\delta}$, then the result is simpler: in that case $\boldsymbol{\delta}\mathbf{\Gamma}^{t-1} = \boldsymbol{\delta}$ for all $t \in \mathbb{N}$, and so

$$\Pr(X_t = x) = \boldsymbol{\delta}\mathbf{P}(x)\mathbf{1}'. \tag{2.4}$$

Bivariate distributions

The calculation of many of the distributions relating to an HMM is most easily done by first noting that, in any directed graphical model, the joint distribution of a set of random variables V_i is given by

$$\Pr(V_1, V_2, \ldots, V_n) = \prod_{i=1}^{n} \Pr(V_i \mid \mathrm{pa}(V_i)), \tag{2.5}$$

where $\mathrm{pa}(V_i)$ denotes the set of all 'parents' of V_i in the set V_1, V_2, $\ldots$, V_n; see, for example, Davison (2003, p. 250) or Jordan (2004).

In the directed graph of the four random variables X_t, X_{t+k}, C_t, C_{t+k} (for positive integer k), C_t has no parents, $\mathrm{pa}(X_t) = \{C_t\}$, $\mathrm{pa}(C_{t+k}) = \{C_t\}$ and $\mathrm{pa}(X_{t+k}) = \{C_{t+k}\}$. It therefore follows that

$$\Pr(X_t, X_{t+k}, C_t, C_{t+k}) = \Pr(C_t) \Pr(X_t|C_t) \Pr(C_{t+k}|C_t) \Pr(X_{t+k}|C_{t+k}),$$

and hence that

$$
\begin{aligned}
\Pr(X_t &= v, X_{t+k} = w) \\
&= \sum_{i=1}^{m} \sum_{j=1}^{m} \Pr(X_t = v, X_{t+k} = w, C_t = i, C_{t+k} = j) \\
&= \sum_{i=1}^{m} \sum_{j=1}^{m} \underbrace{\Pr(C_t = i)}_{u_i(t)} p_i(v) \underbrace{\Pr(C_{t+k} = j \mid C_t = i)}_{\gamma_{ij}(k)} p_j(w) \\
&= \sum_{i=1}^{m} \sum_{j=1}^{m} u_i(t) p_i(v) \gamma_{ij}(k) p_j(w).
\end{aligned}
$$

(Here and elsewhere, $\gamma_{ij}(k)$ denotes the (i, j) element of $\boldsymbol{\Gamma}^k$.) Writing the above double sum as a product of matrices yields

$$\Pr(X_t = v, X_{t+k} = w) = \mathbf{u}(t)\mathbf{P}(v)\boldsymbol{\Gamma}^k\mathbf{P}(w)\mathbf{1}'. \tag{2.6}$$

If the Markov chain is stationary, this reduces to

$$\Pr(X_t = v, X_{t+k} = w) = \boldsymbol{\delta}\mathbf{P}(v)\boldsymbol{\Gamma}^k\mathbf{P}(w)\mathbf{1}'. \tag{2.7}$$

Similarly, one can obtain expressions for the higher-order marginal distributions; in the stationary case, the formula for a trivariate distribution is, for positive integers k and l,

$$\Pr(X_t = v, X_{t+k} = w, X_{t+k+l} = z) = \boldsymbol{\delta}\mathbf{P}(v)\boldsymbol{\Gamma}^k\mathbf{P}(w)\boldsymbol{\Gamma}^l\mathbf{P}(z)\mathbf{1}'.$$

2.2.3 Moments

First, we note that

$$\mathrm{E}(X_t) = \sum_{i=1}^{m} \mathrm{E}(X_t \mid C_t = i) \Pr(C_t = i) = \sum_{i=1}^{m} u_i(t)\mathrm{E}(X_t \mid C_t = i),$$

which, in the stationary case, reduces to

$$E(X_t) = \sum_{i=1}^{m} \delta_i E(X_t \mid C_t = i).$$

More generally, analogous results hold for $E(g(X_t))$ and $E(g(X_t, X_{t+k}))$, for any functions g for which the relevant state-dependent expectations exist. In the stationary case

$$E(g(X_t)) = \sum_{i=1}^{m} \delta_i E(g(X_t) \mid C_t = i); \qquad (2.8)$$

and

$$E(g(X_t, X_{t+k})) = \sum_{i,j=1}^{m} E(g(X_t, X_{t+k}) \mid C_t = i, C_{t+k} = j)\, \delta_i \gamma_{ij}(k). \quad (2.9)$$

Often we shall be interested in a function g which factorizes as

$$g(X_t, X_{t+k}) = g_1(X_t) g_2(X_{t+k}),$$

in which case equation (2.9) becomes

$$E(g(X_t, X_{t+k})) = \sum_{i,j=1}^{m} E(g_1(X_t) \mid C_t = i)\, E(g_2(X_{t+k}) \mid C_{t+k} = j)\, \delta_i \gamma_{ij}(k).$$

$$(2.10)$$

These expressions enable us to find covariances and correlations; convenient explicit expressions exist in many cases. For the case of a stationary two-state Poisson–HMM:

- $E(X_t) = \delta_1 \lambda_1 + \delta_2 \lambda_2$;
- $Var(X_t) = E(X_t) + \delta_1 \delta_2 (\lambda_2 - \lambda_1)^2 \geq E(X_t)$;
- $Cov(X_t, X_{t+k}) = \delta_1 \delta_2 (\lambda_2 - \lambda_1)^2 (1 - \gamma_{12} - \gamma_{21})^k$, for $k \in \mathbb{N}$.

Notice that the resulting formula for the correlation of X_t and X_{t+k} is of the form $\rho(k) = A(1 - \gamma_{12} - \gamma_{21})^k$ with $A \in [0, 1)$, and that $A = 0$ if $\lambda_1 = \lambda_2$. For more details, and for more general results, see Exercises 3 and 4.

2.3 The likelihood

The aim of this section is to develop a convenient formula for the likelihood L_T of T consecutive observations $x_1, x_2, \ldots, x_T$ assumed to be generated by an m-state HMM. That such a formula exists is indeed fortunate, but by no means obvious. We shall see that the computation of the likelihood, consisting as it does of a sum of m^T terms, each of which is a product of $2T$ factors, appears to require $O(Tm^T)$ operations. However, it has long been known in several contexts that the likelihood is easily computable; see, for example, Baum (1972), Lange and Boehnke (1983), and Cosslett and Lee (1985). What we describe here is in fact a special case of a much more general theory; see Smyth, Heckerman and Jordan (1997) or Jordan (2004).

It is our purpose here to demonstrate that L_T can in general be computed relatively simply in $O(Tm^2)$ operations. The way will then be open to estimate parameters by numerical maximization of the likelihood. First the likelihood of a two-state model will be explored, and then the general formula will be presented.

2.3.1 The likelihood of a two-state Bernoulli–HMM

Example. Consider the stationary two-state HMM with t.p.m.

$$\boldsymbol{\Gamma} = \begin{pmatrix} \frac{1}{2} & \frac{1}{2} \\ \frac{1}{4} & \frac{3}{4} \end{pmatrix}$$

and state-dependent distributions given by

$$\begin{aligned} \Pr(X_t = x \mid C_t = 1) &= \frac{1}{2} \quad \text{(for } x = 0, 1), \\ \Pr(X_t = 1 \mid C_t = 2) &= 1. \end{aligned}$$

We call a model of this kind a **Bernoulli–HMM**. The stationary distribution of the Markov chain is $\boldsymbol{\delta} = \frac{1}{3}(1, 2)$. Consider the probability $\Pr(X_1 = X_2 = X_3 = 1)$. First, note that, by equation (2.5),

$$\begin{aligned} &\Pr(X_1, X_2, X_3, C_1, C_2, C_3) \\ &= \Pr(C_1)\Pr(X_1 \mid C_1)\Pr(C_2 \mid C_1)\Pr(X_2 \mid C_2)\Pr(C_3 \mid C_2)\Pr(X_3 \mid C_3); \end{aligned}$$

and then sum over the values assumed by C_1, C_2, C_3. The result is

$$\begin{aligned} &\Pr(X_1 = 1, X_2 = 1, X_3 = 1) \\ &= \sum_{i=1}^{2}\sum_{j=1}^{2}\sum_{k=1}^{2} \Pr(X_1 = 1, X_2 = 1, X_3 = 1, C_1 = i, C_2 = j, C_3 = k) \\ &= \sum_{i=1}^{2}\sum_{j=1}^{2}\sum_{k=1}^{2} \delta_i p_i(1)\gamma_{ij}p_j(1)\gamma_{jk}p_k(1). \end{aligned} \tag{2.11}$$

Notice that the triple sum (2.11) has $m^T = 2^3$ terms, each of which is a product of $2T = 2 \times 3$ factors. To evaluate the required probability, the different possibilities for the values of i, j and k can be listed and the sum (2.11) calculated as in Table 2.1.

Summation of the last column of Table 2.1 tells us that $\Pr(X_1 =$

Table 2.1 *Example of a likelihood computation.*

i	j	k	$p_i(1)$	$p_j(1)$	$p_k(1)$	δ_i	γ_{ij}	γ_{jk}	Product
1	1	1	$\frac{1}{2}$	$\frac{1}{2}$	$\frac{1}{2}$	$\frac{1}{3}$	$\frac{2}{4}$	$\frac{2}{4}$	$\frac{1}{96}$
1	1	2	$\frac{1}{2}$	$\frac{1}{2}$	1	$\frac{1}{3}$	$\frac{2}{4}$	$\frac{2}{4}$	$\frac{1}{48}$
1	2	1	$\frac{1}{2}$	1	$\frac{1}{2}$	$\frac{1}{3}$	$\frac{2}{4}$	$\frac{1}{4}$	$\frac{1}{96}$
1	2	2	$\frac{1}{2}$	1	1	$\frac{1}{3}$	$\frac{2}{4}$	$\frac{3}{4}$	$\frac{1}{16}$
2	1	1	1	$\frac{1}{2}$	$\frac{1}{2}$	$\frac{2}{3}$	$\frac{1}{4}$	$\frac{2}{4}$	$\frac{1}{48}$
2	1	2	1	$\frac{1}{2}$	1	$\frac{2}{3}$	$\frac{1}{4}$	$\frac{2}{4}$	$\frac{1}{24}$
2	2	1	1	1	$\frac{1}{2}$	$\frac{2}{3}$	$\frac{3}{4}$	$\frac{1}{4}$	$\frac{1}{16}$
2	2	2	1	1	1	$\frac{2}{3}$	$\frac{3}{4}$	$\frac{3}{4}$	$\frac{3}{8}$

$$\frac{29}{48}$$

$1, X_2 = 1, X_3 = 1) = \frac{29}{48}$. In passing we note that the largest element in that column is $\frac{3}{8}$; the state sequence that maximizes the joint probability

$$\Pr(X_1 = 1, X_2 = 1, X_3 = 1, C_1 = i, C_2 = j, C_3 = k)$$

is therefore the sequence $i = 2$, $j = 2$, $k = 2$. Equivalently, it maximizes the conditional probability $\Pr(C_1 = i, C_2 = j, C_3 = k \mid X_1 = 1, X_2 = 1, X_3 = 1)$. This is an example of 'global decoding', which will be discussed in Section 5.4.2.

But a more convenient way to present the sum is to use matrix notation. Let $\mathbf{P}(u)$ be defined (as before) as $\mathrm{diag}(p_1(u), p_2(u))$. Then

$$\mathbf{P}(0) = \begin{pmatrix} \frac{1}{2} & 0 \\ 0 & 0 \end{pmatrix} \quad \text{and} \quad \mathbf{P}(1) = \begin{pmatrix} \frac{1}{2} & 0 \\ 0 & 1 \end{pmatrix},$$

and the triple sum (2.11) can be written as a matrix product:

$$\boldsymbol{\delta}\mathbf{P}(1)\boldsymbol{\Gamma}\mathbf{P}(1)\boldsymbol{\Gamma}\mathbf{P}(1)\mathbf{1}'.$$

More generally, the likelihood turns out to be a T-fold sum which can also be written as a matrix product.

2.3.2 The likelihood in general

Here we consider the likelihood of an HMM in general. We suppose there is an observation sequence $x_1, x_2, \ldots, x_T$ generated by such a model. We seek the probability L_T of observing that sequence, as calculated under an m-state HMM which has *initial* distribution $\boldsymbol{\delta}$ and t.p.m. $\boldsymbol{\Gamma}$ for the Markov chain, and state-dependent probability (density) functions p_i. In many of our applications we shall assume that $\boldsymbol{\delta}$ is the stationary distribution implied by $\boldsymbol{\Gamma}$, but it is not necessary to make that assumption in general.

Proposition 1 *The likelihood is given by*

$$L_T = \boldsymbol{\delta}\mathbf{P}(x_1)\boldsymbol{\Gamma}\mathbf{P}(x_2)\boldsymbol{\Gamma}\mathbf{P}(x_3)\cdots\boldsymbol{\Gamma}\mathbf{P}(x_T)\mathbf{1}'. \qquad (2.12)$$

If $\boldsymbol{\delta}$, the distribution of C_1, is the stationary distribution of the Markov chain, then in addition

$$L_T = \boldsymbol{\delta}\boldsymbol{\Gamma}\mathbf{P}(x_1)\boldsymbol{\Gamma}\mathbf{P}(x_2)\boldsymbol{\Gamma}\mathbf{P}(x_3)\cdots\boldsymbol{\Gamma}\mathbf{P}(x_T)\mathbf{1}'. \qquad (2.13)$$

Proof. We present only the case of discrete observations. First, note that

$$L_T = \Pr(\mathbf{X}^{(T)} = \mathbf{x}^{(T)}) = \sum_{c_1,c_2,\ldots,c_T=1}^{m} \Pr(\mathbf{X}^{(T)} = \mathbf{x}^{(T)}, \mathbf{C}^{(T)} = \mathbf{c}^{(T)}),$$

and that, by equation (2.5),

$$\Pr(\mathbf{X}^{(T)}, \mathbf{C}^{(T)}) = \Pr(C_1)\prod_{k=2}^{T}\Pr(C_k \mid C_{k-1})\prod_{k=1}^{T}\Pr(X_k \mid C_k). \qquad (2.14)$$

It follows that

$$
\begin{aligned}
L_T &= \sum_{c_1,\ldots,c_T=1}^{m}\left(\delta_{c_1}\gamma_{c_1,c_2}\gamma_{c_2,c_3}\cdots\gamma_{c_{T-1},c_T}\right)\left(p_{c_1}(x_1)p_{c_2}(x_2)\cdots p_{c_T}(x_T)\right)\\
&= \sum_{c_1,\ldots,c_T=1}^{m}\delta_{c_1}p_{c_1}(x_1)\gamma_{c_1,c_2}p_{c_2}(x_2)\gamma_{c_2,c_3}\cdots\gamma_{c_{T-1},c_T}p_{c_T}(x_T)\\
&= \boldsymbol{\delta}\mathbf{P}(x_1)\boldsymbol{\Gamma}\mathbf{P}(x_2)\boldsymbol{\Gamma}\mathbf{P}(x_3)\cdots\boldsymbol{\Gamma}\mathbf{P}(x_T)\mathbf{1}',
\end{aligned}
$$

which is equation (2.12). The last equality above exploits the fact that a multiple sum of terms having a certain simple multiplicative form can in general be written as a matrix product. Exercise 7(b) provides the detailed justification.

If $\boldsymbol{\delta}$ is the stationary distribution of the Markov chain, we have

$$\boldsymbol{\delta}\mathbf{P}(x_1) = \boldsymbol{\delta}\boldsymbol{\Gamma}\mathbf{P}(x_1),$$

hence equation (2.13), which involves an extra factor of $\boldsymbol{\Gamma}$ but may be slightly simpler to code. $\square$

A very simple but crucial consequence of the matrix expression for the likelihood is the 'forward algorithm' for recursive computation of the likelihood. Such recursive computation plays a key role, not only in likelihood evaluation and hence parameter estimation, but also in forecasting, decoding and model checking. The recursive nature of likelihood evaluation via either (2.12) or (2.13) is computationally much more efficient than brute-force summation over all possible state sequences. The fact that such computationally inexpensive recursive schemes can be used to address various questions of interest is a key feature of HMMs. Recursive evaluation of such multiple sums has been discussed by Lange and Boehnke (1983) and Lange (2002, p. 120).

To state the forward algorithm we define the vector $\boldsymbol{\alpha}_t$, for $t = 1, 2, \ldots, T$, by

$$\boldsymbol{\alpha}_t = \boldsymbol{\delta}\mathbf{P}(x_1)\boldsymbol{\Gamma}\mathbf{P}(x_2)\boldsymbol{\Gamma}\mathbf{P}(x_3)\cdots\boldsymbol{\Gamma}\mathbf{P}(x_t) = \boldsymbol{\delta}\mathbf{P}(x_1)\prod_{s=2}^{t}\boldsymbol{\Gamma}\mathbf{P}(x_s), \quad (2.15)$$

with the convention that an empty product is the identity matrix. It follows immediately from this definition that

$$L_T = \boldsymbol{\alpha}_T\mathbf{1}', \quad \text{and} \quad \boldsymbol{\alpha}_t = \boldsymbol{\alpha}_{t-1}\boldsymbol{\Gamma}\mathbf{P}(x_t) \quad \text{for } t \geq 2.$$

Accordingly, we can conveniently set out as follows the computations involved in the likelihood formula (2.12):

$$\boldsymbol{\alpha}_1 = \boldsymbol{\delta}\mathbf{P}(x_1);$$

$$\boldsymbol{\alpha}_t = \boldsymbol{\alpha}_{t-1}\boldsymbol{\Gamma}\mathbf{P}(x_t) \quad \text{for } t = 2, 3, \ldots, T;$$

$$L_T = \boldsymbol{\alpha}_T\mathbf{1}'.$$

That the number of operations involved is of order Tm^2 can be deduced thus. For each of the values of t in the loop, there are m elements of $\boldsymbol{\alpha}_t$ to be computed, and each of those elements is a sum of m products of three quantities: an element of $\boldsymbol{\alpha}_{t-1}$, a transition probability γ_{ij}, and a state-dependent probability (or density) $p_j(x_t)$.

The corresponding scheme for computation of (2.13) (i.e. if $\boldsymbol{\delta}$, the distribution of C_1, is the stationary distribution of the Markov chain) is

$$\boldsymbol{\alpha}_0 = \boldsymbol{\delta};$$

$$\boldsymbol{\alpha}_t = \boldsymbol{\alpha}_{t-1}\boldsymbol{\Gamma}\mathbf{P}(x_t) \quad \text{for } t = 1, 2, \ldots, T;$$

$$L_T = \boldsymbol{\alpha}_T\mathbf{1}'.$$

The elements of the vector $\boldsymbol{\alpha}_t$ are usually referred to as **forward probabilities**; the reason for this name will be seen in Section 4.1.1, where we show that the jth element of $\boldsymbol{\alpha}_t$ is $\Pr(\mathbf{X}^{(t)} = \mathbf{x}^{(t)}, C_t = j)$.

We show here **R** code that uses the forward algorithm to evaluate the likelihood of observations $x_1, \ldots, x_T$ under a Poisson–HMM with at least two states, t.p.m. $\boldsymbol{\Gamma}$, vector of state-dependent means $\boldsymbol{\lambda}$, and initial distribution $\boldsymbol{\delta}$ (not necessarily the stationary distribution). Note, however, that, unless the series is short, one needs to guard against underflow and evaluate the log-likelihood rather than the likelihood; see p. 49 for code that does so.

```
alpha                <- delta*dpois(x[1],lambda)
for (i in 2:T) alpha <- alpha %*% Gamma*dpois(x[i],lambda)
sum(alpha)
```

In the above discussion we have used the multiple-sum expression for the likelihood in order to arrive at the matrix expression, and then used the matrix expression to arrive at the forward recursion. An alternative route, which anticipates some of the material of Chapter 4, is to *define* the vector of forward probabilities α_t by

$$\alpha_t(j) = \Pr(\mathbf{X}^{(t)} = \mathbf{x}^{(t)}, C_t = j), \quad j = 1, 2, \ldots, m,$$

and then to deduce the forward recursion:

$$\alpha_t = \alpha_{t-1}\boldsymbol{\Gamma}\mathbf{P}(x_t).$$

The matrix expression is then a simple consequence of the forward recursion. This alternative route is described in Exercise 8.

2.3.3 HMMs are not Markov processes

HMMs do not in general satisfy the Markov property. This we can now establish via a simple counterexample. Let X_t and C_t be as defined in the example in Section 2.3.1. We already know that

$$\Pr(X_1 = 1, X_2 = 1, X_3 = 1) = \frac{29}{48},$$

and from equations (2.4) and (2.7) it can be established that $\Pr(X_2 = 1) = \frac{5}{6}$ and that

$$\Pr(X_1 = 1, X_2 = 1) = \Pr(X_2 = 1, X_3 = 1) = \frac{17}{24}.$$

It therefore follows that

$$
\begin{aligned}
\Pr(X_3 = 1 \mid X_1 = 1, X_2 = 1) &= \frac{\Pr(X_1 = 1, X_2 = 1, X_3 = 1)}{\Pr(X_1 = 1, X_2 = 1)} \\
&= \frac{29/48}{17/24} = \frac{29}{34}
\end{aligned}
$$

and that

$$
\begin{aligned}
\Pr(X_3 = 1 \mid X_2 = 1) &= \frac{\Pr(X_2 = 1, X_3 = 1)}{\Pr(X_2 = 1)} \\
&= \frac{17/24}{5/6} = \frac{17}{20}.
\end{aligned}
$$

Hence $\Pr(X_3 = 1 \mid X_2 = 1) \neq \Pr(X_3 = 1 \mid X_1 = 1, X_2 = 1)$; this HMM does not satisfy the Markov property. That some HMMs do satisfy the property, however, is clear. For instance, a two-state Bernoulli–HMM can degenerate in obvious fashion to the underlying Markov chain; one simply identifies each of the two observable values with one of the two underlying states. For the conditions under which an HMM will itself satisfy the Markov property, see Spreij (2001).

2.3.4 The likelihood when data are missing

In a time series context it is potentially awkward if some of the data are missing. In the case of hidden Markov time series models, however, the adjustment that needs to be made to the likelihood computation if data are missing turns out to be a simple one.

Suppose, for example, that one has available the observations x_1, x_2, x_4, x_7, x_8, ..., x_T of an HMM, but x_3, x_5 and x_6 are missing. Then the likelihood of the observations is given by

$$\Pr(X_1 = x_1, X_2 = x_2, X_4 = x_4, X_7 = x_7, \ldots, X_T = x_T)$$

$$= \sum \delta_{c_1} \gamma_{c_1,c_2} \gamma_{c_2,c_4}(2) \gamma_{c_4,c_7}(3) \gamma_{c_7,c_8} \cdots \gamma_{c_{T-1},c_T}$$
$$\times p_{c_1}(x_1) p_{c_2}(x_2) p_{c_4}(x_4) p_{c_7}(x_7) \cdots p_{c_T}(x_T),$$

where (as before) $\gamma_{ij}(k)$ denotes a k-step transition probability, and the sum is taken over all indices c_t other than c_3, c_5 and c_6. But this is just

$$\sum \delta_{c_1} p_{c_1}(x_1) \gamma_{c_1,c_2} p_{c_2}(x_2) \gamma_{c_2,c_4}(2) p_{c_4}(x_4) \gamma_{c_4,c_7}(3) p_{c_7}(x_7)$$

$$\times \cdots \times \gamma_{c_{T-1},c_T} p_{c_T}(x_T)$$
$$= \boldsymbol{\delta} \mathbf{P}(x_1) \boldsymbol{\Gamma} \mathbf{P}(x_2) \boldsymbol{\Gamma}^2 \mathbf{P}(x_4) \boldsymbol{\Gamma}^3 \mathbf{P}(x_7) \cdots \boldsymbol{\Gamma} \mathbf{P}(x_T) \mathbf{1}'.$$

With $L_T^{-(3,5,6)}$ denoting the likelihood of the observations (other than x_3, x_5 and x_6), it follows that

$$L_T^{-(3,5,6)} = \boldsymbol{\delta} \mathbf{P}(x_1) \boldsymbol{\Gamma} \mathbf{P}(x_2) \boldsymbol{\Gamma}^2 \mathbf{P}(x_4) \boldsymbol{\Gamma}^3 \mathbf{P}(x_7) \cdots \boldsymbol{\Gamma} \mathbf{P}(x_T) \mathbf{1}'.$$

In general, in the expression for the likelihood the diagonal matrices $\mathbf{P}(x_t)$ corresponding to missing observations x_t are replaced by the identity matrix; that is, the corresponding state-dependent probabilities $p_i(x_t)$ are replaced by 1 for all states i. If one can assume that the missingness is ignorable, this 'ignorable likelihood' is a reasonable basis for estimating parameters (Little, 2009, p. 411).

The fact that, even if some observations are missing, the likelihood of an HMM can be computed easily is especially useful in the derivation of conditional distributions, as will be shown in Section 5.2.

2.3.5 The likelihood when observations are interval-censored

Suppose that we wish to fit a Poisson–HMM to a series of counts, some of which are interval-censored. For instance, the value of x_t may be known only for $4 \leq t \leq T$, with the information $x_1 \leq 5$, $2 \leq x_2 \leq 3$ and $x_3 > 10$ available about the remaining observations. For simplicity, let us first assume that the Markov chain has only two states. In that case, one replaces the diagonal matrices $\mathbf{P}(x_i)$ $(i = 1, 2, 3)$ in the likelihood expression (2.12) by the matrices

$$\text{diag}(\Pr(X_1 \leq 5 \mid C_1 = 1), \Pr(X_1 \leq 5 \mid C_1 = 2)),$$
$$\text{diag}(\Pr(2 \leq X_2 \leq 3 \mid C_2 = 1), \Pr(2 \leq X_2 \leq 3 \mid C_2 = 2)), \text{ and}$$
$$\text{diag}(\Pr(X_3 > 10 \mid C_3 = 1), \Pr(X_3 > 10 \mid C_3 = 2)).$$

More generally, suppose that $a \leq x_t \leq b$, where a may be $-\infty$ (although that is not relevant to the Poisson case), b may be ∞, and the Markov chain has m states. One replaces $\mathbf{P}(x_t)$ in the likelihood by the $m \times m$ diagonal matrix of which the ith diagonal element is $\Pr(a \leq X_t \leq b \mid C_t = i)$. See Exercise 12. It is worth noting that missing data can be regarded as an extreme case of such interval-censoring.

Exercises

1. Consider a *stationary* two-state Poisson–HMM with parameters

$$\mathbf{\Gamma} = \begin{pmatrix} 0.1 & 0.9 \\ 0.4 & 0.6 \end{pmatrix} \quad \text{and} \quad \boldsymbol{\lambda} = (1, 3).$$

In each of the following ways, compute the probability that the first three observations from this model are 0, 2, 1.

(a) Consider all possible sequences of states of the Markov chain that could have occurred. Compute the probability of each sequence, and the probability of the observations given each sequence.

(b) Apply the formula

$$\Pr(X_1 = 0, X_2 = 2, X_3 = 1) = \boldsymbol{\delta}\mathbf{P}(0)\mathbf{\Gamma}\mathbf{P}(2)\mathbf{\Gamma}\mathbf{P}(1)\mathbf{1}',$$

where

$$\mathbf{P}(s) = \begin{pmatrix} \lambda_1^s e^{-\lambda_1}/s! & 0 \\ 0 & \lambda_2^s e^{-\lambda_2}/s! \end{pmatrix} = \begin{pmatrix} 1^s e^{-1}/s! & 0 \\ 0 & 3^s e^{-3}/s! \end{pmatrix}.$$

2. Consider again the model defined in Exercise 1. In that question you were asked to compute $\Pr(X_1 = 0, X_2 = 2, X_3 = 1)$. Now compute $\Pr(X_1 = 0, X_3 = 1)$ in each of the following ways.

(a) Consider all possible sequences of states of the Markov chain that could have occurred. Compute the probability of each sequence, and the probability of the observations given each sequence.

(b) Apply the formula

$$\Pr(X_1=0, X_3=1) = \boldsymbol{\delta}\mathbf{P}(0)\boldsymbol{\Gamma}\mathbf{I}_2\boldsymbol{\Gamma}\mathbf{P}(1)\mathbf{1}' = \boldsymbol{\delta}\mathbf{P}(0)\boldsymbol{\Gamma}^2\mathbf{P}(1)\mathbf{1}',$$

and check that this probability is equal to your answer in (a).

3. Consider an m-state HMM $\{X_t : t = 1, 2, \ldots\}$, based on a stationary Markov chain with transition probability matrix $\boldsymbol{\Gamma}$ and stationary distribution $\boldsymbol{\delta} = (\delta_1, \delta_2, \ldots, \delta_m)$, and having (univariate) state-dependent distributions $p_i(x)$. Let μ_i and σ_i^2 denote the mean and variance of the distribution p_i, $\boldsymbol{\mu}$ the vector $(\mu_1, \mu_2, \ldots, \mu_m)$, and $\mathbf{M}$ the matrix $\mathrm{diag}(\boldsymbol{\mu})$.

Derive the following results for the moments of $\{X_t\}$.

(a) $\mathrm{E}(X_t) = \sum_{i=1}^{m} \delta_i \mu_i = \boldsymbol{\delta}\boldsymbol{\mu}'$.
(b) $\mathrm{E}(X_t^2) = \sum_{i=1}^{m} \delta_i(\sigma_i^2 + \mu_i^2)$.
(c) $\mathrm{Var}(X_t) = \sum_{i=1}^{m} \delta_i(\sigma_i^2 + \mu_i^2) - (\boldsymbol{\delta}\boldsymbol{\mu}')^2$.
(d) If $m = 2$, $\mathrm{Var}(X_t) = \delta_1\sigma_1^2 + \delta_2\sigma_2^2 + \delta_1\delta_2(\mu_1 - \mu_2)^2$.
(e) For $k \in \mathbb{N}$, i.e. for positive integers k,

$$\mathrm{E}(X_t X_{t+k}) = \sum_{i=1}^{m}\sum_{j=1}^{m} \delta_i\mu_i\gamma_{ij}(k)\mu_j = \boldsymbol{\delta}\mathbf{M}\boldsymbol{\Gamma}^k\boldsymbol{\mu}'.$$

(f) For $k \in \mathbb{N}$,

$$\rho(k) = \mathrm{Corr}(X_t, X_{t+k}) = \frac{\boldsymbol{\delta}\mathbf{M}\boldsymbol{\Gamma}^k\boldsymbol{\mu}' - (\boldsymbol{\delta}\boldsymbol{\mu}')^2}{\mathrm{Var}(X_t)}.$$

Note that, if the eigenvalues of $\boldsymbol{\Gamma}$ are distinct, this is a linear combination of the kth powers of those eigenvalues.

(g) If the state-dependent means μ_i are all equal, X_t and X_{t+k} are uncorrelated for $k \in \mathbb{N}$.

Timmermann (2000) and Frühwirth-Schnatter (2006, pp. 308–312) are useful references for moments. See also Exercise 1 of Chapter 1.

4. (Marginal moments and autocorrelation function of a Poisson–HMM: special case of Exercise 3.) Consider a stationary m-state Poisson–HMM $\{X_t : t = 1, 2, \ldots\}$ with transition probability matrix $\boldsymbol{\Gamma}$ and state-dependent means $\boldsymbol{\lambda} = (\lambda_1, \lambda_2, \ldots, \lambda_m)$. Let $\boldsymbol{\delta} = (\delta_1, \delta_2, \ldots, \delta_m)$ be the stationary distribution of the Markov chain. Let $\boldsymbol{\Lambda} = \mathrm{diag}(\boldsymbol{\lambda})$.

Derive the following results.

(a) $\mathrm{E}(X_t) = \boldsymbol{\delta}\boldsymbol{\lambda}'$.
(b) $\mathrm{E}(X_t^2) = \sum_{i=1}^{m}(\lambda_i^2 + \lambda_i)\delta_i = \boldsymbol{\delta}\boldsymbol{\Lambda}\boldsymbol{\lambda}' + \boldsymbol{\delta}\boldsymbol{\lambda}'$.

(c) $\text{Var}(X_t) = \boldsymbol{\delta}\boldsymbol{\Lambda}\boldsymbol{\lambda}' + \boldsymbol{\delta}\boldsymbol{\lambda}' - (\boldsymbol{\delta}\boldsymbol{\lambda}')^2 = \text{E}(X_t) + \boldsymbol{\delta}\boldsymbol{\Lambda}\boldsymbol{\lambda}' - (\boldsymbol{\delta}\boldsymbol{\lambda}')^2 \geq \text{E}(X_t).$

(d) For $k \in \mathbb{N}$, $\text{E}(X_t X_{t+k}) = \boldsymbol{\delta}\boldsymbol{\Lambda}\boldsymbol{\Gamma}^k\boldsymbol{\lambda}'.$

(e) For $k \in \mathbb{N}$,

$$\rho(k) = \text{Corr}(X_t, X_{t+k}) = \frac{\boldsymbol{\delta}\boldsymbol{\Lambda}\boldsymbol{\Gamma}^k\boldsymbol{\lambda}' - (\boldsymbol{\delta}\boldsymbol{\lambda}')^2}{\boldsymbol{\delta}\boldsymbol{\Lambda}\boldsymbol{\lambda}' + \boldsymbol{\delta}\boldsymbol{\lambda}' - (\boldsymbol{\delta}\boldsymbol{\lambda}')^2}.$$

(f) In the case $m = 2$, $\rho(k) = Aw^k$ for $k \in \mathbb{N}$, where

$$A = \frac{\delta_1\delta_2(\lambda_2 - \lambda_1)^2}{\delta_1\delta_2(\lambda_2 - \lambda_1)^2 + \boldsymbol{\delta}\boldsymbol{\lambda}'}$$

and $w = 1 - \gamma_{12} - \gamma_{21}$. Notice that the extra level of randomness in the HMM, as compared with the underlying Markov chain, has reduced the autocorrelations by the factor $A \in [0, 1)$.

5. (A serially dependent process with zero autocorrelation.) In finance, time-series models consisting of serially uncorrelated but dependent random variables are often of interest. We consider here a stationary HMM $\{X_t\}$, with normal state-dependent distributions, that is such a process. Suppose that

$$\boldsymbol{\Gamma} = \begin{pmatrix} 0.990 & 0.005 & 0.005 \\ 0.010 & 0.980 & 0.010 \\ 0.015 & 0.015 & 0.970 \end{pmatrix}$$

and that, given $C_t = i$, $X_t \sim N(1, \sigma_i^2)$, with $(\sigma_1, \sigma_2, \sigma_3) = (1, 10, 20)$. By Exercise 3(g), X_t and X_{t+k} are uncorrelated for $k \in \mathbb{N}$.

(a) Simulate (say) 10 000 observations $\{x_t\}$ from this model. One way of doing so is to modify the code in Section A.1.5, which applies to the case of Poisson state-dependent distributions.

(b) Using the **R** function `acf`, plot the sample ACF of:
   i. $\{x_t\}$;
   ii. $\{|x_t|\}$;
   iii. $\{x_t^2\}$.

   What can you conclude from these three sample ACFs?

(c) Find the ACF of $\{X_t^2\}$ under the model, and superimpose it on your plot of the ACF of $\{x_t^2\}$. You should get a plot similar to that shown in Figure 2.4.

6. We have the general expression

$$L_T = \boldsymbol{\delta}\textbf{P}(x_1)\boldsymbol{\Gamma}\textbf{P}(x_2)\cdots\boldsymbol{\Gamma}\textbf{P}(x_T)\textbf{1}'$$

for the likelihood of an HMM, e.g. of Poisson type.

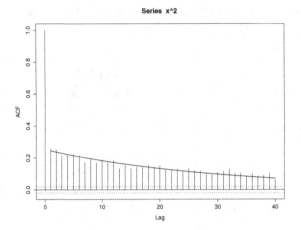

Figure 2.4 *Exercise 5: Sample autocorrelation function of the squares of 10 000 simulated observations, plus ACF of* $\{X_t^2\}$ *under the model (continuous line).*

(a) Consider the special case in which the Markov chain degenerates to a sequence of independent, identically-distributed random variables, i.e. an independent mixture model. Show that, in this case, the likelihood simplifies to the expression given in equation (1.1) for the likelihood of an *independent* mixture.

(b) Suppose instead that, for all i and x, $p_i(x) = p(x)$. What does the likelihood expression

$$\delta \mathbf{P}(x_1) \boldsymbol{\Gamma} \mathbf{P}(x_2) \boldsymbol{\Gamma} \mathbf{P}(x_3) \cdots \boldsymbol{\Gamma} \mathbf{P}(x_T) \mathbf{1}'$$

now reduce to, and what do you conclude?

7. This exercise shows that a sum of m^T terms of a certain simple multiplicative form can (perhaps surprisingly) be computed efficiently, in $O(Tm^2)$ operations.

Consider a multiple sum S of the following general form:

$$S = \sum_{i_1=1}^{m} \sum_{i_2=1}^{m} \cdots \sum_{i_T=1}^{m} h(i_1) \prod_{t=2}^{T} f_t(i_{t-1}, i_t).$$

For $i_1 = 1, 2, \ldots, m$, define the (row) vector $\boldsymbol{\alpha}_1$ by

$$\alpha_1(i_1) \equiv h(i_1);$$

and for $r = 1, 2, \ldots, T - 1$ and $i_{r+1} = 1, 2, \ldots, m$, define

$$\alpha_{r+1}(i_{r+1}) \equiv \sum_{i_r=1}^{m} \alpha_r(i_r) f_{r+1}(i_r, i_{r+1}).$$

That is, the vector $\boldsymbol{\alpha}_{r+1}$ is defined by, and can be computed as, $\boldsymbol{\alpha}_{r+1} = \boldsymbol{\alpha}_r \mathbf{F}_{r+1}$, where the $m \times m$ matrix $\mathbf{F}_t$ has (i, j) element equal to $f_t(i, j)$.

(a) Show by induction on T that $\alpha_T(i_T)$ is precisely the sum over all but i_T, i.e. that

$$\alpha_T(i_T) = \sum_{i_1}\sum_{i_2}\cdots\sum_{i_{T-1}} h(i_1) \prod_{t=2}^{T} f_t(i_{t-1}, i_t).$$

(b) Hence show that $S = \sum_{i_T} \alpha_T(i_T) = \boldsymbol{\alpha}_T \mathbf{1}' = \boldsymbol{\alpha}_1 \mathbf{F}_2 \mathbf{F}_3 \cdots \mathbf{F}_T \mathbf{1}'$.

8. Consider an m-state HMM with the basic dependence structure as depicted in Figure 2.2.

(a) Consider the vector $\boldsymbol{\alpha}_t = (\alpha_t(1), \ldots, \alpha_t(m))$ defined by

$$\alpha_t(j) = \Pr(\mathbf{X}^{(t)} = \mathbf{x}^{(t)}, C_t = j), \quad j = 1, 2, \ldots, m.$$

Use conditional probability and the conditional independence assumptions to show that

$$\alpha_t(j) = \sum_{i=1}^{m} \alpha_{t-1}(i) \gamma_{ij} p_j(x_t).$$

(b) Verify for yourself that the result from (a), written in matrix notation, yields the forward recursion

$$\boldsymbol{\alpha}_t = \boldsymbol{\alpha}_{t-1} \boldsymbol{\Gamma} \mathbf{P}(x_t), \quad t = 2, \ldots, T.$$

(c) Hence derive the matrix expression for the likelihood.

9. Write a function `pois-HMM.moments(m,lambda,gamma,lag.max=10)` that computes the expectation, variance and autocorrelation function (for lags 0 to `lag.max`) of an m-state stationary Poisson–HMM with t.p.m. `gamma` and state-dependent means `lambda`.

Hint: when finding the autocorrelation function, use the **R** package `expm` to compute the necessary powers of the t.p.m.

10. Write the three functions listed below, relating to the marginal distribution of an m-state Poisson–HMM with parameters `lambda`, `gamma`, and possibly `delta`. In each case, if `delta` is specified as NULL, the stationary distribution should be used. You can use your function `statdist` (see Exercise 9(b) of Chapter 1) to provide the stationary distribution.

```
dpois.HMM(x, m, lambda, gamma, delta=NULL)
ppois.HMM(x, m, lambda, gamma, delta=NULL)
qpois.HMM(p, m, lambda, gamma, delta=NULL)
```

The function dpois.HMM computes the probability function at the arguments specified by the vector x, ppois.HMM the distribution function, and qpois.HMM the inverse distribution function.

11. Consider the function pois.HMM.generate_sample in Section A.1.5 that generates observations from a stationary m-state Poisson–HMM. Test the function by generating a long sequence of observations (10 000, say), and then check whether the sample mean, variance, ACF and relative frequencies correspond to what you expect.

12. (Interval-censored observations.)

   (a) Suppose that, in a series of unbounded counts $x_1, \ldots, x_T$, only the observation x_t is interval-censored, and $a \leq x_t \leq b$, where b may be ∞. Prove the statement made in Section 2.3.5 that the likelihood of a Poisson–HMM with m states is obtained by replacing $\mathbf{P}(x_t)$ in the expression (2.12) by the $m \times m$ diagonal matrix of which the ith diagonal element is $\Pr(a \leq X_t \leq b \mid C_t = i)$.

   (b) Extend part (a) to allow for any number of interval-censored observations.

CHAPTER 3

Estimation by direct maximization of the likelihood

3.1 Introduction

We saw in equation (2.12) that the likelihood of an HMM is given by

$$L_T = \Pr\left(\mathbf{X}^{(T)} = \mathbf{x}^{(T)}\right) = \boldsymbol{\delta}\mathbf{P}(x_1)\boldsymbol{\Gamma}\mathbf{P}(x_2)\cdots\boldsymbol{\Gamma}\mathbf{P}(x_T)\mathbf{1}',$$

where $\boldsymbol{\delta}$ is the initial distribution (that of C_1) and $\mathbf{P}(x)$ the $m \times m$ diagonal matrix with ith diagonal element the state-dependent probability or density $p_i(x)$. In principle, we can therefore compute $L_T = \boldsymbol{\alpha}_T\mathbf{1}'$ recursively via

$$\boldsymbol{\alpha}_1 = \boldsymbol{\delta}\mathbf{P}(x_1)$$

and

$$\boldsymbol{\alpha}_t = \boldsymbol{\alpha}_{t-1}\boldsymbol{\Gamma}\mathbf{P}(x_t), \quad \text{for } t = 2, 3, \ldots, T.$$

If the Markov chain is assumed stationary (in which case $\boldsymbol{\delta} = \boldsymbol{\delta}\boldsymbol{\Gamma}$), we can choose to use instead

$$\boldsymbol{\alpha}_0 = \boldsymbol{\delta}$$

and

$$\boldsymbol{\alpha}_t = \boldsymbol{\alpha}_{t-1}\boldsymbol{\Gamma}\mathbf{P}(x_t), \quad \text{for } t = 1, 2, \ldots, T.$$

We shall first consider the stationary case.

The number of operations involved is of order Tm^2, making the evaluation of the likelihood quite feasible even for large T. Parameter estimation can therefore be performed by numerical maximization of the likelihood with respect to the parameters.

But there are several problems that need to be addressed when parameters are estimated in this way. The main problems are numerical underflow, constraints on the parameters, and multiple local maxima in the likelihood function. In this chapter we first discuss how to overcome these problems, in order to arrive at a general strategy for computing MLEs. Then we discuss the estimation of standard errors for parameters. We defer to the next chapter the EM algorithm, which necessitates some discussion of the forward and backward probabilities.

3.2 Scaling the likelihood computation

In the case of discrete state-dependent distributions, the elements of $\boldsymbol{\alpha}_t$, being made up of products of probabilities, become progressively smaller as t increases, and are eventually rounded to zero. In fact, with probability 1 the likelihood approaches 0 (or possibly ∞ in the continuous case) exponentially fast; see Leroux and Puterman (1992). The remedy is, however, the same for over- and underflow, and we confine our attention to underflow.

Since the likelihood is a product of matrices, not of scalars, it is not possible to circumvent numerical underflow simply by computing the log of the likelihood as the sum of logs of its factors. In this respect the computation of the likelihood of an independent mixture model is simpler than that of an HMM.

To solve the problem, Durbin *et al.* (1998, p. 78) suggest (*inter alia*) a method of computation that relies on the following approximation. Suppose we wish to compute $\log(p+q)$, where $p > q$. Write $\log(p+q)$ as

$$\log p + \log(1 + q/p) = \log p + \log(1 + \exp(\tilde{q} - \tilde{p})),$$

where $\tilde{p} = \log p$ and $\tilde{q} = \log q$. The function $\log(1 + e^x)$ is then approximated by interpolation from a table of its values; apparently quite a small table will give a reasonable degree of accuracy.

We prefer to compute the logarithm of L_T by using a strategy of scaling the vector of forward probabilities $\boldsymbol{\alpha}_t$. Effectively we scale the vector $\boldsymbol{\alpha}_t$ at each time t so that its elements add to 1, keeping track of the sum of the logs of the scale factors thus applied.

Define, for $t = 0, 1, \ldots, T$, the vector

$$\boldsymbol{\phi}_t = \boldsymbol{\alpha}_t / w_t,$$

where $w_t = \sum_i \alpha_t(i) = \boldsymbol{\alpha}_t \mathbf{1}'$. First, we note certain immediate consequences of the definitions of $\boldsymbol{\phi}_t$ and w_t:

$$
\begin{aligned}
w_0 = \boldsymbol{\alpha}_0 \mathbf{1}' &= \boldsymbol{\delta} \mathbf{1}' = 1; \\
\boldsymbol{\phi}_0 &= \boldsymbol{\delta}; \\
w_t \boldsymbol{\phi}_t &= w_{t-1} \boldsymbol{\phi}_{t-1} \boldsymbol{\Gamma} \mathbf{P}(x_t); \\
L_T = \boldsymbol{\alpha}_T \mathbf{1}' &= w_T (\boldsymbol{\phi}_T \mathbf{1}') = w_T.
\end{aligned}
\tag{3.1}
$$

Hence $L_T = w_T = \prod_{t=1}^{T}(w_t / w_{t-1})$. From (3.1) it follows that

$$w_t = w_{t-1}(\boldsymbol{\phi}_{t-1} \boldsymbol{\Gamma} \mathbf{P}(x_t) \mathbf{1}'),$$

and so we conclude that

$$\log L_T = \sum_{t=1}^{T} \log(w_t / w_{t-1}) = \sum_{t=1}^{T} \log(\boldsymbol{\phi}_{t-1} \boldsymbol{\Gamma} \mathbf{P}(x_t) \mathbf{1}').$$

The computation of the log-likelihood is summarized below in the form of an algorithm. Note that $\boldsymbol{\Gamma}$ and $\mathbf{P}(x_t)$ are $m \times m$ matrices, $\mathbf{v}$ and $\boldsymbol{\phi}_t$ are vectors of length m, u is a scalar, and l is the scalar in which the log-likelihood is accumulated.

$$\boldsymbol{\phi}_0 \leftarrow \boldsymbol{\delta}; \; l \leftarrow 0$$
$$\text{for } t = 1, 2, \ldots, T$$
$$\mathbf{v} \leftarrow \boldsymbol{\phi}_{t-1}\boldsymbol{\Gamma}\mathbf{P}(x_t)$$
$$u \leftarrow \mathbf{v}\mathbf{1}'$$
$$l \leftarrow l + \log u$$
$$\boldsymbol{\phi}_t \leftarrow \mathbf{v}/u$$
$$\text{return } l$$

The required log-likelihood, $\log L_T$, is then given by the final value of l. This procedure will almost always prevent underflow. Clearly, minor variations of the technique are possible: the scale factor w_t could be chosen instead to be the largest element of the vector being scaled, or the mean of its elements (as opposed to the sum).

The algorithm is easily modified to compute the log-likelihood without assuming stationarity of the Markov chain. With $\boldsymbol{\delta}$ denoting the initial distribution, the more general algorithm is

$$w_1 \leftarrow \boldsymbol{\delta}\mathbf{P}(x_1)\mathbf{1}'; \; \boldsymbol{\phi}_1 \leftarrow \boldsymbol{\delta}\mathbf{P}(x_1)/w_1; \; l \leftarrow \log w_1$$
$$\text{for } t = 2, 3, \ldots, T$$
$$\mathbf{v} \leftarrow \boldsymbol{\phi}_{t-1}\boldsymbol{\Gamma}\mathbf{P}(x_t)$$
$$u \leftarrow \mathbf{v}\mathbf{1}'$$
$$l \leftarrow l + \log u$$
$$\boldsymbol{\phi}_t \leftarrow \mathbf{v}/u$$
$$\text{return } l$$

If the initial distribution happens to be the stationary distribution, the more general algorithm still applies.

The following code implements this last version of the algorithm in order to compute the log-likelihood of observations $x_1, \ldots, x_T$ under a Poisson–HMM with at least two states, transition probability matrix $\boldsymbol{\Gamma}$, vector of state-dependent means $\boldsymbol{\lambda}$, and initial distribution $\boldsymbol{\delta}$.

```
alpha        <- delta*dpois(x[1],lambda)
lscale       <- log(sum(alpha))
alpha        <- alpha/sum(alpha)
for (i in 2:T) {
  alpha        <- alpha %*% Gamma*dpois(x[i],lambda)
  lscale       <- lscale+log(sum(alpha))
  alpha        <- alpha/sum(alpha)
}
lscale
```

This code improves on that shown on p. 39 in that the vector of forward probabilities is scaled to have sum 1 at all times. But it is probably unnecessary to scale the forward probabilities at time 1, and if one omits that part of the scaling, the algorithm and code simplify slightly.

3.3 Maximization of the likelihood subject to constraints

3.3.1 Reparametrization to avoid constraints

The elements of Γ and those of λ, the vector of state-dependent means in a Poisson–HMM, are subject to non-negativity and other constraints. In particular, the row sums of Γ equal 1. Estimates of parameters should also satisfy such constraints. Thus, when maximizing the likelihood we need to solve a constrained optimization problem, not an unconstrained one.

Special-purpose software, such as NPSOL (Gill *et al.*, 1986) or the corresponding NAG routine E04UCF, can be used to maximize a function of several variables which are subject to constraints. The advice of Gill, Murray and Wright (1981, p. 267) is that it is 'rarely appropriate to alter linearly constrained problems'. However, depending on the implementation and the nature of the data, constrained optimization can be slow. For example, the constrained optimizer `constrOptim` available in **R** is acknowledged to be slow if the optimum lies on the boundary of the parameter space. We shall focus on the use of the unconstrained optimizer `nlm`. Exercise 3 explores the use of `constrOptim`, which can minimize a function subject to linear inequality constraints.

In general, there are two groups of constraints: those that apply to the parameters of the state-dependent distributions and those that apply to the parameters of the Markov chain. The first group of constraints depends on which state-dependent distribution(s) are chosen; for example, the 'success probability' of a binomial distribution lies between 0 and 1.

In the case of a Poisson–HMM the relevant constraints are:

- the means λ_i of the state-dependent distributions must, for $i = 1, \ldots, m$, be non-negative;

- the rows of the transition probability matrix Γ must add to 1, and all the parameters γ_{ij} must be non-negative.

Here the constraints can be imposed by making transformations. The transformation of the parameters λ_i is easy. Define $\eta_i = \log \lambda_i$, for $i = 1, \ldots, m$. Then $\eta_i \in \mathbb{R}$. After we have maximized the likelihood with respect to the unconstrained parameters, the constrained parameter estimates can be obtained by transforming back: $\widehat{\lambda}_i = \exp \widehat{\eta}_i$.

The reparametrization of the matrix Γ requires more work, but can be accomplished quite elegantly. Note that Γ has m^2 entries but only

$m(m-1)$ free parameters, as there are m row-sum constraints

$$\sum_{j=1}^{m} \gamma_{ij} = 1 \quad (i = 1, \ldots, m).$$

We shall show one possible transformation between the m^2 constrained probabilities γ_{ij} and $m(m-1)$ unconstrained real numbers $\tau_{ij}, i \neq j$.

For the sake of readability we display the case $m = 3$. We begin by defining the matrix

$$\mathbf{T} = \begin{pmatrix} - & \tau_{12} & \tau_{13} \\ \tau_{21} & - & \tau_{23} \\ \tau_{31} & \tau_{32} & - \end{pmatrix},$$

a matrix with $m(m-1)$ entries $\tau_{ij} \in \mathbb{R}$. Now let $g : \mathbb{R} \to \mathbb{R}^+$ be a strictly increasing function, for example,

$$g(x) = e^x \quad \text{or} \quad g(x) = \begin{cases} e^x & x \leq 0 \\ x+1 & x \geq 0. \end{cases}$$

Define

$$\nu_{ij} = \begin{cases} g(\tau_{ij}) & \text{for } i \neq j \\ 1 & \text{for } i = j. \end{cases}$$

We then set $\gamma_{ij} = \nu_{ij}/\sum_{k=1}^{m} \nu_{ik}$ (for $i, j = 1, 2, \ldots, m$) and $\boldsymbol{\Gamma} = (\gamma_{ij})$. It is left to the reader as an exercise to verify that the resulting matrix $\boldsymbol{\Gamma}$ satisfies the constraints of a t.p.m. We shall refer to the parameters η_i and τ_{ij} as **working parameters**, and to the parameters λ_i and γ_{ij} as **natural parameters**.

Using the above transformations of $\boldsymbol{\Gamma}$ and $\boldsymbol{\lambda}$, we can perform the calculation of the likelihood-maximizing parameters in two steps.

1. Maximize L_T with respect to the working parameters $\mathbf{T} = \{\tau_{ij}\}$ and $\boldsymbol{\eta} = (\eta_1, \ldots, \eta_m)$. These are all unconstrained.

2. Transform the estimates of the working parameters to estimates of the natural parameters:

$$\widehat{\mathbf{T}} \to \widehat{\boldsymbol{\Gamma}}, \quad \widehat{\boldsymbol{\eta}} \to \widehat{\boldsymbol{\lambda}}.$$

Consider $\boldsymbol{\Gamma}$ for the case $g(x) = e^x$ and general m. Here we have

$$\gamma_{ij} = \frac{\exp(\tau_{ij})}{1 + \sum_{k \neq i} \exp(\tau_{ik})}, \quad \text{for } i \neq j,$$

and the diagonal elements of $\boldsymbol{\Gamma}$ follow from the row sums of 1. The transformation in the opposite direction is

$$\tau_{ij} = \log \left(\frac{\gamma_{ij}}{1 - \sum_{k \neq i} \gamma_{ik}} \right) = \log \left(\gamma_{ij}/\gamma_{ii} \right), \quad \text{for } i \neq j.$$

This generalization of the logit and inverse logit transforms has long been used in the context of compositional data; see Aitchison (1982), where several other transforms are described as well.

We now display some relatively simple code that will transform natural parameters to working and vice versa. The code refers to a Poisson–HMM with $m \geq 2$ states, in which the Markov chain may, if appropriate, be assumed stationary. In that case the stationary distribution $\boldsymbol{\delta}$ is not supplied, but is computed when needed from the t.p.m. $\boldsymbol{\Gamma}$ by solving $\boldsymbol{\delta}(\mathbf{I}_m - \boldsymbol{\Gamma} + \mathbf{U}) = \mathbf{1}$; see p. 18 and Exercise 9(b) of Chapter 1. Otherwise $\boldsymbol{\delta}$ is treated as a (natural) parameter and transformed in order to remove the constraints $\delta_i \geq 0$ and $\sum_i \delta_i = 1$ (although there is a simpler route; see Section 4.2.4).

```
# Transform Poisson natural parameters to working parameters
pois.HMM.pn2pw <- function(m,lambda,gamma,delta=NULL,stationary=TRUE)
{
 tlambda <- log(lambda)
 foo     <- log(gamma/diag(gamma))
 tgamma  <- as.vector(foo[!diag(m)])
 if(stationary) {tdelta <-NULL} else {tdelta<-log(delta[-1]/delta[1])}
 parvect <- c(tlambda,tgamma,tdelta)
 return(parvect)
}
```

```
# Transform Poisson working parameters to natural parameters
pois.HMM.pw2pn <- function(m,parvect,stationary=TRUE)
{
 lambda        <- exp(parvect[1:m])
 gamma         <- diag(m)
 gamma[!gamma] <- exp(parvect[(m+1):(m*m)])
 gamma         <- gamma/apply(gamma,1,sum)
 if(stationary) {delta<-solve(t(diag(m)-gamma+1),rep(1,m))} else
                {foo<-c(1,exp(parvect[(m*m+1):(m*m+m-1)]))
                 delta<-foo/sum(foo)}
 return(list(lambda=lambda,gamma=gamma,delta=delta))
}
```

For code which includes and uses these functions, and for some discussion thereof, see Sections A.1.1–A.1.4 and A.2.1.

3.3.2 Embedding in a continuous-time Markov chain

A different reparametrization is discussed by Zucchini and MacDonald (1998). In a continuous-time Markov chain on a finite state space, the transition probability matrix $\mathbf{P}_t$ over t time units is given by $\mathbf{P}_t = \exp(t\mathbf{Q})$, where $\mathbf{Q}$ is the matrix of transition intensities. The row sums of $\mathbf{Q}$ are 0, but the only constraint on the off-diagonal elements of $\mathbf{Q}$ is that they be non-negative. It is not in general the case that a discrete-time Markov chain is embeddable in a continuous-time Markov chain; see Exercise 11. But if one is prepared to assume that the discrete-time

Markov chain of interest is thus embeddable, the one-step transition probabilities of the discrete-time chain can then be parametrized via $\boldsymbol{\Gamma} = \exp(\mathbf{Q})$. This is effectively what one is doing if one uses the $\mathbf{R}$ package msm (Jackson *et al.*, 2003) to fit HMMs.

3.4 Other problems

3.4.1 *Multiple maxima in the likelihood*

The likelihood of an HMM is a complicated function of the parameters and frequently has several local maxima. The goal of course is to find the global maximum, but there is no simple method of determining in general whether a numerical maximization algorithm has reached the global maximum. Depending on the starting values, it can easily happen that the algorithm identifies a local, but not the global, maximum. This applies also to the main alternative method of estimation, the EM algorithm, which is discussed in Chapter 4. A sensible strategy is therefore to use a range of starting values for the maximization, and to see whether the same maximum is identified in each case.

3.4.2 *Starting values for the iterations*

It is often easy to find plausible starting values for some of the parameters of an HMM: for instance, if one seeks to fit a Poisson–HMM with two states, and the sample mean is 10, one could try 8 and 12, or 5 and 15, for the values of the two state-dependent means. More systematic strategies based on the quantiles of the observations are possible, however. For example, if the model has three states, use as the starting values of the state-dependent means the lower quartile, median and upper quartile of the observed counts.

It is less easy to guess values of the transition probabilities γ_{ij}. One strategy is to assign a common starting value (e.g. 0.01 or 0.05) to all the off-diagonal transition probabilities. A consequence of such a choice, perhaps convenient, is that the corresponding stationary distribution is uniform over the states; this follows by symmetry. Choosing good starting values for parameters tends to steer one away from numerical instability.

3.4.3 *Unbounded likelihood*

In the case of HMMs with continuous state-dependent distributions, just as in the case of independent mixtures (see Section 1.2.3), it may happen that the likelihood is unbounded in the vicinity of certain parameter combinations. As before, we suggest that, if this creates difficulties, one

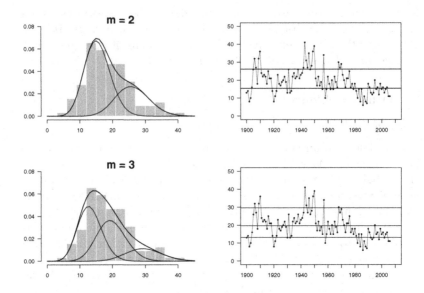

Figure 3.1 *Earthquakes series. Left: marginal distributions of Poisson–HMMs with two and three states, and their components, compared with a histogram of the observations. Right: the state-dependent means (horizontal lines) compared to the observations.*

maximizes the discrete likelihood instead of the joint density. This has the advantage in any case that it applies more generally to interval-censored data. Applications of this kind are described in Sections 17.4 and 17.5.

3.5 Example: earthquakes

Figure 3.1 shows the result of fitting (stationary) Poisson–hidden Markov models with two and three states to the earthquakes series by means of the unconstrained optimizer nlm. The relevant code (with starting values) appears in Section A.2.1. The two-state model is

$$\Gamma = \begin{pmatrix} 0.9340 & 0.0660 \\ 0.1285 & 0.8715 \end{pmatrix},$$

with $\delta = (0.6608, 0.3392)$, $\lambda = (15.472, 26.125)$, and log-likelihood given by $l = -342.3183$. It is clear that the fitted (Markov-dependent) mixture of two Poisson distributions provides a much better fit to the marginal distribution of the observations than does a single Poisson distribution, but the fit can be further improved by using a mixture of three or four Poisson distributions.

The three-state model is

$$\boldsymbol{\Gamma} = \begin{pmatrix} 0.955 & 0.024 & 0.021 \\ 0.050 & 0.899 & 0.051 \\ 0.000 & 0.197 & 0.803 \end{pmatrix},$$

with $\boldsymbol{\delta} = (0.4436, 0.4045, 0.1519)$, $\boldsymbol{\lambda} = (13.146, 19.721, 29.714)$ and $l = -329.4603$. The four-state is

$$\boldsymbol{\Gamma} = \begin{pmatrix} 0.805 & 0.102 & 0.093 & 0.000 \\ 0.000 & 0.976 & 0.000 & 0.024 \\ 0.050 & 0.000 & 0.902 & 0.048 \\ 0.000 & 0.000 & 0.188 & 0.812 \end{pmatrix},$$

with $\boldsymbol{\delta} = (0.0936, 0.3983, 0.3643, 0.1439)$, $\boldsymbol{\lambda} = (11.283, 13.853, 19.695, 29.700)$, and $l = -327.8316$.

The means and variances of the marginal distributions of the four models compare as follows with those of the observations. By a one-state Poisson–HMM we mean a model that assumes that the observations are realizations of independent Poisson random variables with common mean.

	mean	variance
observations:	19.364	51.573
'one-state HMM':	19.364	19.364
two-state HMM:	19.086	44.523
three-state HMM:	18.322	50.709
four-state HMM:	18.021	49.837

As regards the autocorrelation functions of the models, that is, $\rho(k) = \mathrm{Corr}(X_{t+k}, X_t)$, we have the following results, valid for all $k \in \mathbb{N}$, based on the conclusions of Exercise 4 of Chapter 2:

- two states, $\rho(k) = 0.5713 \times 0.8055^k$;

- three states, $\rho(k) = 0.4447 \times 0.9141^k + 0.1940 \times 0.7433^k$;

- four states, $\rho(k) = 0.2332 \times 0.9519^k + 0.3682 \times 0.8174^k + 0.0369 \times 0.7252^k$.

In all these cases the ACF is just a linear combination of the kth powers of the eigenvalues other than 1 of the transition probability matrix.

For model selection, for example, choosing between competing models such as HMMs and independent mixtures, or choosing the number of components in either, see Section 6.1.

A phenomenon that is noticeable when one fits models with three or more states to relatively short series is that the estimates of one or more of the transition probabilities turn out to be very close to zero; see the three-state model above (one such probability, γ_{13}) and the four-state model (six of the 12 off-diagonal transition probabilities).

This phenomenon can be explained as follows. In a stationary Markov chain, the expected number of transitions from state i to state j in a series of T observations is $(T-1)\delta_i\gamma_{ij}$. For $\delta_3 = 0.152$ and $T = 107$ (as in our three-state model), this expectation will be less than 1 if $\gamma_{31} < 0.062$. In such a series, therefore, it is likely that if γ_{31} is fairly small there will be no transitions from state 3 to state 1, and so when we seek to estimate γ_{31} in an HMM the estimate is likely to be effectively zero. As m increases, the probabilities δ_i and γ_{ij} get smaller on average; this makes it increasingly likely that at least one estimated transition probability is effectively zero.

3.6 Standard errors and confidence intervals

Relatively little is known about the properties of the maximum likelihood estimators of HMMs; only asymptotic results are available. To exploit these results one requires estimates of the variance–covariance matrix of the estimators of the parameters. One can estimate the standard errors from the Hessian of the log-likelihood at the maximum, but this approach runs into difficulties when some of the parameters are on the boundary of their parameter space, which occurs quite often when HMMs are fitted. An alternative here is the parametric bootstrap, for which see Section 3.6.2. The algorithm is easy to code (see Section A.1.5), but the computations are time-consuming.

3.6.1 Standard errors via the Hessian

Although the point estimates $\widehat{\boldsymbol{\Theta}} = (\widehat{\boldsymbol{\Gamma}}, \widehat{\boldsymbol{\lambda}})$ are easy to compute, exact interval estimates are not available. Cappé *et al.* (2005, Chapter 12) show that, under certain regularity conditions, the MLEs of HMM parameters are consistent, asymptotically normal and efficient. Thus, if we can estimate the standard errors of the MLEs, then, using the asymptotic normality, we can also compute approximate confidence intervals. However, as pointed out by Frühwirth-Schnatter (2006, p. 53) in the context of independent mixture models, 'The regularity conditions are often violated, including cases of great practical concern, among them small data sets, mixtures with small component weights, and overfitting mixtures with too many components.' Furthermore, McLachlan and Peel (2000, p. 68) warn: 'In particular for mixture models, it is well known that the sample size n has to be very large before the asymptotic theory of maximum likelihood applies.'

With the above caveats in mind we can, in order to estimate the standard errors of the MLEs of an HMM, use the approximate Hessian of minus the log-likelihood at the minimum (e.g. as supplied by `nlm`). We can invert it and so estimate the asymptotic variance–covariance matrix of the estimators of the parameters. A problem with this suggestion is that, if the parameters have been transformed, the Hessian available will be that which refers to the working parameters ϕ_i, not the original, more readily interpretable, natural parameters θ_i ($\boldsymbol{\Gamma}$ and $\boldsymbol{\lambda}$ in the case of a Poisson–HMM).

The situation is therefore that we have, at the minimum of $-l$, the Hessian with respect to the working parameters,

$$\mathbf{H} = -\left(\frac{\partial^2 l}{\partial \phi_i \partial \phi_j}\right),$$

but what we really need is the Hessian with respect to the natural parameters,

$$\mathbf{G} = -\left(\frac{\partial^2 l}{\partial \theta_i \partial \theta_j}\right).$$

There is, however, the following relationship between the two Hessians at the minimum:

$$\mathbf{H} = \mathbf{MGM'} \quad \text{and} \quad \mathbf{G}^{-1} = \mathbf{M'H^{-1}M}, \tag{3.2}$$

where $\mathbf{M}$ is defined by $m_{ij} = \partial\theta_j / \partial\phi_i$. See also Monahan (2011, p. 247) for this relation between the Hessians. (Note that all the derivatives appearing here are as evaluated at the minimum.) In the case of a Poisson–HMM, the elements of $\mathbf{M}$ are quite simple; see Exercise 7 for details.

With $\mathbf{M}$ at our disposal, we can use (3.2) to deduce $\mathbf{G}^{-1}$ from $\mathbf{H}^{-1}$, and use $\mathbf{G}^{-1}$ to find standard errors for the natural parameters, provided such parameters are not on the boundary of the parameter space. An alternative route to the standard errors with respect to the natural parameters which often works well, and is less laborious, is this. First find the MLE by solving the constrained optimization problem, then rerun the optimization without constraints, starting at or very close to the MLE. If the resulting estimate is the same as the MLE already found, the corresponding Hessian then directly supplies the standard errors with respect to the natural parameters. But if one is to make a normality assumption and base a confidence interval on it, such a normality assumption is more likely, but not guaranteed, to be reasonable on the working-parameter scale than on the (constrained) natural-parameter scale.

Furthermore, it is true in many applications that some of the estimated (natural) parameters lie on or very close to the boundary; this limits the usefulness of the above results. As already pointed out on p. 55, for series of moderate length the estimates of some transition probabilities are expected to be close to zero. This is true of $\widehat{\gamma}_{13}$ in the three-state model for the earthquakes series. An additional example of this type can be found in Section 17.3.2. In Section 19.2.1, several of the estimates of the parameters in the state-dependent distributions are practically zero, their lower bound; see Table 19.1. The same phenomenon is apparent in Section 23.9.2; see Table 23.1.

Recursive computation of the Hessian

An alternative method of computing the Hessian is that of Lystig and Hughes (2002). They present the forward algorithm $\alpha_t = \alpha_{t-1}\boldsymbol{\Gamma}\mathbf{P}(x_t)$ in a form which incorporates automatic or 'natural' scaling, and then extend that approach in order to compute (in a single pass, along with the log-likelihood) its Hessian and gradient with respect to the natural parameters, those we have denoted above by θ_i. Turner (2008) has used this approach in order to find the analytical derivatives needed to maximize HMM likelihoods directly by the Levenberg–Marquardt algorithm.

Although this may be a more efficient and more accurate method of computing the Hessian than the use of (3.2), it does not solve the fundamental problem that the use of the Hessian to compute standard errors (and thence confidence intervals) is unreliable if some of the parameters are on or near the boundary of their parameter space.

3.6.2 Bootstrap standard errors and confidence intervals

As an alternative to the techniques described in Section 3.6.1 one may use the **parametric bootstrap** (Efron and Tibshirani, 1993). Roughly speaking, the idea of the parametric bootstrap is to assess the properties of the model with parameters $\boldsymbol{\Theta}$ by using those of the model with parameters $\widehat{\boldsymbol{\Theta}}$. The following steps are performed to estimate the variance–covariance matrix of $\widehat{\boldsymbol{\Theta}}$.

1. Fit the model, i.e. compute $\widehat{\boldsymbol{\Theta}}$.
2. (a) Generate a sample, called a bootstrap sample, of observations from the fitted model, i.e. the model with parameters $\widehat{\boldsymbol{\Theta}}$. The length should be the same as the original number of observations.
   (b) Estimate the parameters $\boldsymbol{\Theta}$ by $\widehat{\boldsymbol{\Theta}}^*$ for the bootstrap sample.
   (c) Repeat steps (a) and (b) B times (with B 'large') and record the values $\widehat{\boldsymbol{\Theta}}^*$.

The variance–covariance matrix of $\widehat{\boldsymbol{\Theta}}$ is then estimated by the sample variance–covariance matrix of the bootstrap estimates $\widehat{\boldsymbol{\Theta}}^*(b)$, $b = 1, 2, \ldots, B$:

$$\widehat{\text{Var-Cov}}(\widehat{\boldsymbol{\Theta}}) = \frac{1}{B-1} \sum_{b=1}^{B} \left(\widehat{\boldsymbol{\Theta}}^*(b) - \widehat{\boldsymbol{\Theta}}^*(\cdot)\right)' \left(\widehat{\boldsymbol{\Theta}}^*(b) - \widehat{\boldsymbol{\Theta}}^*(\cdot)\right),$$

where $\widehat{\boldsymbol{\Theta}}^*(\cdot) = B^{-1} \sum_{b=1}^{B} \widehat{\boldsymbol{\Theta}}^*(b)$.

EXAMPLE: PARAMETRIC BOOTSTRAP 59

The parametric bootstrap requires code to generate realizations from a fitted model; for a Poisson–HMM this is given in Section A.1.5. Since code to fit models is available, that same code can be used to fit models to the bootstrap sample.

The bootstrap method can be used to estimate confidence intervals directly. In the example given in the next section we use the well-known 'percentile method' (Efron and Tibshirani, 1993); other options are available.

3.7 Example: the parametric bootstrap applied to the three-state model for the earthquakes data

Table 3.1 *Earthquakes data: bootstrap confidence intervals for the parameters of the three-state HMM.*

Parameter	MLE	90% conf. limits	
λ_1	13.146	11.463	14.253
λ_2	19.721	13.708	21.142
λ_3	29.714	20.929	33.160
γ_{11}	0.954	0.750	0.988
γ_{12}	0.024	0.000	0.195
γ_{13}	0.021	0.000	0.145
γ_{21}	0.050	0.000	0.179
γ_{22}	0.899	0.646	0.974
γ_{23}	0.051	0.000	0.228
γ_{31}	0.000	0.000	0.101
γ_{32}	0.197	0.000	0.513
γ_{33}	0.803	0.481	0.947
δ_1	0.444	0.109	0.716
δ_2	0.405	0.139	0.685
δ_3	0.152	0.042	0.393

A bootstrap sample of size 500 was generated from the three-state model for the earthquakes data, which appears on p. 55. In fitting models to the bootstrap samples, we noticed that, in two cases out of the 500, the starting values which we were in general using caused numerical instability or convergence problems. By choosing better starting values for these two cases we were able to fit models successfully and complete the exercise. The resulting sample of parameter values then produced the 90% confidence intervals for the parameters that are displayed in Table 3.1, and the estimated parameter correlations that are displayed in Table 3.2. What is noticeable is that the intervals for the state-dependent

Table 3.2 *Earthquakes data: bootstrap estimates of the correlations of the estimators of λ_i, for $i = 1, 2, 3$.*

	λ_1	λ_2	λ_3
λ_1	1.000	0.483	0.270
λ_2		1.000	0.688
λ_3			1.000

means λ_i overlap, the intervals for the stationary probabilities δ_i are very wide, and the estimators $\widehat{\lambda}_i$ are quite strongly correlated.

These results, in particular the correlations shown in Table 3.2, should make one wary of over-interpreting a model with nine parameters based on only 107 (dependent) observations. In particular, they suggest that the states are not well defined, and one should be cautious of attaching a substantive interpretation to them.

Exercises

1. Consider the following parametrization of the t.p.m. of an m-state Markov chain. Let $\tau_{ij} \in \mathbb{R}$ $(i, j = 1, 2, \ldots, m;\ i \neq j)$ be $m(m - 1)$ arbitrary real numbers. Let $g : \mathbb{R} \to \mathbb{R}^+$ be some strictly increasing function, e.g. $g(x) = e^x$. Define ν_{ij} and γ_{ij} as on p. 51.

   (a) Show that the matrix $\boldsymbol{\Gamma}$ with entries γ_{ij} that are constructed in this way is a t.p.m., i.e. show that $0 \leq \gamma_{ij} \leq 1$ for all i and j, and that the row sums of $\boldsymbol{\Gamma}$ are equal to 1.

   (b) Given an $m \times m$ t.p.m. $\boldsymbol{\Gamma} = (\gamma_{ij})$, derive an expression for the parameters τ_{ij}, for $i, j = 1, 2, \ldots, m;\ i \neq j$.

2. The purpose of this exercise is to investigate the numerical behaviour of an 'unscaled' evaluation of the likelihood of an HMM, and to compare this with the behaviour of an alternative algorithm that applies scaling.

   Consider the stationary two-state Poisson–HMM with parameters

   $$\boldsymbol{\Gamma} = \begin{pmatrix} 0.9 & 0.1 \\ 0.2 & 0.8 \end{pmatrix}, \quad (\lambda_1, \lambda_2) = (1, 5).$$

   Compute the likelihood, L_{10}, of the following sequence of ten observations in two ways: 2, 8, 6, 3, 6, 1, 0, 0, 4, 7.

   (a) Use the unscaled method $L_{10} = \boldsymbol{\alpha}_{10}\mathbf{1}'$, where $\boldsymbol{\alpha}_0 = \boldsymbol{\delta}$ and $\boldsymbol{\alpha}_t =$

$$\boldsymbol{\alpha}_{t-1}\mathbf{B}_t;$$

$$\mathbf{B}_t = \boldsymbol{\Gamma}\begin{pmatrix} p_1(x_t) & 0 \\ 0 & p_2(x_t) \end{pmatrix};$$

and

$$p_i(x_t) = \lambda_i^{x_t} e^{-\lambda_i}/x_t!, \quad i = 1,2; \; t = 1,2,\ldots,10.$$

Examine the numerical values of the vectors $\boldsymbol{\alpha}_0, \boldsymbol{\alpha}_1, \ldots, \boldsymbol{\alpha}_{10}$.

(b) Use the first algorithm given in Section 3.2 to compute $\log L_{10}$. Examine the numerical values of the vectors $\boldsymbol{\phi}_0, \boldsymbol{\phi}_1, \ldots, \boldsymbol{\phi}_{10}$. (It is easiest to store these vectors as rows in an 11×2 matrix.)

3. Use the **R** function `constrOptim` to fit HMMs with two to four states to the earthquakes data, and compare your models with those given in Section 3.5.

4. Another approach to the non-negativity and row-sum constraints on $\boldsymbol{\Gamma}$ is to convert them into 'box constraints', i.e. constraints of the form $a \le \theta_i \le b$. A box-constrained optimizer, such as `optim` in **R** with method `L-BFGS-B`, can then be used.

Consider therefore the following transformation:

$$\begin{aligned} w_1 &= \sin^2 \theta_1, \\ w_i &= \left(\textstyle\prod_{j=1}^{i-1} \cos^2 \theta_j\right) \sin^2 \theta_i, \quad i = 2,\ldots,m-1, \\ w_m &= \textstyle\prod_{i=1}^{m-1} \cos^2 \theta_i. \end{aligned}$$

Show how this transformation can be used to convert the constraints

$$\sum_{i=1}^{m} w_i = 1, \quad w_i \ge 0; \; i = 1,\ldots,m,$$

into box constraints.

5. (a) Consider a stationary Markov chain, with t.p.m. $\boldsymbol{\Gamma}$ and stationary distribution $\boldsymbol{\delta}$. Show that the expected number of transitions from state i to state j in a series of T observations (i.e. in $T-1$ transitions) is $(T-1)\delta_i\gamma_{ij}$.

Hint: this expectation is $\sum_{t=2}^{T} \Pr(X_{t-1} = i, X_t = j)$.

(b) Show that, for $\delta_3 = 0.152$ and $T = 107$, this expectation is less than 1 if $\gamma_{31} < 0.062$.

6. Prove the relation (3.2) between the Hessian $\mathbf{H}$ of $-l$ with respect to the working parameters and the Hessian $\mathbf{G}$ of $-l$ with respect to the natural parameters, both being evaluated at the minimum of $-l$.

7. (See Section 3.6.1.) Consider an m-state Poisson–HMM, with natural parameters γ_{ij} and λ_i, and working parameters τ_{ij} and η_i defined as in Section 3.3.1, with $g(x) = e^x$.

(a) Show that

$$\begin{aligned}
\partial \gamma_{ij}/\partial \tau_{ij} &= \gamma_{ij}(1 - \gamma_{ij}), &&\text{for all } i, j; \\
\partial \gamma_{ij}/\partial \tau_{il} &= -\gamma_{ij}\gamma_{il}, &&\text{for } j \neq l; \\
\partial \gamma_{ij}/\partial \tau_{kl} &= 0, &&\text{for } i \neq k; \\
\partial \lambda_i/\partial \eta_i &= e^{\eta_i} = \lambda_i, &&\text{for all } i.
\end{aligned}$$

(b) Hence find the matrix $\mathbf{M}$ in this case.

8. Modify the **R** code in Sections A.1.1–A.1.4 in order to fit a Poisson–HMM to interval-censored observations. (Assume that the observations are available as a $T \times 2$ matrix of which the first column contains the lower bound of the observation and the second the upper bound, possibly **Inf**.)

9. Verify the autocorrelation functions given on p. 55 for the two-, three-and four-state models for the earthquakes data. (Hint: use the **R** function **eigen** to find the eigenvalues and eigenvectors of the relevant transition probability matrices.)

10. Consider again the soap sales series introduced in Exercise 6 of Chapter 1.

   (a) Fit stationary Poisson–HMMs with two, three and four states to these data.

   (b) Find the marginal means and variances, and the ACFs, of these models, and compare them with their sample equivalents.

11. (Embeddability of discrete-time Markov chain in continuous-time.) It is not always possible to embed a discrete-time Markov chain uniquely in a continuous-time chain. That is, given a t.p.m. $\mathbf{\Gamma}$, there does not always exist a unique generator matrix $\mathbf{Q}$ such that $\mathbf{\Gamma} = \exp(\mathbf{Q})$. The following examples show that there may, even in simple cases, be more than one corresponding generator matrix, or there may be none.

   (a) (Example taken from Israel, Rosenthal and Wei (2001, p. 256).) Consider the matrices $\mathbf{\Gamma}$, $\mathbf{Q}_1$ and $\mathbf{Q}_2$ given by

$$\mathbf{\Gamma} = \frac{1}{5}\begin{pmatrix} 2 & 2 & 1 \\ 2 & 2 & 1 \\ 2 & 2 & 1 \end{pmatrix} - \frac{e^{-4\pi}}{5}\begin{pmatrix} -3 & 2 & 1 \\ 2 & -3 & 1 \\ 2 & 2 & -4 \end{pmatrix},$$

$$\mathbf{Q}_1 = 2\pi \begin{pmatrix} -1 & 1 & 0 \\ 0 & -1 & 1 \\ 2 & 0 & -2 \end{pmatrix}, \quad \mathbf{Q}_2 = \frac{4\pi}{5}\begin{pmatrix} -3 & 2 & 1 \\ 2 & -3 & 1 \\ 2 & 2 & -4 \end{pmatrix}.$$

   Verify that $\exp(\mathbf{Q}_1) = \exp(\mathbf{Q}_2) = \mathbf{\Gamma}$.

(b) Theorem 3.1 of Israel *et al.* (2001, p. 249) states the following.

Let **P** *be a transition* [*probability*] *matrix, and suppose that*

i. $\det(\mathbf{P}) \leq 0$, *or*

ii. $\det(\mathbf{P}) > \prod_i p_{ii}$, *or*

iii. *there are states i and j such that j is accessible from i, but*
$p_{ij} = 0$.

Then there does not exist an exact generator for **P**.

Use this theorem to conclude that there is no corresponding generator matrix for the following t.p.m.s:

$$\mathbf{\Gamma} = \begin{pmatrix} 0.4 & 0.6 \\ 0.5 & 0.5 \end{pmatrix}, \qquad \mathbf{\Gamma} = \begin{pmatrix} 0.9 & 0.1 & 0 \\ 0.1 & 0.8 & 0.1 \\ 0.1 & 0.1 & 0.8 \end{pmatrix}.$$

## CHAPTER 4

Estimation by the EM algorithm

I know many statisticians are deeply in love with the EM algorithm.

Speed (2008)

A commonly used method of finding maximum-likelihood estimates of HMMs is the EM algorithm, which we shall describe in Section 4.2, the crux of this chapter. The tools we need to do so are the forward and the backward probabilities, which are also used for decoding and state prediction in Chapter 5. In establishing some useful propositions concerning the forward and backward probabilities we invoke several properties of HMMs, the proofs of which are given in Appendix B.

In the context of HMMs the EM algorithm is known as the Baum–Welch algorithm. The Baum–Welch algorithm is designed to estimate the parameters of an HMM whose Markov chain is homogeneous but not assumed stationary, that is, it is not assumed that $\delta\Gamma = \delta$. Indeed, the method has to be modified if this assumption is made; see Section 4.2.5. If the stationarity assumption is not made, the initial distribution δ is also estimated, in addition to the parameters of the state-dependent distributions and the t.p.m. Γ.

4.1 Forward and backward probabilities

In Section 2.3.2 we defined the (row) vector α_t, for $t = 1, 2, \ldots, T$, as follows:

$$\alpha_t = \delta\mathbf{P}(x_1)\Gamma\mathbf{P}(x_2)\cdots\Gamma\mathbf{P}(x_t) = \delta\mathbf{P}(x_1)\prod_{s=2}^{t}\Gamma\mathbf{P}(x_s), \qquad (4.1)$$

with δ denoting the initial distribution of the Markov chain. We have referred to the elements of α_t as **forward probabilities**, but we have not yet justified their description as probabilities. One of the purposes of this section is to show that $\alpha_t(j)$, the jth component of α_t, is indeed a probability, the joint probability $\Pr(X_1 = x_1, X_2 = x_2, \ldots, X_t = x_t, C_t = j)$.

We shall also need the vector of **backward probabilities** β_t which, for $t = 1, 2, \ldots, T$, is defined by

$$\boldsymbol{\beta}'_t = \boldsymbol{\Gamma}\mathbf{P}(x_{t+1})\boldsymbol{\Gamma}\mathbf{P}(x_{t+2}) \cdots \boldsymbol{\Gamma}\mathbf{P}(x_T)\mathbf{1}' = \left(\prod_{s=t+1}^{T} \boldsymbol{\Gamma}\mathbf{P}(x_s) \right) \mathbf{1}', \quad (4.2)$$

with the convention that an empty product is the identity matrix; the case $t = T$ therefore yields $\boldsymbol{\beta}_T = \mathbf{1}$. We shall show that $\beta_t(j)$, the jth component of $\boldsymbol{\beta}_t$, can be identified as the *conditional* probability $\Pr(X_{t+1} = x_{t+1}, X_{t+2} = x_{t+2}, \ldots, X_T = x_T \mid C_t = j)$. It will then follow that, for $t = 1, 2, \ldots, T$,

$$\alpha_t(j)\beta_t(j) = \Pr(\mathbf{X}^{(T)} = \mathbf{x}^{(T)}, C_t = j).$$

4.1.1 Forward probabilities

It follows immediately from the definition of $\boldsymbol{\alpha}_t$ that, for $t = 1, 2, \ldots, T - 1$, $\boldsymbol{\alpha}_{t+1} = \boldsymbol{\alpha}_t\boldsymbol{\Gamma}\mathbf{P}(x_{t+1})$ or, in scalar form,

$$\alpha_{t+1}(j) = \left(\sum_{i=1}^{m} \alpha_t(i)\gamma_{ij} \right) p_j(x_{t+1}). \quad (4.3)$$

We shall now use the above recursion, and equation (B.1) in Appendix B, to prove the following result by induction.

> **Proposition 2** *For $t = 1, 2, \ldots, T$ and $j = 1, 2, \ldots, m$,*
> $$\alpha_t(j) = \Pr(\mathbf{X}^{(t)} = \mathbf{x}^{(t)}, C_t = j).$$

Proof. Since $\boldsymbol{\alpha}_1 = \boldsymbol{\delta}\mathbf{P}(x_1)$, we have

$$\alpha_1(j) = \delta_j\, p_j(x_1) = \Pr(C_1 = j)\Pr(X_1 = x_1 \mid C_1 = j),$$

hence $\alpha_1(j) = \Pr(X_1 = x_1, C_1 = j)$; that is, the proposition holds for $t = 1$. We now show that, if the proposition holds for some $t \in \mathbb{N}$, then it also holds for $t + 1$:

$$
\begin{aligned}
\alpha_{t+1}(j) &= \sum_{i=1}^{m} \alpha_t(i)\gamma_{ij}p_j(x_{t+1}) \quad \text{(see (4.3))} \\
&= \sum_{i} \Pr(\mathbf{X}^{(t)} = \mathbf{x}^{(t)}, C_t = i)\Pr(C_{t+1} = j \mid C_t = i) \\
&\qquad\qquad\qquad \times \Pr(X_{t+1} = x_{t+1} \mid C_{t+1} = j) \\
&= \sum_{i} \Pr(\mathbf{X}^{(t+1)} = \mathbf{x}^{(t+1)}, C_t = i, C_{t+1} = j) \quad (4.4) \\
&= \Pr(\mathbf{X}^{(t+1)} = \mathbf{x}^{(t+1)}, C_{t+1} = j),
\end{aligned}
$$

as required. The crux is the line numbered (4.4); equation (B.1) provides the justification thereof. □

4.1.2 Backward probabilities

It follows immediately from the definition of β_t that $\beta'_t = \mathbf{\Gamma P}(x_{t+1})\beta'_{t+1}$, for $t = 1, 2, \ldots, T-1$.

Proposition 3 *For* $t = 1, 2, \ldots, T-1$ *and* $i = 1, 2, \ldots, m$,

$$\beta_t(i) = \Pr(X_{t+1} = x_{t+1}, X_{t+2} = x_{t+2}, \ldots, X_T = x_T \mid C_t = i),$$

provided that $\Pr(C_t = i) > 0$. *In a more compact notation,*

$$\beta_t(i) = \Pr(\mathbf{X}_{t+1}^T = \mathbf{x}_{t+1}^T \mid C_t = i),$$

where $\mathbf{X}_a^b$ *denotes the vector* $(X_a, X_{a+1}, \ldots, X_b)$.

This proposition identifies $\beta_t(i)$ as a conditional probability: the probability of the observations being $x_{t+1}, \ldots, x_T$, given that the Markov chain is in state i at time t. (Recall that the forward probabilities are joint probabilities, not conditional.)

Proof. The proof is by induction, but essentially comes down to equations (B.5) and (B.6) of Appendix B. These are

$$\Pr(X_{t+1} \mid C_{t+1})\Pr(\mathbf{X}_{t+2}^T \mid C_{t+1}) = \Pr(\mathbf{X}_{t+1}^T \mid C_{t+1}) \qquad \text{(B.5)}$$

and

$$\Pr(\mathbf{X}_{t+1}^T \mid C_{t+1}) = \Pr(\mathbf{X}_{t+1}^T \mid C_t, C_{t+1}). \qquad \text{(B.6)}$$

To establish validity for $t = T-1$, note that, since $\beta'_{T-1} = \mathbf{\Gamma P}(x_T)\mathbf{1}'$,

$$\beta_{T-1}(i) = \sum_j \Pr(C_T = j \mid C_{T-1} = i)\Pr(X_T = x_T \mid C_T = j). \qquad \text{(4.5)}$$

But, by (B.6),

$$
\begin{aligned}
\Pr(C_T \mid C_{T-1})\Pr(X_T \mid C_T) &= \Pr(C_T \mid C_{T-1})\Pr(X_T \mid C_{T-1}, C_T) \\
&= \Pr(X_T, C_{T-1}, C_T)/\Pr(C_{T-1}). \qquad \text{(4.6)}
\end{aligned}
$$

Substitute from (4.6) into (4.5), and the result is

$$
\begin{aligned}
\beta_{T-1}(i) &= \frac{1}{\Pr(C_{T-1} = i)}\sum_j \Pr(X_T = x_T, C_{T-1} = i, C_T = j) \\
&= \Pr(X_T = x_T, C_{T-1} = i)/\Pr(C_{T-1} = i) \\
&= \Pr(X_T = x_T \mid C_{T-1} = i),
\end{aligned}
$$

as required.

To show that validity for $t+1$ implies validity for t, first note that the recursion for β_t, and the inductive hypothesis, establish that

$$\beta_t(i) = \sum_j \gamma_{ij} \Pr(X_{t+1} = x_{t+1} \mid C_{t+1} = j) \Pr(\mathbf{X}_{t+2}^T = \mathbf{x}_{t+2}^T \mid C_{t+1} = j).$$

$$(4.7)$$

But (B.5) and (B.6) imply that

$$\Pr(X_{t+1} \mid C_{t+1}) \Pr(\mathbf{X}_{t+2}^T \mid C_{t+1}) = \Pr(\mathbf{X}_{t+1}^T \mid C_t, C_{t+1}). \qquad (4.8)$$

Substitute from (4.8) into (4.7), and the result is

$$
\begin{aligned}
\beta_t(i) &= \sum_j \Pr(C_{t+1} = j \mid C_t = i) \Pr(\mathbf{X}_{t+1}^T = \mathbf{x}_{t+1}^T \mid C_t = i, C_{t+1} = j) \\
&= \frac{1}{\Pr(C_t = i)} \sum_j \Pr(\mathbf{X}_{t+1}^T = \mathbf{x}_{t+1}^T, C_t = i, C_{t+1} = j) \\
&= \Pr(\mathbf{X}_{t+1}^T = \mathbf{x}_{t+1}^T, C_t = i) / \Pr(C_t = i),
\end{aligned}
$$

which is the required conditional probability. $\qquad\qquad\square$

Note that the backward probabilities require a backward pass through the data for their evaluation, just as the forward probabilities require a forward pass; hence the names.

4.1.3 Properties of forward and backward probabilities

We now establish a result relating the forward and backward probabilities $\alpha_t(i)$ and $\beta_t(i)$ to the probabilities $\Pr(\mathbf{X}^{(T)} = \mathbf{x}^{(T)}, C_t = i)$. This we shall use in applying the EM algorithm to HMMs, and in local decoding; see Section 5.4.1.

Proposition 4 *For $t = 1, 2, \ldots, T$ and $i = 1, 2, \ldots, m$,*

$$\alpha_t(i)\beta_t(i) = \Pr(\mathbf{X}^{(T)} = \mathbf{x}^{(T)}, C_t = i), \qquad (4.9)$$

and consequently $\boldsymbol{\alpha}_t\boldsymbol{\beta}_t' = \Pr(\mathbf{X}^{(T)} = \mathbf{x}^{(T)}) = L_T$, for each such t.

Proof. By the previous two propositions,

$$
\begin{aligned}
\alpha_t(i)\beta_t(i) &= \Pr(\mathbf{X}_1^t, C_t = i) \Pr(\mathbf{X}_{t+1}^T \mid C_t = i) \\
&= \Pr(C_t = i) \Pr(\mathbf{X}_1^t \mid C_t = i) \Pr(\mathbf{X}_{t+1}^T \mid C_t = i).
\end{aligned}
$$

Now apply the conditional independence of $\mathbf{X}_1^t$ and $\mathbf{X}_{t+1}^T$ given C_t (see equation (B.7) of Appendix B), and the result is that

$$\alpha_t(i)\beta_t(i) = \Pr(C_t = i) \Pr(\mathbf{X}_1^t, \mathbf{X}_{t+1}^T \mid C_t = i) = \Pr(\mathbf{X}^{(T)}, C_t = i).$$

Summation of this equation over i yields the second conclusion. $\qquad\square$

The second conclusion also follows immediately from the matrix expression for the likelihood and the definitions of $\boldsymbol{\alpha}_t$ and $\boldsymbol{\beta}_t$:

$$
\begin{aligned}
L_T &= \left(\boldsymbol{\delta}\mathbf{P}(x_1)\boldsymbol{\Gamma}\mathbf{P}(x_2)\ldots\boldsymbol{\Gamma}\mathbf{P}(x_t)\right)\left(\boldsymbol{\Gamma}\mathbf{P}(x_{t+1})\ldots\boldsymbol{\Gamma}\mathbf{P}(x_T)\mathbf{1}'\right) \\
&= \boldsymbol{\alpha}_t\boldsymbol{\beta}_t'.
\end{aligned}
$$

Note that we now have available T routes to the computation of the likelihood L_T, one for each possible value of t. But the route we have used so far (the case $t = T$, yielding $L_T = \boldsymbol{\alpha}_T\mathbf{1}'$) seems the most convenient, as it requires the computation of forward probabilities only, and only a single pass (forward) through the data.

In applying the EM algorithm to HMMs we shall also need the following two properties.

Proposition 5 *Firstly, for $t = 1, \ldots, T$,*

$$
\Pr(C_t = j \mid \mathbf{X}^{(T)} = \mathbf{x}^{(T)}) = \alpha_t(j)\beta_t(j)/L_T; \qquad (4.10)
$$

and secondly, for $t = 2, \ldots, T$,

$$
\Pr(C_{t-1} = j, C_t = k \mid \mathbf{X}^{(T)} = \mathbf{x}^{(T)}) = \alpha_{t-1}(j)\,\gamma_{jk}\,p_k(x_t)\,\beta_t(k)/L_T. \qquad (4.11)
$$

Proof. The first assertion follows immediately from (4.9) above. The second is an application of equations (B.4) and (B.5) of Appendix B, and the proof proceeds as follows.

$$
\begin{aligned}
&\Pr(C_{t-1} = j, C_t = k \mid \mathbf{X}^{(T)} = \mathbf{x}^{(T)}) \\
&= \Pr(\mathbf{X}^{(T)}, C_{t-1} = j, C_t = k)/L_T \\
&= \Pr(\mathbf{X}^{(t-1)}, C_{t-1} = j)\,\Pr(C_t = k \mid C_{t-1} = j)\,\Pr(\mathbf{X}_t^T \mid C_t = k)/L_T \\
&\qquad\qquad\qquad\qquad \text{(by (B.4))} \\
&= \alpha_{t-1}(j)\,\gamma_{jk}\left(\Pr(X_t \mid C_t = k)\,\Pr(\mathbf{X}_{t+1}^T \mid C_t = k)\right)/L_T \\
&\qquad\qquad\qquad\qquad \text{(by (B.5))} \\
&= \alpha_{t-1}(j)\,\gamma_{jk}\,p_k(x_t)\,\beta_t(k)/L_T. \qquad\qquad\qquad\qquad \square
\end{aligned}
$$

4.2 The EM algorithm

Since the sequence of states occupied by the Markov-chain component of an HMM is not observed, a very natural approach to parameter estimation in HMMs is to treat those states as missing data and to employ the EM algorithm (Dempster, Laird and Rubin, 1977) in order to find maximum likelihood estimates of the parameters. Indeed, the pioneering work of Leonard Baum and his colleagues (see Baum *et al.*, 1970; Baum,

1972; Welch, 2003) on what were later called HMMs was an important precursor of the work of Dempster *et al.*

4.2.1 EM in general

The EM algorithm is an iterative method for performing maximum likelihood estimation when some of the data are missing, and exploits the fact that the complete-data log-likelihood (CDLL) may be straightforward to maximize even if the likelihood of the observed data is not. By 'complete-data log-likelihood' we mean the log-likelihood of the parameters of interest $\boldsymbol{\theta}$, based on both the observed data and the missing data.

The algorithm may be described informally as follows (see, for example, Little and Rubin, 2002, pp. 166–168). Choose starting values for the parameters $\boldsymbol{\theta}$ you wish to estimate. Then repeat the following steps until some convergence criterion has been satisfied, for example, until the resulting change in $\boldsymbol{\theta}$ is less than some threshold.

- **E step** Compute the conditional expectations of the missing data given the observations and given the current estimate of $\boldsymbol{\theta}$. More precisely, compute the conditional expectations of *those functions of* the missing data that appear in the CDLL.
- **M step** Maximize, with respect to $\boldsymbol{\theta}$, the CDLL with the functions of the missing data replaced in it by their conditional expectations.

The resulting value of $\boldsymbol{\theta}$ is then a stationary point of the likelihood of the observed data. In some cases, however, the stationary point reached can be a local (as opposed to global) maximum or a saddle point.

Little and Rubin (p. 168) stress the point that it is not (necessarily) the missing data themselves that are replaced in the CDLL by their conditional expectations, but those functions of the missing data that appear in the CDLL; they describe this as the 'key idea of EM'.

4.2.2 EM for HMMs

In the case of an HMM it is convenient to represent the sequence of states $c_1, c_2, \ldots, c_T$ followed by the Markov chain by the zero–one random variables defined as follows:

$$u_j(t) = 1 \text{ if and only if } c_t = j \quad (t = 1, 2, \ldots, T)$$

and

$$v_{jk}(t) = 1 \text{ if and only if } c_{t-1} = j \text{ and } c_t = k \quad (t = 2, 3, \ldots, T).$$

With this notation, the CDLL of an HMM (i.e. the log-likelihood of the observations $x_1, x_2, \ldots, x_T$ plus the missing data $c_1, c_2, \ldots, c_T$) is given by

$$\log\left(\Pr(\mathbf{x}^{(T)},\mathbf{c}^{(T)})\right) = \log\left(\delta_{c_1}\prod_{t=2}^{T}\gamma_{c_{t-1},c_t}\prod_{t=1}^{T}p_{c_t}(x_t)\right)$$

$$= \log\delta_{c_1} + \sum_{t=2}^{T}\log\gamma_{c_{t-1},c_t} + \sum_{t=1}^{T}\log p_{c_t}(x_t).$$

Hence the CDLL is

$$\log\left(\Pr(\mathbf{x}^{(T)},\mathbf{c}^{(T)})\right)$$

$$= \sum_{j=1}^{m}u_j(1)\log\delta_j + \sum_{j=1}^{m}\sum_{k=1}^{m}\left(\sum_{t=2}^{T}v_{jk}(t)\right)\log\gamma_{jk}$$

$$+ \sum_{j=1}^{m}\sum_{t=1}^{T}u_j(t)\log p_j(x_t) \qquad (4.12)$$

$$= \text{term 1} + \text{term 2} + \text{term 3}.$$

Here δ is to be understood as the *initial* distribution of the Markov chain, the distribution of C_1, not necessarily the stationary distribution. Of course it is not reasonable to try to estimate the initial distribution from just one observation at time 1, especially as the state of the Markov chain itself is not observed. It is therefore interesting to see how the EM algorithm responds to this unreasonable request; see Section 4.2.4. We shall later (Section 4.2.5) add the assumption that the Markov chain is stationary; δ will then denote the stationary distribution implied by Γ, and the question of estimating δ will fall away.

The EM algorithm for HMMs proceeds as follows.

• **E step** Replace all the quantities $v_{jk}(t)$ and $u_j(t)$ by their conditional expectations given the observations $\mathbf{x}^{(T)}$ (and given the current parameter estimates):

$$\widehat{u}_j(t) = \Pr(C_t{=}j \mid \mathbf{x}^{(T)}) = \alpha_t(j)\beta_t(j)/L_T \qquad (4.13)$$

and

$$\widehat{v}_{jk}(t) = \Pr(C_{t-1}{=}j, C_t{=}k \mid \mathbf{x}^{(T)}) = \alpha_{t-1}(j)\,\gamma_{jk}\,p_k(x_t)\,\beta_t(k)/L_T. \qquad (4.14)$$

(See Section 4.1.3, equations (4.10) and (4.11), for justification of the above equalities.) Note that in this context we need the forward probabilities as computed for an HMM that does *not* assume stationarity of the underlying Markov chain $\{C_t\}$; the backward probabilities, however, are not affected by the stationarity or otherwise of $\{C_t\}$.

• **M step** Having replaced $v_{jk}(t)$ and $u_j(t)$ by $\widehat{v}_{jk}(t)$ and $\widehat{u}_j(t)$, maximize the CDLL, expression (4.12), with respect to the three sets of

parameters: the initial distribution $\boldsymbol{\delta}$, the t.p.m. $\boldsymbol{\Gamma}$, and the parameters of the state-dependent distributions (e.g. $\lambda_1, \ldots, \lambda_m$ in the
case of a simple Poisson–HMM).

Examination of (4.12) reveals that the M step splits neatly into three
separate maximizations, since (of the parameters) term 1 depends only
on the initial distribution $\boldsymbol{\delta}$, term 2 on the t.p.m. $\boldsymbol{\Gamma}$, and term 3 on the
'state-dependent parameters'. We must therefore maximize:

1. $\sum_{j=1}^{m} \widehat{u}_j(1) \log \delta_j$ with respect to $\boldsymbol{\delta}$;

2. $\sum_{j=1}^{m} \sum_{k=1}^{m} \left(\sum_{t=2}^{T} \widehat{v}_{jk}(t) \right) \log \gamma_{jk}$ with respect to $\boldsymbol{\Gamma}$; and

3. $\sum_{j=1}^{m} \sum_{t=1}^{T} \widehat{u}_j(t) \log p_j(x_t)$ with respect to the state-dependent parameters. Notice here that the only parameters on which the term
   $\sum_{t=1}^{T} \widehat{u}_j(t) \log p_j(x_t)$ depends are those of the jth state-dependent
   distribution, p_j; this further simplifies the problem.

The solution is as follows.

1. Set $\delta_j = \widehat{u}_j(1) / \sum_{j=1}^{m} \widehat{u}_j(1) = \widehat{u}_j(1)$. (See Exercise 12 of Chapter 1
   for justification.)

2. Set $\gamma_{jk} = f_{jk} / \sum_{k=1}^{m} f_{jk}$, where $f_{jk} = \sum_{t=2}^{T} \widehat{v}_{jk}(t)$. (Apply the result
   of Exercise 12 of Chapter 1 to each row.)

3. The maximization of the third term may be easy or difficult, depending on the nature of the state-dependent distributions assumed. It is
   essentially the standard problem of maximum likelihood estimation
   for the distributions concerned. In the case of Poisson and normal
   distributions, closed-form solutions are available; see Section 4.2.3. In
   some other cases, such as the gamma distributions and the negative
   binomial, numerical maximization will be necessary to carry out this
   part of the M step.

From point 2 in the solution, we see that it is not the quantities $\widehat{v}_{jk}(t)$
themselves that are needed, but their sums f_{jk}. It is worth noting that
the computation of the forward and backward probabilities is susceptible
to under- or overflow error, as are the computation and summation of
the quantities $\widehat{v}_{jk}(t)$. In applying EM as described here, precautions (e.g.
scaling) therefore have to be taken in order to prevent, or at least reduce
the risk of, such error. Code for computing the logarithms of the forward
and backward probabilities of a Poisson–HMM appears in Sections A.1.7
and A.1.8.

4.2.3 M step for Poisson– and normal–HMMs

Here we give part 3 of the M step explicitly for the cases of Poisson and
normal state-dependent distributions. The state-dependent part of the

CDLL (term 3 of expression (4.12)) is

$$\sum_{j=1}^{m}\sum_{t=1}^{T}\widehat{u}_j(t)\log p_j(x_t).$$

For a Poisson–HMM, $p_j(x) = e^{-\lambda_j}\lambda_j^x/x!$, so in that case term 3 is maximized by setting

$$0 = \sum_t \widehat{u}_j(t)(-1 + x_t/\lambda_j);$$

that is, by

$$\widehat{\lambda}_j = \sum_{t=1}^{T}\widehat{u}_j(t)x_t \bigg/ \sum_{t=1}^{T}\widehat{u}_j(t).$$

For a normal–HMM the state-dependent density is of the form $p_j(x) = (2\pi\sigma_j^2)^{-1/2}\exp\left(-\frac{1}{2\sigma_j^2}(x - \mu_j)^2\right)$, and the maximizing values of the state-dependent parameters μ_j and σ_j^2 are

$$\widehat{\mu}_j = \sum_{t=1}^{T}\widehat{u}_j(t)x_t \bigg/ \sum_{t=1}^{T}\widehat{u}_j(t),$$

and

$$\widehat{\sigma}_j^2 = \sum_{t=1}^{T}\widehat{u}_j(t)(x_t - \widehat{\mu}_j)^2 \bigg/ \sum_{t=1}^{T}\widehat{u}_j(t).$$

4.2.4 Starting from a specified state

It is known (Levinson, Rabiner and Sondhi, 1983, p. 1055) that the maximum likelihood estimator of δ is a unit vector: one entry is 1 and the others 0. (See also Exercise 1.) This suggests an alternative method of applying the EM algorithm, used notably by Leroux and Puterman (1992). One maximizes the likelihood of the observations (i.e. the 'incomplete-data' likelihood) conditional on the Markov chain starting in state i, for each of $i = 1, 2, \ldots, m$, and then chooses the largest result. But we shall in the examples that follow treat δ as a vector of parameters requiring estimation, as it is instructive to see what emerges.

4.2.5 EM for the case in which the Markov chain is stationary

Now assume in addition that the underlying Markov chain is stationary, not merely homogeneous. This is often a desirable assumption in time-series applications. The initial distribution δ is then such that $\delta = \delta\Gamma$

and $\delta \mathbf{1}' = 1$, or equivalently,

$$\delta = \mathbf{1}(\mathbf{I}_m - \mathbf{\Gamma} + \mathbf{U})^{-1},$$

with $\mathbf{U}$ being a square matrix of ones. In this case, δ is completely determined by the transition probabilities $\mathbf{\Gamma}$, and the question of estimating δ falls away. However, the M step then gives rise to the following optimization problem: maximize, with respect to $\mathbf{\Gamma}$, the sum of terms 1 and 2 of expression (4.12), that is, maximize

$$\sum_{j=1}^{m} \widehat{u}_j(1) \log \delta_j + \sum_{j=1}^{m} \sum_{k=1}^{m} \left(\sum_{t=2}^{T} \widehat{v}_{jk}(t) \right) \log \gamma_{jk}. \qquad (4.15)$$

Notice that here term 1 also depends on $\mathbf{\Gamma}$. Even in the case of only two states, analytic maximization would require the solution of a pair of equations quadratic in two variables, namely, two of the transition probabilities; see Exercise 3.

Except possibly in special cases, numerical optimization is needed for this part of the M step if stationarity is assumed, or else some modification of EM designed to circumvent such numerical optimization. This is a slight disadvantage of the use of EM, as the stationary version of the models is important in a time-series context.

4.3 Examples of EM applied to Poisson–HMMs

4.3.1 Earthquakes

We now present two- and three-state models fitted by the EM algorithm, as described above, to the earthquakes data. For the two-state model, the starting values of the off-diagonal transition probabilities are taken to be 0.1, and the starting value of δ, the initial distribution, is $(0.5, 0.5)$. Since 19.36 is the sample mean, 10 and 30 are plausible starting values for the state-dependent means λ_1 and λ_2.

In the tables shown, 'iteration 0' refers to the starting values, and 'stationary model' to the parameter values and log-likelihood of the comparable stationary model fitted via nlm by direct numerical maximization.

Several features of Table 4.1 are worth noting. Firstly, the likelihood value of the stationary model is slightly lower than that fitted here by EM (i.e. $-l$ is higher). This is to be expected, as constraining the initial distribution δ to be the stationary distribution can only decrease the maximal value of the likelihood. Secondly, the estimates of the transition probabilities and the state-dependent means are not identical for the two models, but close; this, too, is to be expected. Thirdly, although we know from Section 4.2.4 that δ will approach a unit vector, it is noticeable just how quickly, starting from $(0.5, 0.5)$, it approaches $(1, 0)$.

Table 4.1 *Two-state model for earthquakes, fitted by EM.*

Iteration	γ_{12}	γ_{21}	λ_1	λ_2	δ_1	$-l$
0	0.100000	0.10000	10.000	30.000	0.50000	413.27542
1	0.138816	0.11622	13.742	24.169	0.99963	343.76023
2	0.115510	0.10079	14.090	24.061	1.00000	343.13618
30	0.071653	0.11895	15.419	26.014	1.00000	341.87871
50	0.071626	0.11903	15.421	26.018	1.00000	341.87870
convergence	0.071626	0.11903	15.421	26.018	1.00000	341.87870
stationary model	0.065961	0.12851	15.472	26.125	0.66082	342.31827

Table 4.2 *Three-state model for earthquakes, fitted by EM.*

Iteration	λ_1	λ_2	λ_3	δ_1	δ_2	$-l$
0	10.000	20.000	30.000	0.33333	0.33333	342.90781
1	11.699	19.030	29.741	0.92471	0.07487	332.12143
2	12.265	19.078	29.581	0.99588	0.00412	330.63689
30	13.134	19.713	29.710	1.00000	0.00000	328.52748
convergence	13.134	19.713	29.710	1.00000	0.00000	328.52748
stationary model	13.146	19.721	29.714	0.44364	0.40450	329.46028

Table 4.2 displays similar information for the corresponding three-state models. In this case as well, the starting values of the off-diagonal transition probabilities are all taken to be 0.1 and the starting δ is uniform over the states.

We now present more fully the 'EM' and the stationary versions of the three-state model, which are only summarized in Table 4.2.

• Three-state model with initial distribution $(1, 0, 0)$, fitted by EM:

$$\Gamma = \begin{pmatrix} 0.9393 & 0.0321 & 0.0286 \\ 0.0404 & 0.9064 & 0.0532 \\ 0.0000 & 0.1903 & 0.8097 \end{pmatrix},$$

$$\lambda = (13.134, 19.713, 29.710).$$

• Three-state model based on stationary Markov chain, fitted by direct numerical maximization:

$$\Gamma = \begin{pmatrix} 0.9546 & 0.0244 & 0.0209 \\ 0.0498 & 0.8994 & 0.0509 \\ 0.0000 & 0.1966 & 0.8034 \end{pmatrix},$$

$$\delta = (0.4436, 0.4045, 0.1519), \quad \lambda = (13.146, 19.721, 29.714).$$

Table 4.3 *Six two-state models fitted to the foetal movement counts time series.*

	DNM (stat.)	DNM (state 2)	EM (this work)	EM (L&P)	SEM	Prior feedback
λ_1	3.1148	3.1007	3.1007	3.1006	2.93	2.84
λ_2	0.2564	0.2560	0.2560	0.2560	0.26	0.25
γ_{12}	0.3103	0.3083	0.3083	0.3083	0.28	0.32
γ_{21}	0.0113	0.0116	0.0116	0.0116	0.01	0.015
$-l^*$	–	150.7007	150.7007	150.70	–	–
$-l$	177.5188	177.4833	177.4833	(177.48)	(177.58)	(177.56)
$10^{78}L$	8.0269	8.3174	8.3174	(8.3235)	7.539	7.686

Key to notation and abbreviations:

DNM (stat.)	model starting from stationary distribution of the Markov chain, fitted by direct numerical maximization
DNM (state 2)	model starting from state 2 of the Markov chain, fitted by direct numerical maximization
EM (this work)	model fitted by EM, as described in Sections 4.2.2–4.2.3
EM (L&P)	model fitted by EM by Leroux and Puterman (1992)
SEM	model fitted by stochastic EM algorithm of Chib (1996); from Table 2 of Robert and Titterington (1998)
prior feedback	model fitted by maximum likelihood via 'prior feedback'; from Table 2 of Robert and Titterington (1998)
l^*	log-likelihood omitting constant terms
l	log-likelihood
L	likelihood
–	not needed; deliberately omitted

Figures appearing in brackets in the last three columns of this table have been deduced from figures published in the works cited, but do not themselves appear there; they are at best as accurate as the figures on which they are based. For instance, $10^{78}L = 8.3235$ is based on $-l^* = 150.70$. The figures 150.70, 7.539 and 7.686 in those three columns are exactly as published.

4.3.2 Foetal movement counts

Leroux and Puterman (1992) used EM to fit (among other models) Markov-dependent mixtures to a time series of unbounded counts, a series which has subsequently been analysed by Chib (1996), Robert and Titterington (1998), Robert and Casella (1999, p. 432) and Scott (2002). The series consists of counts of movements by a foetal lamb in 240 consecutive 5-second intervals, and was taken from Wittmann, Rurak and Taylor (1984).

We present here a two-state HMM fitted by EM, and two HMMs fitted by direct numerical maximization of the likelihood, and we compare these with the models in Table 4 of Leroux and Puterman and Table 2 of Robert and Titterington. Robert and Titterington omit any comparison with the results of Leroux and Puterman, a comparison which might have been informative: the estimates of λ_1 differ, and Leroux and Puterman's likelihood value is somewhat higher, as are our values.

The models of Leroux and Puterman start with probability 1 from one of the states; that is, the initial distribution of the Markov chain is a unit vector. But it is of course easy to fit such models by direct numerical maximization if one so wishes. In any code which takes the initial distribution of the Markov chain to be the stationary distribution, one merely replaces that stationary distribution by the relevant unit vector. Our Table 4.3 presents (*inter alia*) three models we have fitted, the second of which is of this kind. The first we have fitted by direct numerical maximization of the likelihood based on the initial distribution being the stationary distribution, the second by direct numerical maximization of the likelihood starting from state 2, and the third by EM. In Table 4.3 they are compared to three models appearing in the published literature.

The results displayed in the table suggest that, if one wishes to fit models by maximum likelihood, then EM and direct numerical maximization are superior to the other methods considered. Robert and Titterington suggest that Chib's algorithm may not have converged, or may have converged to a different local optimum of the likelihood; their prior feedback results appear also to be suboptimal, however.

4.4 Discussion

As Bulla and Berzel (2008) point out, researchers and practitioners tend to use either EM or direct numerical maximization, but not both, to perform maximum likelihood estimation in HMMs, and each approach has its merits. However, one of the merits claimed for EM in some generality turns out to be illusory in the context of HMMs. McLachlan and Krishnan (2008, p. 29) state of the EM algorithm in general (i.e. not specifically in the context of HMMs) that it is 'easy to program, since no evaluation of the likelihood nor its derivatives is involved'. If one applies EM as described above, one has to compute both the forward and the backward probabilities; on the other hand, one needs only the forward probabilities to compute the likelihood, which can then be maximized numerically. In effect, EM does all that is needed to compute the likelihood via the forward probabilities, and then does more. Especially if one has available an optimization routine, such as `nlm`, `optim` or `constrOptim` in **R**, which does not demand the specification of derivatives, ease of programming seems in the present context to be more a characteristic of direct numerical maximization than of EM.

In our experience, it is a major advantage of direct numerical maximization without analytical derivatives that one can, with a minimum of programming effort, repeatedly modify a model in an interactive search for the best model. Often all that is needed is a small change to the code that evaluates the likelihood. It is also usually straightforward to replace one optimizer by another if an optimizer fails, or if one wishes to check in any way the output of a particular optimizer.

Note the experience reported by Altman and Petkau (2005). In their applications direct maximization of the likelihood produced the MLEs far more quickly than did the EM algorithm. See also Turner (2008), who provides a detailed study of direct maximization by the Levenberg–Marquardt algorithm. In two examples he finds that this algorithm is much faster (in the sense of CPU time) than is EM, and it is also clearly faster than optim, both with and without the provision of analytical derivatives. In our opinion the disadvantage of using analytical derivatives in exploratory modelling is the work involved in recoding those derivatives, and checking the code, when one alters a model. Of course for standard models such as the Poisson–HMM, which are likely to be used repeatedly, the advantage of having efficient code would make such labour worthwhile. Cappé et al. (2005, p. 358) provide a discussion of the relative merits of EM and direct maximization of the likelihood of an HMM by gradient-based methods.

Exercises

1. (a) Suppose $L_i > 0$ for $i = 1, 2, \ldots, m$. Maximize $L = \sum_{i=1}^{m} a_i L_i$ over the region $a_i \geq 0$, $\sum_{i=1}^{m} a_i = 1$.

   (b) Consider an HMM with initial distribution $\boldsymbol{\delta}$, and consider the (observed-data) likelihood as a function of $\boldsymbol{\delta}$. Show that, at a maximum of the likelihood, $\boldsymbol{\delta}$ is a unit vector.

2. Consider the example of Visser, Raijmakers and Molenaar (2002, pp. 186–187). There a series of length 1000 is simulated from an HMM with states S_1 and S_2 and the three observation symbols 1, 2 and 3. The transition probability matrix is

$$\mathbf{A} = \begin{pmatrix} 0.9 & 0.1 \\ 0.3 & 0.7 \end{pmatrix},$$

the initial probabilities are $\boldsymbol{\pi} = (0.5, 0.5)$, and the state-dependent distribution in state i is row i of the matrix

$$\mathbf{B} = \begin{pmatrix} 0.7 & 0.0 & 0.3 \\ 0.0 & 0.4 & 0.6 \end{pmatrix}.$$

The parameters $\mathbf{A}$, $\mathbf{B}$ and $\boldsymbol{\pi}$ are then estimated by EM; the estimates of $\mathbf{A}$ and $\mathbf{B}$ are close to $\mathbf{A}$ and $\mathbf{B}$, but that of $\boldsymbol{\pi}$ is $(1,0)$. This estimate for $\boldsymbol{\pi}$ is explained as follows: 'The reason for this is that the sequence of symbols that was generated actually starts with the symbol 1 which can only be produced from state S_1.'

Do you agree with the above statement? What if the probability of symbol 1 in state S_2 had been (say) 0.1 rather than 0.0?

3. Consider the fitting by EM of a two-state HMM based on a *stationary* Markov chain. In the M step, the sum of terms 1 and 2 must be maximized with respect to $\boldsymbol{\Gamma}$; see the expression labelled (4.15).

   Write term 1 + term 2 as a function of γ_{12} and γ_{21}, the off-diagonal transition probabilities, and differentiate to find the equations satisfied by these probabilities at a stationary point. (You should find that the equations are quadratic in both γ_{12} and γ_{21}.)

4. Consider again the soap sales series introduced in Exercise 6 of Chapter 1.

   Use the EM algorithm to fit Poisson–HMMs with two, three and four states to these data.

5. Let $\{X_t\}$ be an HMM on m states.

   (a) Suppose the state-dependent distributions are binomial. More precisely, assume that

   $$\Pr(X_t = x \mid C_t = j) = \binom{n_t}{x} p_j^x (1 - p_j)^{n_t - x}.$$

   Find the value for p_j that will maximize the third term of equation (4.12). (This is needed in order to carry out the M step of EM for a binomial–HMM.)

   (b) Now suppose instead that the state-dependent distributions are exponential, with means $1/\lambda_j$. Find the value for λ_j that will maximize the third term of equation (4.12).

6. Write code to fit normal–, binomial–, and exponential–HMMs by EM.

## CHAPTER 5

Forecasting, decoding and state prediction

Main results of this chapter:

conditional distributions   p. 83

$$\Pr(X_t = x \mid \mathbf{X}^{(-t)} = \mathbf{x}^{(-t)}) = \sum_i w_i(t) p_i(x)$$

forecast distributions   p. 85

$$\Pr(X_{T+h} = x \mid \mathbf{X}^{(T)} = \mathbf{x}^{(T)}) = \sum_i \xi_i(h) p_i(x)$$

state probabilities and local decoding   p. 87

$$\Pr(C_t = i \mid \mathbf{X}^{(T)} = \mathbf{x}^{(T)}) = \alpha_t(i)\beta_t(i)/L_T$$

global decoding maximize over $\mathbf{c}^{(T)}$:   p. 88

$$\Pr(\mathbf{C}^{(T)} = \mathbf{c}^{(T)} \mid \mathbf{X}^{(T)} = \mathbf{x}^{(T)})$$

state prediction   p. 93

$$\Pr(C_{T+h} = i \mid \mathbf{X}^{(T)} = \mathbf{x}^{(T)}) = \boldsymbol{\alpha}_T \boldsymbol{\Gamma}^h \mathbf{e}_i'/L_T$$

5.1 Introduction

Convenient expressions for conditional distributions and forecast distributions are available for HMMs. This makes it easy, for example, to check for outliers or to make interval forecasts. In this chapter, we first show (in Section 5.2) how to compute the conditional distribution of the observation at time t given the observations at all other times. In Section 5.3 we derive the forecast distribution of an HMM. Then, in Section 5.4, we demonstrate how, given the HMM and the observations, one can deduce information about the states occupied by the underlying Markov chain. Such inference is known as decoding. We continue to use the earthquakes series as our illustrative example. Our results are stated for the case of discrete observations X_t; if the observations are continuous, probabilities will need to be replaced by densities.

Note that in this chapter we do not assume stationarity of the Markov chain $\{C_t\}$, only homogeneity: here the row vector $\boldsymbol{\delta}$ denotes the *initial* distribution, that of C_1. Of course the results also hold in the special case in which the Markov chain is stationary, in which case $\boldsymbol{\delta}$ is both the initial and the stationary distribution.

5.2 Conditional distributions

We now derive a formula for the distribution of X_t conditioned on all the other observations of the HMM. We use the notation $\mathbf{x}^{(-t)}$ for the observations at all times other than t; that is, we define

$$\mathbf{x}^{(-t)} \equiv (x_1, \ldots, x_{t-1}, x_{t+1}, \ldots, x_T),$$

and similarly $\mathbf{X}^{(-t)}$.

Using the likelihood of an HMM as discussed in Section 2.3.2, and the definition of the forward and backward probabilities as in Section 4.1, it follows immediately, for $t = 2, 3, \ldots, T$, that

$$
\begin{aligned}
\Pr\left(X_t = x \mid \mathbf{X}^{(-t)} = \mathbf{x}^{(-t)}\right) &= \frac{\boldsymbol{\delta}\mathbf{P}(x_1)\mathbf{B}_2 \cdots \mathbf{B}_{t-1}\boldsymbol{\Gamma}\mathbf{P}(x)\mathbf{B}_{t+1} \cdots \mathbf{B}_T\mathbf{1}'}{\boldsymbol{\delta}\mathbf{P}(x_1)\mathbf{B}_2 \cdots \mathbf{B}_{t-1}\boldsymbol{\Gamma}\mathbf{B}_{t+1} \cdots \mathbf{B}_T\mathbf{1}'} \\
&\propto \boldsymbol{\alpha}_{t-1}\boldsymbol{\Gamma}\mathbf{P}(x)\boldsymbol{\beta}_t'. \qquad (5.1)
\end{aligned}
$$

Here we have used the compact notation $\mathbf{B}_t = \boldsymbol{\Gamma}\mathbf{P}(x_t)$. With this notation, $\boldsymbol{\alpha}_t = \boldsymbol{\delta}\mathbf{P}(x_1)\mathbf{B}_2 \cdots \mathbf{B}_t$ and $\boldsymbol{\beta}_t' = \mathbf{B}_{t+1} \cdots \mathbf{B}_T\mathbf{1}'$.

The result for the case $t = 1$ is

$$
\begin{aligned}
\Pr\left(X_1 = x \mid \mathbf{X}^{(-1)} = \mathbf{x}^{(-1)}\right) &= \frac{\boldsymbol{\delta}\mathbf{P}(x)\mathbf{B}_2 \cdots \mathbf{B}_T\mathbf{1}'}{\boldsymbol{\delta}\mathbf{I}\mathbf{B}_2 \cdots \mathbf{B}_T\mathbf{1}'} \\
&\propto \boldsymbol{\delta}\mathbf{P}(x)\boldsymbol{\beta}_1'. \qquad (5.2)
\end{aligned}
$$

The above conditional distributions are ratios of two likelihoods of an HMM: the numerator is the likelihood of the observations except that the observation x_t is replaced by x, and the denominator is the likelihood of the observations except that x_t is treated as missing.

We now show that these conditional probabilities can be expressed as mixtures of the m state-dependent probability distributions. In both equations (5.1) and (5.2) the required conditional probability has the following form: a row vector multiplied by the $m \times m$ diagonal matrix $\mathbf{P}(x) = \mathrm{diag}(p_1(x), \ldots, p_m(x))$, multiplied by a column vector. It follows, for $t = 1, 2, \ldots, T$, that

$$\Pr\left(X_t = x \mid \mathbf{X}^{(-t)} = \mathbf{x}^{(-t)}\right) \propto \sum_{i=1}^{m} d_i(t)p_i(x),$$

where, in the case of (5.1), $d_i(t)$ is the product of the ith entry of the vector $\boldsymbol{\alpha}_{t-1}\boldsymbol{\Gamma}$ and the ith entry of the vector $\boldsymbol{\beta}_t$; and in the case of (5.2), it is the product of the ith entry of the vector $\boldsymbol{\delta}$ and the ith entry of the vector $\boldsymbol{\beta}_1$. Hence

$$\Pr\left(X_t = x \mid \mathbf{X}^{(-t)} = \mathbf{x}^{(-t)}\right) = \sum_{i=1}^{m} w_i(t)p_i(x), \qquad (5.3)$$

where the mixing probabilities $w_i(t) = d_i(t)/\sum_{j=1}^{m} d_j(t)$ are functions of the observations $\mathbf{x}^{(-t)}$ and of the model parameters. The **R** code for such conditional distributions is given in Section A.1.9.

In Figure 5.1 we present the full array of conditional distributions for the earthquakes data. It is clear that each of the conditional distributions has a different shape, and the shape may change sharply from one time point to the next. In addition, it is striking that some of the observed counts (e.g. those for 1957 and 1958) are extreme relative to their conditional distributions. This observation suggests using the conditional distributions for outlier checking, which will be demonstrated in Section 6.2.

5.3 Forecast distributions

We turn now to another type of conditional distribution, the forecast distribution of an HMM. Specifically we derive two expressions for the conditional distribution of X_{T+h} given $\mathbf{X}^{(T)} = \mathbf{x}^{(T)}$; h is termed the forecast horizon. Again we shall focus on the discrete case; the formulae for the continuous case are the same but with the probability functions replaced by density functions.

For discrete-valued observations the forecast distribution $\Pr(X_{T+h} = x \mid \mathbf{X}^{(T)} = \mathbf{x}^{(T)})$ of an HMM is very similar to the conditional distribution $\Pr(X_t = x \mid \mathbf{X}^{(-t)} = \mathbf{x}^{(-t)})$ just discussed, and can be computed in essentially the same way, as a ratio of likelihoods:

$$\begin{aligned}
\Pr(X_{T+h} = x \mid \mathbf{X}^{(T)} = \mathbf{x}^{(T)}) &= \frac{\Pr(\mathbf{X}^{(T)} = \mathbf{x}^{(T)}, X_{T+h} = x)}{\Pr(\mathbf{X}^{(T)} = \mathbf{x}^{(T)})} \\
&= \frac{\boldsymbol{\delta}\mathbf{P}(x_1)\mathbf{B}_2\mathbf{B}_3 \cdots \mathbf{B}_T\boldsymbol{\Gamma}^h\mathbf{P}(x)\mathbf{1}'}{\boldsymbol{\delta}\mathbf{P}(x_1)\mathbf{B}_2\mathbf{B}_3 \cdots \mathbf{B}_T\mathbf{1}'} \\
&= \frac{\boldsymbol{\alpha}_T\boldsymbol{\Gamma}^h\mathbf{P}(x)\mathbf{1}'}{\boldsymbol{\alpha}_T\mathbf{1}'}.
\end{aligned}$$

Writing $\boldsymbol{\phi}_T = \boldsymbol{\alpha}_T/\boldsymbol{\alpha}_T\mathbf{1}'$ (see Section 3.2), we have

$$\Pr(X_{T+h} = x \mid \mathbf{X}^{(T)} = \mathbf{x}^{(T)}) = \boldsymbol{\phi}_T\boldsymbol{\Gamma}^h\mathbf{P}(x)\mathbf{1}'. \qquad (5.4)$$

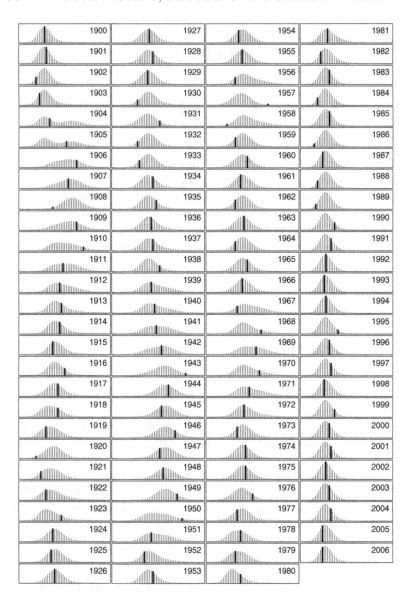

Figure 5.1 *Earthquakes data, three-state HMM: conditional distribution of the number of earthquakes in each year, given all the other observations. The bold bar corresponds to the actual number observed in that year.*

Expressions for joint distributions of several forecasts can be derived along the same lines. (See Exercise 5.)

The forecast distribution can therefore be written as a mixture of the state-dependent probability distributions:

$$\Pr(X_{T+h} = x \mid \mathbf{X}^{(T)} = \mathbf{x}^{(T)}) = \sum_{i=1}^{m} \xi_i(h)p_i(x), \qquad (5.5)$$

where the weight $\xi_i(h)$ is the ith entry of the vector $\boldsymbol{\phi}_T\boldsymbol{\Gamma}^h$. The **R** code for forecasts is given in Section A.1.14.

Since the entire probability distribution of the forecast is known, it is possible to make interval forecasts, and not only point forecasts. This is illustrated in Table 5.1, which lists statistics of some forecast distributions for the earthquake series fitted with a three-state Poisson HMM.

As the forecast horizon h increases, the forecast distribution converges to the marginal distribution of the stationary HMM, that is,

$$\lim_{h\to\infty} \Pr(X_{T+h} = x \mid \mathbf{X}^{(T)} = \mathbf{x}^{(T)}) = \lim_{h\to\infty} \boldsymbol{\phi}_T\boldsymbol{\Gamma}^h\mathbf{P}(x)\mathbf{1}' = \boldsymbol{\delta}^*\mathbf{P}(x)\mathbf{1}',$$

where we temporarily use $\boldsymbol{\delta}^*$ to denote the stationary distribution of the Markov chain (in order to distinguish it from $\boldsymbol{\delta}$, the initial distribution). The limit follows from the observation that, for any non-negative (row) vector $\boldsymbol{\eta}$ whose entries add to 1, the vector $\boldsymbol{\eta}\boldsymbol{\Gamma}^h$ approaches the stationary distribution of the Markov chain as $h \to \infty$, provided the Markov chain satisfies the usual regularity conditions of irreducibility and aperiodicity; see, for example, Feller (1968, p. 394). Sometimes the forecast distribution approaches its limiting distribution only slowly; see Figure 5.2, which displays six of the forecast distributions for the earthquakes series, compared with the limiting distribution. In other cases the approach can be relatively fast; for a case in point, consider a three-state HMM for the soap sales series given in Exercise 6 of Chapter 1. The rate of approach is determined by the size of the largest eigenvalue other than 1 of the t.p.m. $\boldsymbol{\Gamma}$.

5.4 Decoding

In speech recognition and other applications – see, for example, Fredkin and Rice (1992), or Guttorp (1995, p. 101) – it is of interest to determine the states of the Markov chain that are most likely (under the fitted model) to have given rise to the observation sequence. In the context of speech recognition this is known as the decoding problem; see Juang and Rabiner (1991). 'Local decoding' of the state at time t refers to the determination of that state which is most likely at that time. In contrast, 'global decoding' refers to the determination of the most likely *sequence* of states.

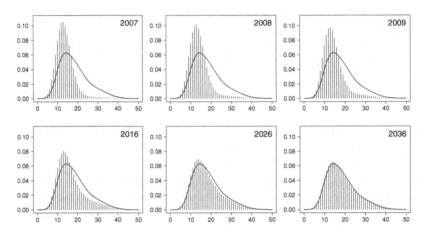

Figure 5.2 *Earthquakes data, three-state Poisson–HMM: forecast distributions for 1 to 30 years ahead, compared to limiting distribution, which is shown as a continuous line.*

Table 5.1 *Earthquakes data, three-state Poisson–HMM: forecasts.*

year:	2007	2008	2009	2016	2026	2036
horizon:	1	2	3	10	20	30
forecast mode:	13	13	13	13	14	14
forecast median:	12.7	12.9	13.1	14.4	15.6	16.2
forecast mean:	13.7	14.1	14.5	16.4	17.5	18.0
nominal 90%						
forecast interval:	[8,21]	[8,23]	[8,25]	[8,30]	[8,32]	[9,32]
exact coverage:	0.908	0.907	0.907	0.918	0.932	0.910

5.4.1 State probabilities and local decoding

Consider again the vectors of forward and backward probabilities, α_t and β_t, as discussed in Section 4.1. For the derivation of the most likely state of the Markov chain at time $t \in \{1, \ldots, T\}$, we shall use the following result, which appears there as equation (4.9):

$$\alpha_t(i)\beta_t(i) = \Pr(\mathbf{X}^{(T)} = \mathbf{x}^{(T)}, C_t = i).$$

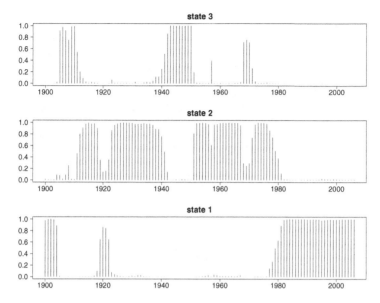

Figure 5.3 *Earthquakes data: state probabilities for fitted three-state HMM.*

Hence the conditional distribution of C_t given the observations can be obtained, for $i = 1, 2, \ldots, m$, as

$$
\Pr(C_t = i \mid \mathbf{X}^{(T)} = \mathbf{x}^{(T)}) = \frac{\Pr(C_t = i, \mathbf{X}^{(T)} = \mathbf{x}^{(T)})}{\Pr(\mathbf{X}^{(T)} = \mathbf{x}^{(T)})}
$$

$$
= \frac{\alpha_t(i)\beta_t(i)}{L_T}. \tag{5.6}
$$

Here L_T can be computed by the scaling method described in Section 3.2. Scaling is also necessary in order to prevent numerical underflow in the evaluation of the product $\alpha_t(i)\beta_t(i)$. The **R** code given in Section A.1.11 implements one method of doing this.

For each time $t \in \{1, \ldots, T\}$ one can therefore determine the distribution of the state C_t, given the observations $\mathbf{x}^{(T)}$, which for m states is a discrete probability distribution with support $\{1, \ldots, m\}$. In Figures 5.3 and 5.4 we display the state probabilities for the earthquakes series, based on the fitted three- and four-state Poisson–HMMs, respectively.

For each $t \in \{1, \ldots, T\}$ the most probable state i_t^*, given the observations, is defined as

$$
i_t^* = \operatorname*{argmax}_{i=1,\ldots,m} \Pr(C_t = i \mid \mathbf{X}^{(T)} = \mathbf{x}^{(T)}). \tag{5.7}
$$

This approach determines the most likely state separately for each t by

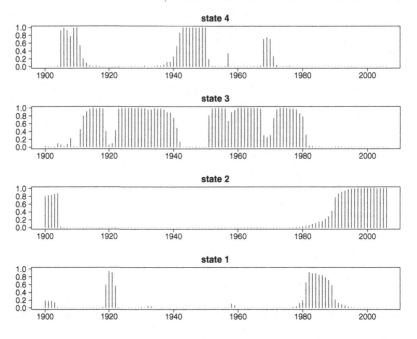

Figure 5.4 *Earthquakes data: state probabilities for fitted four-state HMM.*

maximizing the conditional probability $\Pr(C_t = i \mid \mathbf{X}^{(T)} = \mathbf{x}^{(T)})$ and is therefore called **local decoding**. In Figure 5.5 we display the results of applying local decoding to the earthquakes series, for the fitted three- and four-state models. The relevant **R** code is given in Sections A.1.11 and A.1.13.

5.4.2 Global decoding

In many applications, such as speech recognition, one is interested not so much in the most likely state for each separate time t (as provided by local decoding) as in the most likely *sequence* of (hidden) states. Instead of maximizing $\Pr(C_t = i \mid \mathbf{X}^{(T)} = \mathbf{x}^{(T)})$ over i for each t, one seeks that sequence of states $c_1, c_2, \ldots, c_T$ which maximizes the conditional probability

$$\Pr(\mathbf{C}^{(T)} = \mathbf{c}^{(T)} \mid \mathbf{X}^{(T)} = \mathbf{x}^{(T)}); \tag{5.8}$$

or equivalently, and more conveniently, the joint probability

$$\Pr(\mathbf{C}^{(T)}, \mathbf{X}^{(T)}) = \delta_{c_1} \prod_{t=2}^{T} \gamma_{c_{t-1}, c_t} \prod_{t=1}^{T} p_{c_t}(x_t).$$

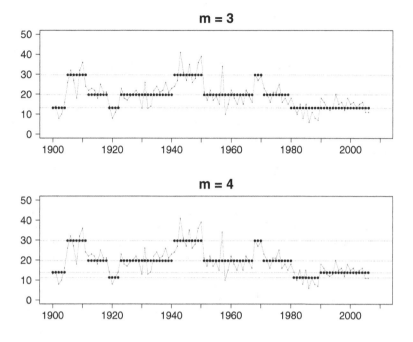

Figure 5.5 *Earthquakes data: local decoding according to three- and four-state HMMs. The horizontal lines indicate the state-dependent means.*

This is a subtly different maximization problem from that of local decoding, and is called **global decoding**. The results of local and global decoding are often very similar but not identical.

Maximizing (5.8) over all possible state sequences c_1, c_2, ..., c_T by brute force would involve m^T function evaluations, which is clearly not feasible except for very small T. Fortunately one can use instead an efficient dynamic programming algorithm to determine the most likely sequence of states, namely the Viterbi algorithm (Viterbi, 1967; Forney, 1973).

We begin by defining

$$\xi_{1i} = \Pr(C_1 = i, X_1 = x_1) = \delta_i \, p_i(x_1),$$

and, for $t = 2, 3, \ldots, T$,

$$\xi_{ti} = \max_{c_1, c_2, \ldots, c_{t-1}} \Pr(\mathbf{C}^{(t-1)} = \mathbf{c}^{(t-1)}, C_t = i, \mathbf{X}^{(t)} = \mathbf{x}^{(t)}).$$

It can then be shown (see Exercise 1) that the probabilities ξ_{tj} satisfy the following recursion, for $t = 2, 3, \ldots, T$ and $j = 1, 2, \ldots, m$:

$$\xi_{tj} = \left(\max_i (\xi_{t-1,i}\, \gamma_{ij}) \right) p_j(x_t). \tag{5.9}$$

This provides an efficient means of computing the $T \times m$ matrix of values ξ_{tj}, as the computational effort is linear in T. The required maximizing sequence of states $i_1, i_2, \ldots, i_T$ can then be determined recursively from

$$i_T = \operatorname*{argmax}_{i=1,\ldots,m} \xi_{Ti} \tag{5.10}$$

and, for $t = T-1, T-2, \ldots, 1$, from

$$i_t = \operatorname*{argmax}_{i=1,\ldots,m} (\xi_{ti}\, \gamma_{i,i_{t+1}}). \tag{5.11}$$

Note that, since the quantity to be maximized in global decoding is simply a product of probabilities (as opposed to a sum of such products), one can choose to maximize its logarithm, in order to prevent numerical underflow; the Viterbi algorithm can easily be rewritten in terms of the logarithms of the probabilities. Alternatively a scaling similar to that used in the likelihood computation can be employed: in that case one scales each of the T rows of the matrix $\{\xi_{ti}\}$ to have row sum 1. The Viterbi algorithm is applicable to both stationary and non-stationary underlying Markov chains; there is no necessity to assume that the initial distribution δ is the stationary distribution.

We now present an **R** function (essentially the same as that in Section A.1.6) which implements the Viterbi algorithm as described above, for a Poisson–HMM. The key lines of the code are

```
foo<-apply(xi[i-1,]*mod$gamma,2,max)*dpois(x[i],mod$lambda)
```

which corresponds to equation (5.9), and

```
iv[i]<-which.max(mod$gamma[,iv[i+1]]*xi[i,])
```

which corresponds to equation (5.11).

```
# Global decoding - Viterbi algorithm
pois.HMM.viterbi<-function(x,mod)
{
 n        <- length(x)
 xi       <- matrix(0,n,mod$m)
 foo      <- mod$delta*dpois(x[1],mod$lambda)
 xi[1,] <- foo/sum(foo)
 for (i in 2:n){
   foo    <- apply(xi[i-1,]*mod$gamma,2,max)*dpois(x[i],mod$lambda)
   xi[i,] <- foo/sum(foo)
   }
 iv       <- numeric(n)
 iv[n]    <- which.max(xi[n,])
 for (i in (n-1):1){
   iv[i] <- which.max(mod$gamma[,iv[i+1]]*xi[i,])
   }
```

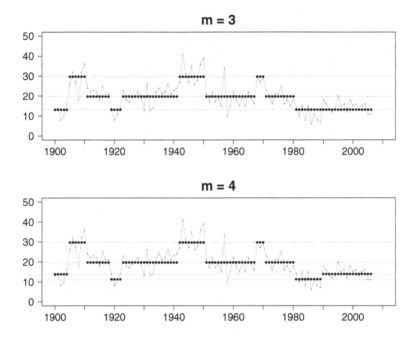

Figure 5.6 *Earthquakes: global decoding according to three- and four-state HMMs. The horizontal lines indicate the state-dependent means.*

```
 return(iv)
 }
```

Figure 5.6 displays, for the fitted three- and four-state models for the earthquakes series, the paths obtained by the Viterbi algorithm. The paths are very similar to those obtained by local decoding: compare with Figure 5.5. But they do differ. In the case of the three-state model, the years 1911, 1941 and 1980 differ. In the case of the four-state model, 1911 and 1941 differ. Notice also the nature of the difference between the upper and lower panels of Figure 5.6: allowing for a fourth state has the effect of splitting one of the states of the three-state model, that with the lowest mean. When four states are allowed, the 'Viterbi path' moves along the lowest state in the years 1919–1922 and 1981–1989 only.

Global decoding is the main objective in many applications, especially when there are substantive interpretations for the states. It is therefore of interest to investigate the performance of global decoding in identifying the correct states. This can be done by simulating a series from an HMM, applying the algorithm in order to decode the simulated observations, and then comparing the Viterbi path with the (known) series of simulated states. We present here an example of such a comparison, based on a simulated sequence of length 100 000 from the three-state (stationary) model for the earthquakes given on p. 55.

The 3×3 table displayed below with its marginals gives the simulated joint distribution of the true state i (rows) and the Viterbi estimate j of the state (columns). The row totals are close to (0.444, 0.405, 0.152), the stationary distribution of the model; this provides a partial check of the simulation. The column totals (the distribution of the state as inferred by Viterbi) are also close to this stationary distribution.

	$j = 1$	2	3	
$i = 1$	0.431	0.017	0.000	0.448
2	0.018	0.366	0.013	0.398
3	0.000	0.020	0.134	0.154
	0.449	0.403	0.147	1.000

From this table one may conclude, for instance, that the estimated probability that the state inferred is 2, if the true state is 1, is $0.017/0.448 = 0.038$. More generally, the left-hand table below gives $\Pr(\text{inferred state} = j \mid \text{true state} = i)$ and the right-hand table gives $\Pr(\text{true state} = i \mid \text{inferred state} = j)$.

	$j = 1$	2	3		$j = 1$	2	3
$i = 1$	0.961	0.038	0.000	$i = 1$	0.958	0.043	0.001
2	0.046	0.921	0.033	2	0.041	0.908	0.088
3	0.002	0.130	0.868	3	0.001	0.050	0.910

Ideally all the diagonal elements of the above two tables would be 1; here they range from 0.868 to 0.961. Such a simulation exercise quantifies the expected accuracy of the Viterbi path and is therefore particularly recommended in applications in which the interpretation of that path is an important objective of the analysis.

5.5 State prediction

In Section 5.4.1 we derived an expression for the conditional distribution of the state C_t, for $t = 1, 2, \ldots, T$, given the observations $\mathbf{x}^{(T)}$. In so doing we considered only present or past states. However, it is also possible to provide the conditional distribution of the state C_t for $t > T$, that is, to perform 'state prediction'.

Given the observations $x_1, \ldots, x_T$, the following set of statements can

Table 5.2 *Earthquakes data. State prediction using a three-state Poisson–HMM: the probability that the Markov chain will be in a given state in the specified year.*

Year	2007	2008	2009	2016	2026	2036
state=1	0.951	0.909	0.871	0.674	0.538	0.482
2	0.028	0.053	0.077	0.220	0.328	0.373
3	0.021	0.038	0.052	0.107	0.134	0.145

be made about future, present and past states (respectively):

$$L_T \Pr(C_t = i \mid \mathbf{X}^{(T)} = \mathbf{x}^{(T)})$$

$$= \begin{cases} \boldsymbol{\alpha}_T \boldsymbol{\Gamma}^{t-T} \mathbf{e}'_i & \text{for } t > T \qquad \text{state prediction} \\ \alpha_T(i) & \text{for } t = T \qquad \text{filtering} \\ \alpha_t(i)\beta_t(i) & \text{for } 1 \le t < T \quad \text{smoothing,} \end{cases}$$

where $\mathbf{e}_i = (0, \ldots, 0, 1, 0, \ldots, 0)$ has a one in the ith position only. The 'filtering' and 'smoothing' parts (for present or past states) are identical to the state probabilities as described in Section 5.4.1, and indeed could here be combined, since $\beta_T(i) = 1$ for all i. The 'state prediction' part is simply a generalization to $t > T$, the future, and can be restated as follows (see Exercise 6): for $i = 1, 2, \ldots, m$,

$$\Pr(C_{T+h} = i \mid \mathbf{X}^{(T)} = \mathbf{x}^{(T)}) = \boldsymbol{\alpha}_T \boldsymbol{\Gamma}^h \mathbf{e}'_i / L_T = \boldsymbol{\phi}_T \boldsymbol{\Gamma}^h \mathbf{e}'_i, \qquad (5.12)$$

with $\boldsymbol{\phi}_T = \boldsymbol{\alpha}_T / \boldsymbol{\alpha}_T \mathbf{1}'$. Note that, as $h \to \infty$, $\boldsymbol{\phi}_T \boldsymbol{\Gamma}^h$ converges to the stationary distribution of the Markov chain and so $\Pr(C_{T+h} = i \mid \mathbf{X}^{(T)})$ converges to the ith element thereof.

Table 5.2 gives, for a range of years, the state predictions based on the three-state model for the earthquake series. The **R** code for state prediction is given in Section A.1.12.

5.6 HMMs for classification

This book focuses on the use of HMMs as general-purpose models for time series, that is, models that encapsulate the main features of the observed series and enable one to extract meaningful information and to forecast. The unobserved states in the model are regarded as artefacts that are useful to accommodate unexplained heterogeneity and serial dependence, but they need not have substantive interpretations. (See also the discussion in Section 6.4.)

In contrast, primarily in the engineering literature, HMMs are often applied solely for the purpose of classifying observations on the state-dependent process into 'real' states. The classic example is in speech recognition (Rabiner, 1989), where the states are phonemes that need to be decoded from recorded speech. Other examples include face, handwriting, gesture and activity recognition. Ecologists may want to determine the behavioural states of an animal based on its remotely observed movements (see, for example, Grünewälder *et al.*, 2012). In such applications each state has an *a priori* well-defined meaning. The objective is to determine the most likely states. The observations are not *per se* of interest; their function is merely to provide information about the states.

In such applications one usually begins by taking a training sample in which both the states and the state-dependent observations are recorded. One then estimates the parameters of the model by maximizing the complete-data log-likelihood, analogously as in Section 4.2.2. The fitted model is used to determine the states that are most likely for new series whose states are unobserved. That step involves decoding states for the (unlabelled) series given the 'trained' HMM. In the machine-learning literature, such a procedure is referred to as *supervised learning*. In contrast, in the methods and applications covered in this book the role of the states is almost always completely data-driven (*unsupervised learning*).

Exercises

1. Prove the recursion (5.9):

$$\xi_{tj} = \left(\max_i(\xi_{t-1,i}\,\gamma_{ij}) \right) p_j(x_t).$$

2. Apply local and global decoding to a three-state model for the soap sales series introduced in Exercise 6 of Chapter 1, and compare the results to see how much the conclusions differ.

3. Compute the h-step-ahead state predictions for the soap sales series, for $h = 1, \ldots, 5$. How close are these distributions to the stationary distribution of the Markov chain?

4. (a) Using the same sequence of random numbers in each case, generate sequences of length 1000 from the Poisson–HMMs with

$$\Gamma = \begin{pmatrix} 0.8 & 0.1 & 0.1 \\ 0.1 & 0.8 & 0.1 \\ 0.1 & 0.1 & 0.8 \end{pmatrix},$$

and (i) $\lambda = (10, 20, 30)$, and (ii) $\lambda = (15, 20, 25)$. Keep a record of the sequence of states, which should be the same in (i) and (ii).

   (b) Use the Viterbi algorithm to infer the most likely sequence of states in each case, and compare these two sequences to the 'true' underlying sequence, i.e. the generated one.

(c) What conclusions do you draw about the accuracy of the Viterbi algorithm in this model?

5. (Bivariate forecast distributions for HMMs.)

   (a) Find the joint distribution of X_{T+1} and X_{T+2}, given $\mathbf{X}^{(T)}$, in as simple a form as you can.

   (b) For the earthquakes data, find $\Pr(X_{T+1} \leq 10, X_{T+2} \leq 10 \mid \mathbf{X}^{(T)})$.

6. Prove equation (5.12):

$$\Pr(C_{T+h} = i \mid \mathbf{X}^{(T)} = \mathbf{x}^{(T)}) = \boldsymbol{\alpha}_T \boldsymbol{\Gamma}^h \mathbf{e}_i' / L_T = \boldsymbol{\phi}_T \boldsymbol{\Gamma}^h \mathbf{e}_i'.$$

## CHAPTER 6

Model selection and checking

In the basic HMM with m states, increasing m always improves the fit of the model (as judged by the likelihood). But along with the improvement comes a quadratic increase in the number of parameters, and the improvement in fit has to be traded off against this increase. A criterion for model selection is therefore needed.

In some cases, it is sensible to reduce the number of parameters by making assumptions on the state-dependent distributions or on the t.p.m. of the Markov chain. For an example of the former, see p. 252, where, in order to model a series of categorical observations with 16 circular categories, von Mises distributions (with two parameters) are used as the state-dependent distributions. For an example of the latter, see Section 20.2, where we describe discrete state-space stochastic volatility models which are m-state HMMs with only three or four parameters. Notice that in this case the number of parameters does not increase at all with increasing m.

In this chapter we give a brief account of model selection in HMMs (Section 6.1), and then describe the use of pseudo-residuals in order to check for deficiencies in the selected model (Section 6.2).

6.1 Model selection by AIC and BIC

A problem which arises naturally when one uses hidden Markov (or other) models is that of selecting an appropriate model – for example, of choosing the appropriate number of states m, sometimes described as the 'order' of the HMM, or of choosing between competing state-dependent distributions such as Poisson and negative binomial. Although the question of order estimation for an HMM is neither trivial nor settled (see Cappé *et al.*, 2005, Chapter 15), we need some criterion for model comparison. The material outlined below is based on Zucchini (2000), which gives an introductory account of model selection. Celeux and Durand (2008) present and discuss several model selection techniques for selecting the number of states in an HMM.

Assume that the observations $x_1, \ldots, x_T$ were generated by the unknown 'true' or 'operating' model f and that one fits models from two different approximating families, $\{g_1 \in G_1\}$ and $\{g_2 \in G_2\}$. The goal of model selection is to identify the model which is in some sense the best.

We outline the two most popular approaches to model selection. In the frequentist approach one selects the family estimated to be closest to the operating model. For that purpose one defines a discrepancy (a measure of 'lack of fit') between the operating and the fitted models, $\Delta(f,\widehat{g}_1)$ and $\Delta(f,\widehat{g}_2)$. These discrepancies depend on the operating model f, which is unknown, and so it is not possible to determine which of the two discrepancies is smaller, that is, which model should be selected. Instead one bases selection on estimators of the expected discrepancies, namely $\widehat{\mathrm{E}}_f(\Delta(f,\widehat{g}_1))$ and $\widehat{\mathrm{E}}_f(\Delta(f,\widehat{g}_2))$, which are referred to as model selection criteria. By choosing the Kullback–Leibler discrepancy, and under the conditions listed in Appendix A of Linhart and Zucchini (1986), the model selection criterion simplifies to the Akaike information criterion (AIC):

$$\mathrm{AIC} = -2\log L + 2p,$$

where $\log L$ is the log-likelihood of the fitted model and p denotes the number of parameters of the model. The first term is a measure of fit, and decreases with increasing number of states m. The second term is a penalty term, and increases with increasing m.

The Bayesian approach to model selection is to select the family which is estimated to be most likely to be true. In a first step, before considering the observations, one specifies the priors, which are the probabilities $\Pr(f \in G_1)$ and $\Pr(f \in G_2)$ that f stems from the approximating family. In a second step one computes and compares the posteriors, which are the probabilities that f belongs to the approximating family, given the observations, $\Pr(f \in G_1 \mid \mathbf{x}^{(T)})$ and $\Pr(f \in G_2 \mid \mathbf{x}^{(T)})$. Under certain conditions (see, for example, Wasserman, 2000), this approach results in the Bayesian information criterion (BIC) which differs from AIC in the penalty term:

$$\mathrm{BIC} = -2\log L + p\log T,$$

where $\log L$ and p are as for the AIC, and T is the number of observations. Compared to the AIC, the penalty term of the BIC has more weight for $T > e^2$, which holds in most applications. Thus the BIC often favours models with fewer parameters than does the AIC.

For the earthquakes series, the AIC and BIC both select three states; see Figure 6.1, which plots the AIC and BIC against the number of states m of the HMM. The values of the two criteria are provided in Table 6.1. These values are also compared to those of the independent mixture models of Section 1.2.4. Although the HMMs demand more parameters than the comparable independent mixtures, the resulting values of the AIC and BIC are lower than those obtained for the independent mixtures.

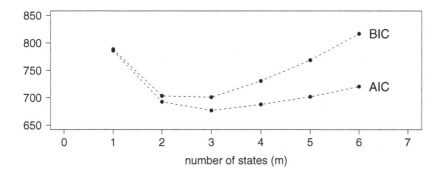

Figure 6.1 *Earthquakes series: model selection criteria AIC and BIC.*

Table 6.1 *Earthquakes data: comparison of (stationary) hidden Markov and independent mixture models by AIC and BIC.*

Model	k	$-\log L$	AIC	BIC
'1-state HM'	1	391.9189	785.8	788.5
2-state HM	4	342.3183	692.6	703.3
3-state HM	9	329.4603	**676.9**	**701.0**
4-state HM	16	327.8316	687.7	730.4
5-state HM	25	325.9000	701.8	768.6
6-state HM	36	324.2270	720.5	816.7
indep. mixture (2)	3	360.3690	726.7	734.8
indep. mixture (3)	5	356.8489	723.7	737.1
indep. mixture (4)	7	356.7337	727.5	746.2

Several comments arise from Table 6.1. Firstly, given the serial dependence manifested in Figure 2.1, it is not surprising that the independent mixture models do not perform well relative to the HMMs. Secondly, although it is perhaps obvious *a priori* that one should not even try to fit a model with as many as 25 or 36 parameters to 107 observations, and dependent observations at that, it is interesting to explore the likelihood functions in the case of HMMs with five and six states. The likelihood appears to be highly multimodal in these cases, and it is easy to find several local maxima by using different starting values. A strategy that seems to succeed in these cases is to start all the off-diagonal transition probabilities at small values (such as 0.01) and to space out the

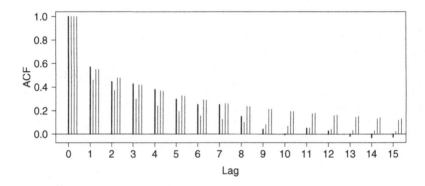

Figure 6.2 *Earthquakes data: sample ACF and ACF of three models. The bold bars on the left represent the sample ACF, and the other bars those of the HMMs with (from left to right) two, three and four states.*

state-dependent means over a range somewhat less than the range of the observations.

According to both AIC and BIC, the model with three states is the most appropriate. But, more generally, the model selected may depend on the selection criterion adopted. The selected model is displayed on p. 55, and the state-dependent distributions, together with the resulting marginal, are displayed in Figure 3.1 on p. 54.

It is also useful to compare the autocorrelation functions of the HMMs with two to four states with the sample ACF. The ACFs of the models can be found by using the results of Exercise 4 of Chapter 2, and appeared on p. 55. In tabular form the ACFs are as follows:

k:	1	2	3	4	5	6	7	8
observations	0.570	0.444	0.426	0.379	0.297	0.251	0.251	0.149
2-state model	0.460	0.371	0.299	0.241	0.194	0.156	0.126	0.101
3-state model	0.551	0.479	0.419	0.370	0.328	0.292	0.261	0.235
4-state model	0.550	0.477	0.416	0.366	0.324	0.289	0.259	0.234

In Figure 6.2 the sample ACF is juxtaposed to those of the models with two, three and four states. It is clear that the ACFs of the models with three and four states correspond well to the sample ACF up to about lag 7. However, one can apply more systematic diagnostics, as will now be shown.

6.2 Model checking with pseudo-residuals

Even when one has selected what is by some criterion the 'best' model, there still remains the problem of deciding whether the model is indeed adequate; one needs tools to assess the general goodness of fit of the model, and to identify outliers relative to the model. In the simpler context of normal-theory regression models (for instance), the role of residuals as a tool for model checking is very well established. In this section we describe quantities we call pseudo-residuals (but also known as quantile residuals) which are intended to fulfil this role much more generally, and which are useful in the context of HMMs. We consider two versions of these pseudo-residuals (in Sections 6.2.2 and 6.2.3); both rely on being able to perform likelihood computations routinely, which is certainly the case for HMMs. A detailed account, in German, of the construction and application of pseudo-residuals is provided by Stadie (2002). See also Zucchini and MacDonald (1999).

6.2.1 Introducing pseudo-residuals

To motivate pseudo-residuals we need the following simple result. Let X be a random variable with continuous distribution function F. Then $U \equiv F(X)$ is uniformly distributed on the unit interval, which we write

$$U \sim \mathrm{U}(0,1).$$

The proof is left to the reader as Exercise 1.

The **uniform pseudo-residual** of an observation x_t from a continuous random variable X_t is defined as the probability, under the fitted model, of obtaining an observation less than or equal to x_t:

$$u_t = \Pr(X_t \le x_t) = F_{X_t}(x_t).$$

That is, u_t is the observation x_t transformed by its distribution function under the model. If the model is correct, this type of pseudo-residual is distributed $\mathrm{U}(0,1)$, with residuals for extreme observations close to 0 or 1. With the help of these uniform pseudo-residuals, observations from different distributions can be compared. If we have observations $x_1, \ldots, x_T$ and a model $X_t \sim F_t$, for $t = 1, \ldots, T$ (i.e. each x_t has its own distribution function F_t), then the x_t-values cannot be compared directly. However, the pseudo-residuals u_t are identically $\mathrm{U}(0,1)$ (if the model is true), and can sensibly be compared. If a histogram or quantile–quantile plot ('qq-plot') of the uniform pseudo-residuals u_t casts doubt on the conclusion that they are $\mathrm{U}(0,1)$, one can deduce that the model is not valid.

Although the uniform pseudo-residual is useful in this way, it has a drawback if used for outlier identification. For example, if one considers

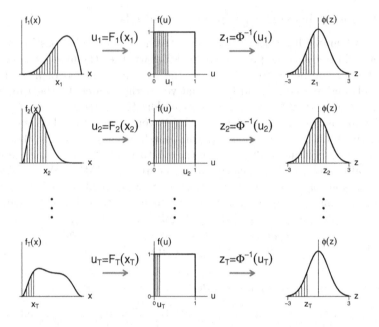

Figure 6.3 *Construction of normal pseudo-residuals in the continuous case.*

the values lying close to 0 or 1 on an index plot, it is hard to see whether a value is very unlikely or not. A value of 0.999, for instance, is difficult to distinguish from a value of 0.97, and such an index plot not a useful device for detecting outliers.

This deficiency of uniform pseudo-residuals can, however, easily be remedied by using the following result. Let Φ be the distribution function of the standard normal distribution and X a random variable with distribution function F. Then $Z \equiv \Phi^{-1}(F(X))$ is distributed standard normal. (For the proof we refer again to Exercise 1.) We now define the **normal pseudo-residual** as

$$z_t = \Phi^{-1}(u_t) = \Phi^{-1}(F_{X_t}(x_t)).$$

If the fitted model is valid, these normal pseudo-residuals are distributed standard normal, with the value of the residual equal to 0 when the observation coincides with the median. Note that, by their definition, normal pseudo-residuals measure the deviation from the median, and not from the expectation. The construction of normal pseudo-residuals is illustrated in Figure 6.3. If the observations $x_1, \ldots, x_T$ were indeed generated by the model $X_t \sim F_t$, the normal pseudo-residuals z_t would follow a standard normal distribution. One can therefore check the model either

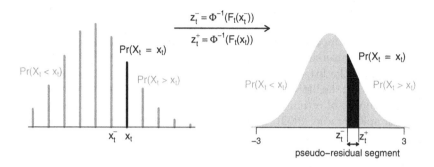

Figure 6.4 *Construction of normal pseudo-residuals in the discrete case.*

by visually analysing the histogram or qq-plot of the normal pseudo-residuals, or by performing tests for normality.

This normal version of pseudo-residuals has the advantage that the absolute value of the residual increases with increasing deviation from the median and that extreme observations can be identified more easily on a normal scale. This becomes obvious if one compares index plots of uniform and normal pseudo-residuals.

Note that the theory of pseudo-residuals as outlined so far can be applied to continuous distributions only. In the case of discrete observations, the pseudo-residuals can, however, be modified to allow for the discreteness. The pseudo-residuals are no longer defined as points, but as intervals. Thus, for a discrete random variable X_t with distribution function F_{X_t} we define the uniform pseudo-residual segments as

$$[u_t^-; u_t^+] = [F_{X_t}(x_t^-); F_{X_t}(x_t)],\qquad(6.1)$$

with x_t^- denoting the greatest realization possible that is strictly less than x_t, and we define the normal pseudo-residual segments as

$$[z_t^-; z_t^+] = [\Phi^{-1}(u_t^-); \Phi^{-1}(u_t^+)] = [\Phi^{-1}(F_{X_t}(x_t^-)); \Phi^{-1}(F_{X_t}(x_t))].\ (6.2)$$

The construction of the normal pseudo-residual segment of a discrete random variable is illustrated in Figure 6.4.

Both versions of pseudo-residual segments (uniform and normal) contain information on how extreme and how rare the observations are, although the uniform version represents the rarity or otherwise more directly, as the length of the segment is the corresponding probability. For example, the lower limit u_t^- of the uniform pseudo-residual interval specifies the probability of observing a value strictly less than x_t, $1 - u_t^+$ gives the probability of a value strictly greater than x_t, and the difference $u_t^+ - u_t^-$ is equal to the probability of the observation x_t under the fitted model. The pseudo-residual segments can be interpreted as interval-censored realizations of a uniform (or standard normal) distribution, if the fitted model is valid. Though this is correct only if the parameters of the fitted model are known, it is still approximately correct if the number of estimated parameters is small compared to the size of the sample (Stadie, 2002). Diagnostic plots of pseudo-residual segments of discrete random variables necessarily look rather different from those of continuous random variables.

It is easy to construct an index plot of pseudo-residual segments or to plot these against any independent or dependent variable. However, in order to construct a qq-plot of the pseudo-residual segments one has to specify an ordering of the pseudo-residual segments. One possibility is to sort on the so-called 'mid-pseudo-residuals' which are defined as

$$z_t^m = \Phi^{-1}\left(\frac{u_t^- + u_t^+}{2}\right). \qquad (6.3)$$

Furthermore, the mid-pseudo-residuals can themselves be used for checking for normality, for example via a histogram of mid-pseudo-residuals. But we can claim no more than approximate normality for such mid-pseudo-residuals.

Now, having outlined the properties of pseudo-residuals, we can consider the use of pseudo-residuals in the context of HMMs. The analysis of the pseudo-residuals of an HMM serves two purposes: the assessment of the general fit of a selected model, and the detection of outliers. Depending on the aspects of the model that are to be analysed, one can distinguish two kinds of pseudo-residual that are useful for an HMM: those that are based on the conditional distribution given all other observations, which we call **ordinary pseudo-residuals**, and those based on the conditional distribution given all preceding observations, which we call **forecast pseudo-residuals**.

That the pseudo-residuals of a set of observations are (approximately) identically distributed, either $U(0, 1)$ or standard normal, is their crucial property. But for our purposes it is not important whether such pseudo-residuals are independent of each other; indeed, we shall see in Section 6.3.2 that it would be wrong to assume of ordinary pseudo-residuals that they are independent.

Note that Dunn and Smyth (1996) discuss (under the name 'quantile residual') what we have called normal pseudo-residuals, and point out that they are a case of Cox–Snell residuals (Cox and Snell, 1968).

6.2.2 Ordinary pseudo-residuals

The first technique considers the observations one at a time and seeks those which, relative to the model and *all* other observations in the series, are sufficiently extreme to suggest that they differ in nature or origin from the others. This means that one computes a pseudo-residual z_t from the conditional distribution of X_t, given $\mathbf{X}^{(-t)}$: a 'full conditional distribution', in the terminology used in Markov chain Monte Carlo. For continuous observations the normal pseudo-residual is

$$z_t = \Phi^{-1}\left(\Pr(X_t \leq x_t \mid \mathbf{X}^{(-t)} = \mathbf{x}^{(-t)})\right).$$

If the model is correct, z_t is a realization of a standard normal random variable. For discrete observations the normal pseudo-residual segment is $[z_t^-; z_t^+]$, where

$$z_t^- = \Phi^{-1}\left(\Pr(X_t < x_t \mid \mathbf{X}^{(-t)} = \mathbf{x}^{(-t)})\right)$$

and

$$z_t^+ = \Phi^{-1}\left(\Pr(X_t \leq x_t \mid \mathbf{X}^{(-t)} = \mathbf{x}^{(-t)})\right).$$

In the discrete case the conditional probabilities $\Pr(X_t = x \mid \mathbf{X}^{(-t)} = \mathbf{x}^{(-t)})$ are given by equations (5.1) and (5.2) in Section 5.2; the continuous case is similar, with probabilities replaced by densities.

Section 6.3.1 applies ordinary pseudo-residuals to HMMs with one to four states fitted to the earthquakes data, and a further example of their use appears in Figure 15.3 and the corresponding text.

6.2.3 Forecast pseudo-residuals

The second technique for outlier detection seeks observations that are extreme relative to the model and all *preceding* observations (as opposed to all other observations). In this case the relevant conditional distribution is that of X_t given $\mathbf{X}^{(t-1)}$. The corresponding (normal) pseudo-residuals are

$$z_t = \Phi^{-1}\left(\Pr(X_t \leq x_t \mid \mathbf{X}^{(t-1)} = \mathbf{x}^{(t-1)})\right)$$

for continuous observations; and $[z_t^-; z_t^+]$ for discrete, where

$$z_t^- = \Phi^{-1}\left(\Pr(X_t < x_t \mid \mathbf{X}^{(t-1)} = \mathbf{x}^{(t-1)})\right)$$

and

$$z_t^+ = \Phi^{-1}\left(\Pr(X_t \leq x_t \mid \mathbf{X}^{(t-1)} = \mathbf{x}^{(t-1)})\right).$$

In the discrete case the required conditional probability $\Pr(X_t = x_t \mid \mathbf{X}^{(t-1)} = \mathbf{x}^{(t-1)})$ is given by the ratio of the likelihood of the first t

observations to that of the first $t - 1$:

$$\Pr(X_t = x \mid \mathbf{X}^{(t-1)} = \mathbf{x}^{(t-1)}) = \frac{\boldsymbol{\alpha}_{t-1} \boldsymbol{\Gamma} \mathbf{P}(x) \mathbf{1}'}{\boldsymbol{\alpha}_{t-1} \mathbf{1}'} .$$

The pseudo-residuals of this second type are described as forecast pseudo-residuals because they measure the deviation of an observation from the median of the corresponding one-step-ahead forecast. If a forecast pseudo-residual is extreme, this indicates that the observation concerned is an outlier, or that the model no longer provides an acceptable description of the series. This provides a method for the continuous monitoring of the behaviour of a time series. An example of such monitoring is given at the end of Section 22.4; see Figure 22.5.

The idea of forecast pseudo-residual appears – as 'conditional quantile residual' – in Dunn and Smyth (1996); in the last paragraph on p. 243 they point out that the quantile residuals they describe can be extended to serially dependent data. The basic idea of (uniform) forecast pseudo-residuals goes back to Rosenblatt (1952), however. Both Brockwell (2007) and Dunn and Smyth describe a way of extending what we call forecast pseudo-residuals to distributions other than continuous. Instead of using a segment of positive length to represent the residual if the observations are not continuous, they choose a point distributed uniformly on that segment. The use of a segment of positive length has the advantage, we believe, of explicitly displaying the discreteness of the observation, and indicating both its extremeness and its rarity.

Another example of the use of forecast pseudo-residuals appears in Figure 15.4 and the corresponding text.

6.3 Examples

6.3.1 Ordinary pseudo-residuals for the earthquakes

In Figure 6.5 we show several types of residual plot for the fitted models of the earthquakes series, using the first definition of pseudo-residual, that based on the conditional distribution relative to all other observations, $\Pr(X_t = x \mid \mathbf{X}^{(-t)} = \mathbf{x}^{(-t)})$. The relevant code appears in Section A.1.10. It is interesting to compare the pseudo-residuals of the selected three-state model to those of the models with one, two and four states.

As regards the residual plots provided in Figure 6.5, it is clear that the selected three-state model provides an acceptable fit while, for example, the normal pseudo-residuals of the one-state model (a single Poisson distribution) deviate strikingly from the standard normal distribution. If, however, we consider only the residual plots (other than perhaps the qq-plot), and not the model selection criteria, we might even accept the two-state model as an adequate alternative.

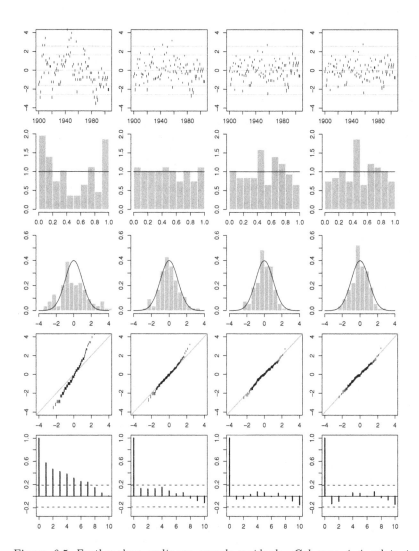

Figure 6.5 *Earthquakes: ordinary pseudo-residuals. Columns 1–4 relate to HMMs with (respectively) 1, 2, 3, 4 states. The top row shows index plots of the normal pseudo-residuals, with horizontal lines at 0, ±1.96, ±2.58. The second and third rows show histograms of the uniform and the normal pseudo-residuals. The fourth row shows quantile–quantile plots of the normal pseudo-residuals, with the theoretical quantiles on the horizontal axis. The last row shows the autocorrelation functions of the normal pseudo-residuals.*

Looking at the last row of Figure 6.5, one might be tempted to conjecture that, if a model is 'true', the ordinary pseudo-residuals will be independent, or at least uncorrelated. From the example in the next section we shall see that that would be an incorrect conclusion.

6.3.2 Dependent ordinary pseudo-residuals

Consider the stationary Gaussian AR(1) process $X_t = \phi X_{t-1} + \varepsilon_t$, with the innovations ε_t independent standard normal. It follows that $|\phi| < 1$ and $\text{Var}(X_t) = 1/(1 - \phi^2)$.

Let t lie strictly between 1 and T. The conditional distribution of X_t given $\mathbf{X}^{(-t)}$ is that of X_t given only X_{t-1} and X_{t+1}. This latter conditional distribution can be found by noting that the joint distribution of X_t, X_{t-1} and X_{t+1} (in that order) is normal with mean vector $\mathbf{0}$ and covariance matrix

$$\Sigma = \frac{1}{1 - \phi^2} \begin{pmatrix} 1 & \phi & \phi \\ \phi & 1 & \phi^2 \\ \phi & \phi^2 & 1 \end{pmatrix}.$$

The required conditional distribution then turns out to be normal with mean $\phi(X_{t-1} + X_{t+1})/(1 + \phi^2)$ and variance $(1 + \phi^2)^{-1}$. Hence the corresponding uniform pseudo-residual is

$$\Phi\left(\left(X_t - \frac{\phi}{1 + \phi^2}(X_{t-1} + X_{t+1})\right)\sqrt{1 + \phi^2}\right),$$

and the normal pseudo-residual is

$$z_t = \left(X_t - \frac{\phi}{1 + \phi^2}(X_{t-1} + X_{t+1})\right)\sqrt{1 + \phi^2}.$$

By the properties of ordinary normal pseudo-residuals (see Section 6.2.2), z_t is unconditionally standard normal; this can also be verified directly.

Whether (for example) the pseudo-residuals z_t and z_{t+1} are independent is not immediately obvious. The answer is that they are not independent; the correlation of z_t and z_{t+1} can (for $t > 1$ and $t < T - 1$) be obtained by routine manipulation, and turns out to be

$$-\phi/(1 + \phi^2);$$

see Exercise 3. This is opposite in sign to ϕ and smaller in modulus. For instance, if $\phi = 1/\sqrt{2}$, the correlation of z_t and z_{t+1} is $-2\phi/3$. (In contrast, the corresponding *forecast* pseudo-residuals do have zero correlation at lag 1.)

One can, also by routine manipulation, show that $\text{Cov}(z_t, z_{t+2}) = 0$ and that, for all integers $k \geq 3$, $\text{Cov}(z_t, z_{t+k}) = \phi \, \text{Cov}(z_t, z_{t+k-1})$. Consequently, for all integers $k \geq 2$, $\text{Cov}(z_t, z_{t+k}) = 0$.

6.4 Discussion

This may be an appropriate point at which to stress the dangers of over-interpretation. Although our earthquakes model seems adequate in important respects, this does not imply that it can be interpreted substantively. Nor, indeed, are we aware of any convincing seismological interpretation of the three states we propose. But models need not have a substantive interpretation to be useful; many useful statistical models are merely empirical models, in the sense in which Cox (1990) uses that term.

Latent-variable models of all kinds, including independent mixtures and HMMs, seem to be particularly prone to over-interpretation, and we would caution against the error of reification: the tendency to regard as physically real anything that has been given a name. In this spirit we can do no better than to follow Gould (1997, p. 350) in quoting John Stuart Mill:

> The tendency has always been strong to believe that whatever received a name must be an entity or being, having an independent existence of its own. And if no real entity answering to the name could be found, men did not for that reason suppose that none existed, but imagined that it was something peculiarly abstruse and mysterious.*

Exercises

1. (a) Let X be a continuous random variable with distribution function F. Show that the random variable $U = F(X)$ is uniformly distributed on the interval $[0, 1]$, i.e. $U \sim U(0, 1)$.

   (b) Suppose that $U \sim U(0, 1)$ and let F be the distribution function of a continuous random variable. Show that the random variable $X = F^{-1}(U)$ has the distribution function F.

   (c) i. Give the explicit expression for F^{-1} for the exponential distribution, i.e. the distribution with density function $f(x) = \lambda e^{-\lambda x}$, $x \geq 0$.

      ii. Verify your result by generating 1000 uniformly distributed random numbers, transforming these by applying F^{-1}, and then examining the histogram of the resulting values.

   (d) Show that for a continuous random variable X with distribution function F, the random variable $Z = \Phi^{-1}(F(X))$ is distributed standard normal.

2. Consider the AR(1) process in the example of Section 6.3.2. That

---

* This appears to be a slightly inaccurate quotation from a footnote added by J.S. Mill to Volume II, Chapter XIV of the second edition of a work of his father, James Mill: *Analysis of the Phenomena of the Human Mind*.

the conditional distribution of X_t given $\mathbf{X}^{(-t)}$ depends only on X_{t-1} and X_{t+1} is fairly obvious because X_t depends on $X_1, \ldots, X_{t-2}$ only through X_{t-1}, and on $X_{t+2}, \ldots, X_T$ only through X_{t+1}.

Establish this more formally by writing the densities of $\mathbf{X}^{(T)}$ and $\mathbf{X}^{(-t)}$ in terms of conditional densities $p(x_u \mid x_{u-1})$ and noting that, in their ratio, many of the factors cancel.

3. Verify, in the AR(1) example of Section 6.3.2, that for appropriate ranges of t-values:

   (a) given X_{t-1} and X_{t+1},

   $$X_t \sim \mathrm{N}\left(\phi(X_{t-1} + X_{t+1})/(1+\phi^2), (1+\phi^2)^{-1}\right);$$

   (b) $\mathrm{Var}(z_t) = 1$;
   (c) $\mathrm{Corr}(z_t, z_{t+1}) = -\phi/(1+\phi^2)$;
   (d) $\mathrm{Cov}(z_t, z_{t+2}) = 0$; and
   (e) for all integers $k \geq 3$, $\mathrm{Cov}(z_t, z_{t+k}) = \phi\,\mathrm{Cov}(z_t, z_{t+k-1})$.

4. Let z_t be as in the AR(1) example of Section 6.3.2. Show as follows that $\{z_t\}$ is a moving-average process of order 1:

   $$z_t = (1+\phi^2)^{-\frac{1}{2}}(\varepsilon_t - \phi\varepsilon_{t+1}).$$

   (This is probably the simplest way of arriving at conclusions (c) and (d) in Exercise 3.)

5. Generate a two-state stationary Poisson–HMM $\{X_t\}$ with t.p.m. $\boldsymbol{\Gamma} = \begin{pmatrix} 0.6 & 0.4 \\ 0.4 & 0.6 \end{pmatrix}$, and with state-dependent means $\lambda_1 = 2$ and $\lambda_2 = 5$ for $t = 1, \ldots, 100$, and $\lambda_1 = 2$ and $\lambda_2 = 7$ for $t = 101, \ldots, 120$. Fit a model to the first 80 observations, and use forecast pseudo-residuals to monitor the next 40 observations for evidence of a change.

6. Consider again the soap sales series introduced in Exercise 6 of Chapter 1.

   (a) Use the AIC and BIC to decide how many states are needed in a Poisson–HMM for these data.

   (b) Compute the pseudo-residuals relative to Poisson–HMMs with 1–4 states, and use plots similar to those in Figure 6.5 to decide how many states are needed.

## CHAPTER 7

Bayesian inference for Poisson–hidden Markov models

As alternative to the frequentist approach, one can also consider Bayesian estimation. There are several approaches to Bayesian inference in hidden Markov models; see, for instance, Chib (1996), Robert and Titterington (1998), Robert, Rydén and Titterington (2000), Scott (2002), Cappé *et al.* (2005), Frühwirth-Schnatter (2006) and Rydén (2008). Here we follow Scott (2002) and Congdon (2006).

Our purpose is to demonstrate an application of Bayesian inference to Poisson–HMMs. There are obstacles to be overcome, such as label switching and the difficulty of estimating m, the number of states, and some of these are model specific.

7.1 Applying the Gibbs sampler to Poisson–HMMs

7.1.1 Introduction and outline

We consider here a Poisson–HMM $\{X_t\}$ on m states, with underlying Markov chain $\{C_t\}$. We denote the state-dependent means, as usual, by $\boldsymbol{\lambda} = (\lambda_1, \ldots, \lambda_m)$, and the transition probability matrix of the Markov chain by $\boldsymbol{\Gamma}$.

Given a sequence of observations $x_1, x_2, \ldots, x_T$, a fixed m, and prior distributions on the parameters $\boldsymbol{\lambda}$ and $\boldsymbol{\Gamma}$, our objective in this section is to estimate the posterior distribution of these parameters by means of the Gibbs sampler. We shall later (in Section 7.2) drop the assumption that m is known, and also consider the Bayesian estimation thereof.

The prior distributions we assume for the parameters are of the following forms. The rth row $\boldsymbol{\Gamma}_r$ of the t.p.m. $\boldsymbol{\Gamma}$ is assumed to have a Dirichlet distribution with parameter vector $\boldsymbol{\nu}_r$, and the increment $\tau_j = \lambda_j - \lambda_{j-1}$ (with $\lambda_0 \equiv 0$) to have a gamma distribution with shape parameter a_j and rate parameter b_j. Furthermore, the rows of $\boldsymbol{\Gamma}$ and the quantities τ_j are assumed mutually independent in their prior distributions.

Our notation and terminology are as follows. Random variables Y_1, $\ldots, Y_m$ are here said to have a Dirichlet distribution with parameter vector $(\nu_1, \ldots, \nu_m)$ if their joint density is proportional to

$$y_1^{\nu_1-1} y_2^{\nu_2-1} \cdots y_m^{\nu_m-1}.$$

More precisely, this expression, with y_m replaced by $1 - \sum_{i=1}^{m-1} y_i$, is (up to proportionality) the joint density of $Y_1, \ldots, Y_{m-1}$ on the unit simplex in dimension $m - 1$, that is, on the subspace of $\mathbb{R}^{m-1}$ defined by $\sum_{i=1}^{m-1} y_i \leq 1$, $y_i \geq 0$. A random variable X is said to have a gamma distribution with shape parameter a and rate parameter b if its density is (for positive x)

$$f(x) = \frac{b^a}{\Gamma(a)} x^{a-1} e^{-bx}.$$

With this parametrization, X has mean a/b, variance a/b^2 and coefficient of variation (c.v.) $1/\sqrt{a}$.

If it were possible to observe the Markov chain, updating the transition probabilities $\boldsymbol{\Gamma}$ would be straightforward. Here, however, we have to generate sample paths of the Markov chain in order to update $\boldsymbol{\Gamma}$.

An important part of Scott's model structure, which we copy, is this. Each observed count x_t is considered to be the sum $\sum_j x_{jt}$ of contributions from up to m regimes, the contribution of regime j to x_t being x_{jt}. Note that, if the Markov chain is in state i at a given time, regimes $1, \ldots, i$ are *all* said to be active at that time, and regimes $i+1, \ldots, m$ to be inactive. This is an unusual use of the word 'regime', but convenient here.

Instead of parametrizing the model in terms of the m state-dependent means λ_i, we parametrize it in terms of the non-negative increments $\boldsymbol{\tau} = (\tau_1, \ldots, \tau_m)$, where $\tau_j = \lambda_j - \lambda_{j-1}$ (with $\lambda_0 \equiv 0$). Equivalently,

$$\lambda_i = \sum_{j=1}^{i} \tau_j.$$

This has the effect of placing the λ_j in increasing order, which is useful in order to prevent the technical problem known as label switching. For an account of this problem, see, for example, Frühwirth-Schnatter (2006, Section 3.5.5). The random variable τ_j can be described as the mean contribution of regime j, if active, to the count observed at a given time.

In outline, we proceed as follows.

- Given the observed counts $\mathbf{x}^{(T)}$ and the current values of the parameters $\boldsymbol{\Gamma}$ and $\boldsymbol{\lambda}$, we generate a sample path of the Markov chain (MC).

- We use this sample path to decompose the observed counts into (simulated) regime contributions.

- With the MC sample path available, and the regime contributions, we can now update $\boldsymbol{\Gamma}$ and $\boldsymbol{\tau}$, hence $\boldsymbol{\lambda}$.

The above steps are repeated a large number of times and, after a 'burn-in period', the resulting samples of values of $\boldsymbol{\Gamma}$ and $\boldsymbol{\lambda}$ provide the required estimates of their posterior distributions. In what follows, we use $\boldsymbol{\theta}$ to represent both $\boldsymbol{\Gamma}$ and $\boldsymbol{\lambda}$.

7.1.2 Generating sample paths of the Markov chain

Given the observations $\mathbf{x}^{(T)}$ and the current values of the parameters $\boldsymbol{\theta}$, we wish to simulate a sample path $\mathbf{C}^{(T)}$ of the Markov chain, from its conditional distribution

$$\Pr(\mathbf{C}^{(T)} \mid \mathbf{x}^{(T)}, \boldsymbol{\theta}) = \Pr(C_T \mid \mathbf{x}^{(T)}, \boldsymbol{\theta}) \times \prod_{t=1}^{T-1} \Pr(C_t \mid \mathbf{x}^{(T)}, \mathbf{C}_{t+1}^T, \boldsymbol{\theta}).$$

We shall be drawing values for $C_T, C_{T-1}, \ldots, C_1$, in that order, and some of the quantities that we shall need in order to do so are the probabilities

$$\Pr(C_t \mid \mathbf{x}^{(t)}, \boldsymbol{\theta}) = \frac{\Pr(C_t, \mathbf{x}^{(t)} \mid \boldsymbol{\theta})}{\Pr(\mathbf{x}^{(t)} \mid \boldsymbol{\theta})} = \frac{\alpha_t(C_t)}{L_t} \propto \alpha_t(C_t), \quad \text{for } t = 1, \ldots, T. \tag{7.1}$$

As before (see p. 65), $\boldsymbol{\alpha}_t = (\alpha_t(1), \ldots, \alpha_t(m))$ denotes the vector of forward probabilities

$$\alpha_t(i) = \Pr(\mathbf{x}^{(t)}, C_t = i),$$

which can be computed from the recursion $\boldsymbol{\alpha}_t = \boldsymbol{\alpha}_{t-1} \boldsymbol{\Gamma} \mathbf{P}(x_t)$ ($t = 2, \ldots, T$), with $\boldsymbol{\alpha}_1 = \boldsymbol{\delta} \mathbf{P}(x_1)$. L_t is the likelihood of the first t observations.

We start the simulation by drawing C_T, the state of the Markov chain at the final time T, from $\Pr(C_T \mid \mathbf{x}^{(T)}, \boldsymbol{\theta}) \propto \alpha_T(C_T)$, (i.e. case $t = T$ of equation (7.1)). We then simulate the states C_t (in the order $t = T - 1$, $T - 2$, $\ldots$, 1) by making use of the following proportionality argument, as in Chib (1996):

$$\Pr(C_t \mid \mathbf{x}^{(T)}, \mathbf{C}_{t+1}^T, \boldsymbol{\theta})$$
$$\propto \Pr(C_t \mid \mathbf{x}^{(t)}, \boldsymbol{\theta}) \Pr(\mathbf{x}_{t+1}^T, \mathbf{C}_{t+1}^T \mid \mathbf{x}^{(t)}, C_t, \boldsymbol{\theta})$$
$$\propto \Pr(C_t \mid \mathbf{x}^{(t)}, \boldsymbol{\theta}) \Pr(C_{t+1} \mid C_t, \boldsymbol{\theta}) \Pr(\mathbf{x}_{t+1}^T, \mathbf{C}_{t+2}^T \mid \mathbf{x}^{(t)}, C_t, C_{t+1}, \boldsymbol{\theta})$$
$$\propto \alpha_t(C_t) \Pr(C_{t+1} \mid C_t, \boldsymbol{\theta}). \tag{7.2}$$

The third factor appearing in the second-last line is independent of C_t, hence the simplification. (See Exercise 4.) Expression (7.2) is easily available, since the second factor in it is simply a one-step transition probability in the Markov chain. We are therefore in a position to simulate sample paths of the Markov chain, given observations $\mathbf{x}^{(T)}$ and parameters $\boldsymbol{\theta}$.

7.1.3 Decomposing the observed counts into regime contributions

Suppose we have a sample path $\mathbf{C}^{(T)}$ of the Markov chain, generated as described in Section 7.1.2, and suppose that $C_t = i$, so that regimes 1, $\ldots, i$ are active at time t. Our next step is to decompose each observation x_t ($t = 1, 2, \ldots, T$) into regime contributions $x_{1t}, \ldots, x_{it}$ such that $\sum_{j=1}^{i} x_{jt} = x_t$. We therefore need the joint distribution of $X_{1t}, \ldots, X_{it}$, given $C_t = i$ and $X_t = x_t$ (and given $\boldsymbol{\theta}$). This is multinomial with total x_t and probability vector proportional to $(\tau_1, \ldots, \tau_i)$; see Exercise 1.

7.1.4 Updating the parameters

The t.p.m. $\boldsymbol{\Gamma}$ can now be updated, that is, new estimates produced. This we do by drawing $\boldsymbol{\Gamma}_r$, the rth row of $\boldsymbol{\Gamma}$, from the Dirichlet distribution with parameter vector $\boldsymbol{\nu}_r + \mathbf{T}_r$, where $\mathbf{T}_r$ is the rth row of the (simulated) matrix of transition counts; see Section 7.1.2. (Recall that the prior for $\boldsymbol{\Gamma}_r$ is Dirichlet($\boldsymbol{\nu}_r$), and see Exercise 2.)

Similarly, the vector $\boldsymbol{\lambda}$ of state-dependent means is updated by drawing τ_j ($j = 1, \ldots, m$) from a gamma distribution with parameters $a_j + \sum_{t=1}^{T} x_{jt}$ and $b_j + N_j$; here N_j denotes the number of times regime j was active in the simulated sample path of the Markov chain, and x_{jt} the contribution of regime j to x_t. (Recall that the prior for τ_j is a gamma distribution with shape parameter a_j and rate parameter b_j, and see Exercise 3.)

7.2 Bayesian estimation of the number of states

In the Bayesian approach to model selection, the number of states, m, is a parameter whose value is assessed from its posterior distribution, $p(m \mid \mathbf{x}^{(T)})$. Computing this posterior distribution is, however, not an easy problem; indeed, it has been described as 'notoriously difficult to calculate' (Scott, James and Sugar, 2005).

Using p as a general symbol for probability mass or density functions, one has

$$p(m \mid \mathbf{x}^{(T)}) = p(m)\,p(\mathbf{x}^{(T)} \mid m)/p(\mathbf{x}^{(T)}) \propto p(m)\,p(\mathbf{x}^{(T)} \mid m), \quad (7.3)$$

where $p(\mathbf{x}^{(T)} \mid m)$ is called the **integrated likelihood**. If only two models are being compared, the posterior odds are equal to the product of the 'Bayes factor' and the prior odds:

$$\frac{p(m_2 \mid \mathbf{x}^{(T)})}{p(m_1 \mid \mathbf{x}^{(T)})} = \frac{p(\mathbf{x}^{(T)} \mid m_2)}{p(\mathbf{x}^{(T)} \mid m_1)} \times \frac{p(m_2)}{p(m_1)}. \quad (7.4)$$

7.2.1 Use of the integrated likelihood

In order to use (7.3) or (7.4) we need to estimate the integrated likelihood

$$p(\mathbf{x}^{(T)} \mid m) = \int p(\boldsymbol{\theta}_m, \mathbf{x}^{(T)} \mid m) \, \mathrm{d}\boldsymbol{\theta}_m = \int p(\mathbf{x}^{(T)} \mid m, \boldsymbol{\theta}_m) \, p(\boldsymbol{\theta}_m \mid m) \, \mathrm{d}\boldsymbol{\theta}_m.$$

One way of doing so would be to simulate from $p(\boldsymbol{\theta}_m \mid m)$, the prior distribution of the parameters $\boldsymbol{\theta}_m$ of the m-state model. But it is convenient and – especially if the prior is diffuse – more efficient to use a method that requires instead a sample from the posterior distribution, $p(\boldsymbol{\theta}_m \mid \mathbf{x}^{(T)}, m)$. Such a method is as follows.

Write the integrated likelihood as

$$\int p(\mathbf{x}^{(T)} \mid m, \boldsymbol{\theta}_m) \, \frac{p(\boldsymbol{\theta}_m \mid m)}{p^*(\boldsymbol{\theta}_m)} \, p^*(\boldsymbol{\theta}_m) \, \mathrm{d}\boldsymbol{\theta}_m;$$

that is, write it in a form suitable for the use of a sample from some convenient density $p^*(\boldsymbol{\theta}_m)$ for the parameters $\boldsymbol{\theta}_m$. Since we have available a sample $\boldsymbol{\theta}_m^{(j)}$ $(j = 1, 2, \ldots, B)$ from the posterior distribution, we can use that sample; that is, we can take $p^*(\boldsymbol{\theta}_m) = p(\boldsymbol{\theta}_m \mid \mathbf{x}^{(T)}, m)$. Newton and Raftery (1994) therefore suggest *inter alia* that the integrated likelihood can be estimated by

$$\widehat{I} = \sum_{j=1}^{B} w_j p(\mathbf{x}^{(T)} \mid m, \boldsymbol{\theta}_m^{(j)}) \bigg/ \sum_{j=1}^{B} w_j, \qquad (7.5)$$

where

$$w_j = \frac{p(\boldsymbol{\theta}_m^{(j)} \mid m)}{p(\boldsymbol{\theta}_m^{(j)} \mid \mathbf{x}^{(T)}, m)}.$$

After some manipulation this simplifies to the harmonic mean of the likelihood values of a sample from the posterior,

$$\widehat{I} = \left(B^{-1} \sum_{j=1}^{B} \left(p(\mathbf{x}^{(T)} \mid m, \boldsymbol{\theta}_m^{(j)}) \right)^{-1} \right)^{-1} ; \qquad (7.6)$$

see Exercise 5 for the details. This is, however, not the only route one can follow in deriving the estimator (7.6); see Exercise 6 for another possibility.

Newton and Raftery state that, under quite general conditions, $\widehat{I}$ is a simulation-consistent estimator of $p(\mathbf{x}^{(T)} \mid m)$. But there is a major drawback to this harmonic mean estimator, its infinite variance, and the question of which estimator to use for $p(\mathbf{x}^{(T)} \mid m)$ does not seem to have been settled. Raftery *et al.* (2007) suggest two alternatives to the harmonic mean estimator, but no clear recommendation emerges, and one of the discussants of that paper (Draper, 2007) bemoans the disheartening 'ad-hockery' of the many proposals that have over the years been made for coping with the instability of expectations with respect to (often diffuse) priors.

7.2.2 Model selection by parallel sampling

However, it is possible to estimate $p(m \mid \mathbf{x}^{(T)})$ relatively simply by 'parallel sampling' of the competing models, provided that the set of competing models is sufficiently small; see Congdon (2006) and Scott (2002). Denote by $\boldsymbol{\theta}$ the vector $(\boldsymbol{\theta}_1, \boldsymbol{\theta}_2, \ldots, \boldsymbol{\theta}_K)$, and similarly $\boldsymbol{\theta}^{(j)}$; K is the maximum number of states. Make the assumption that

$$p(m, \boldsymbol{\theta}) = p(\boldsymbol{\theta}_m \mid m)\, p(m);$$

that is, assume that the model with m states does not depend on the parameters of the j-state model, for $j \neq m$.

We wish to estimate $p(m \mid \mathbf{x}^{(T)})$ (for $m = 1, \ldots, K$) by

$$B^{-1} \sum_{j=1}^{B} p(m \mid \mathbf{x}^{(T)}, \boldsymbol{\theta}^{(j)}). \tag{7.7}$$

We use the fact that, with the above assumption,

$$p(m \mid \mathbf{x}^{(T)}, \boldsymbol{\theta}^{(j)}) \propto G_m^{(j)},$$

where

$$G_m^{(j)} \equiv p(\mathbf{x}^{(T)} \mid \boldsymbol{\theta}_m^{(j)}, m)\, p(\boldsymbol{\theta}_m^{(j)} \mid m)\, p(m). \tag{7.8}$$

(See Appendix 1 of Congdon (2006).) Hence

$$p(m \mid \mathbf{x}^{(T)}, \boldsymbol{\theta}^{(j)}) = G_m^{(j)} \Big/ \sum_{k=1}^{K} G_k^{(j)}.$$

This expression for $p(m \mid \mathbf{x}^{(T)}, \boldsymbol{\theta}^{(j)})$ can then be inserted in (7.7) to complete the estimate of $p(m \mid \mathbf{x}^{(T)})$.

7.3 Example: earthquakes

We apply the techniques described above to the series of annual counts of major earthquakes. The prior distributions used are as follows. The gamma distributions used as priors for the λ-increments (i.e. for the quantities τ_j) all have mean $50/(m+1)$ and c.v. 1 in one analysis, and 2 in a second. The Dirichlet distributions used as priors for the rows of $\boldsymbol{\Gamma}$ all have all parameters equal to 1. The prior distribution for m, the number of states, assigns probability $\frac{1}{6}$ to each of the values $1, 2, \ldots, 6$. The number of iterations used was $B = 100\,000$, with a burn-in period of 5000.

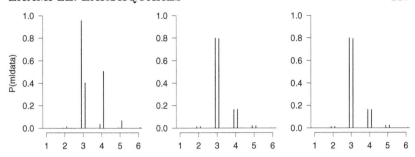

Figure 7.1 *Earthquakes data, posterior distributions of m given a uniform prior on* $\{1, 2, \ldots, 6\}$. *Each panel shows two posterior distributions from independent runs. In the left and centre panels the c.v. in the gamma prior is 1. Left panel: harmonic mean estimator. Centre panel: parallel sampling estimator. Right panel: parallel sampling estimator with c.v. = 1 (left bars) and c.v. = 2 (right bars).*

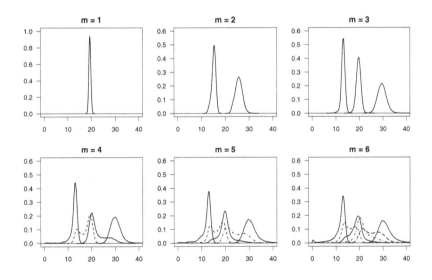

Figure 7.2 *Earthquakes data: posterior distributions of the state-dependent means for one- to six-state Poisson–HMMs.*

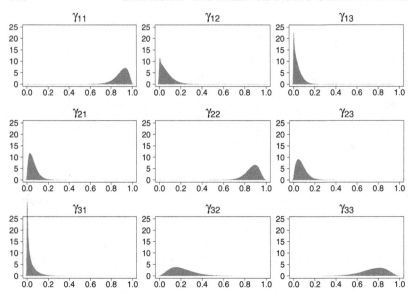

Figure 7.3 *Three-state Poisson–HMM (c.v. = 1) for earthquakes data: posterior distribution of transition probability matrix* **Γ**.

Figure 7.1 displays three comparisons of estimates of the posterior distribution of m. It is clear from the comparison of two independent runs of the harmonic mean estimator (left panel) that it is indeed very unstable. In the first run it would have chosen a three-state model by a large margin, and in the second run a four-state model. In contrast, the parallel sampling estimator (centre and right panels) produces very consistent results even if the c.v. of the prior distributions for the λ-increments is changed from 1 to 2, and clearly identifies $m = 3$ as the posterior mode.

The posterior distributions of the Poisson means for $m = 1, \ldots, 6$ are displayed in Figure 7.2. The posterior distributions of the entries of **Γ** for the three-state model (with c.v. = 1) are displayed in Figure 7.3. Table 7.1 lists posterior statistics for the three-state model. The posterior modes are generally quite close to the maximum likelihood estimates given on p. 55. In particular, the values of $\boldsymbol{\lambda}$ are almost the same, but the posterior modes for the entries of **Γ** are mostly closer to 0.5 than are the corresponding MLEs.

Table 7.1 *Earthquakes data: posterior statistics for the three-state model (c.v. = 1).*

Parameter	Min	Q1	Mode	Median	Mean	Q3	Max
λ_1	6.21	12.62	13.12	13.15	13.12	13.68	16.85
λ_2	13.53	19.05	19.79	19.74	19.71	20.42	27.12
λ_3	22.08	28.33	29.87	29.59	29.64	30.88	43.88
γ_{11}	0.001	0.803	0.882	0.861	0.843	0.905	0.998
γ_{12}	0.000	0.047	0.056	0.085	0.104	0.139	0.964
γ_{13}	0.000	0.020	0.011	0.042	0.053	0.075	0.848
γ_{21}	0.000	0.043	0.050	0.070	0.083	0.108	0.979
γ_{22}	0.009	0.784	0.858	0.837	0.824	0.880	0.992
γ_{23}	0.000	0.052	0.060	0.082	0.093	0.122	0.943
γ_{31}	0.000	0.021	0.011	0.049	0.068	0.096	0.758
γ_{32}	0.000	0.144	0.180	0.213	0.229	0.296	0.918
γ_{33}	0.010	0.627	0.757	0.718	0.703	0.795	0.986

7.4 Discussion

From the above example it seems clear that the Bayesian approach is demanding computationally and in certain other respects. The model needs to be parametrized in a way that avoids label switching. In HMMs the labels associated with the states are arbitrary; the model is invariant under permutation of the labels. This is irrelevant to maximum likelihood estimation, but it is a problem in the context of MCMC estimation of posterior distributions. One must ensure that only one of the $m!$ permutations of the labels is used in the simulation. In the algorithm for Poisson–HMMs discussed above, label switching was avoided by ordering the states according to the means of the state-dependent distributions. An analogous reparametrization has been used for the normal case by Robert and Titterington (1998). Prior distributions need to be specified for the parameters and, as a rule, the choice of prior distribution is driven by mathematical convenience rather than by prior information. The above difficulties are model specific; the derivations, priors and hence the computer code all change substantially if the state-dependent distribution changes.

Although the computational demands of MCMC are high in comparison with those of ML *point estimation*, this is not a fair comparison. Interval estimation using the parametric bootstrap, though easy to implement (see Section 3.6.2), is comparably time-consuming. Interval estimates based instead on approximate standard errors obtained from the Hessian (as in Section 3.6.1) require potentially demanding parametrization-specific derivations.

A warning note regarding MCMC has been sounded by Celeux, Hurn and Robert (2000), and echoed by Chopin (2007, p. 282): 'we consider that almost the entirety of Markov chain Monte Carlo samplers implemented for mixture models has failed to converge!' This statement was made of independent mixture models but is presumably also applicable to HMMs.

Another Bayesian approach to model selection in HMMs is the use of reversible jump Markov chain Monte Carlo (RJMCMC) techniques, in which m, the number of states, is treated as a parameter of the model and, like the other parameters, updated in each iteration. This has the advantage over parallel sampling that relatively few iterations are 'wasted' on unpromising values of m, thereby reducing the total number of iterations needed to achieve a given degree of accuracy. Such an advantage might be telling for very long time series and a potentially large number of states, but for typical applications and sample sizes, models with $m > 5$ are rarely feasible. The disadvantage for users who have to write their own code is the complexity of the algorithm.

Robert *et al.* (2000) describe in detail the use of that approach in HMMs with normal distributions as the state-dependent distributions, and provide several examples of its application. As far as we are aware, this approach has not been extended to HMMs with other possible state-dependent distributions, such as Poisson, and we refer the reader interested in RJMCMC to Robert *et al.* and the references therein, especially Green (1995) and Richardson and Green (1997).

In Section 8.4 we describe the use of the package `R2OpenBUGS` to run the Gibbs sampler on a Poisson–HMM.

Exercises

1. Consider u defined by $u = \sum_{j=1}^{i} u_j$, where the variables u_j are independent Poisson random variables with means τ_j.

   Show that, conditional on u, the joint distribution of $u_1, u_2, \ldots, u_i$ is multinomial with total u and probability vector $(\tau_1, \ldots, \tau_i)/\sum_{j=1}^{i} \tau_j$.

2. (Updating of Dirichlet distributions.) Let $\mathbf{w} = (w_1, w_2, \ldots, w_m)$ be an observation from a multinomial distribution with probability vector $\mathbf{y}$, which has a Dirichlet distribution with parameter vector $\mathbf{d} = (d_1, d_2, \ldots, d_m)$.

   Show that the posterior distribution of $\mathbf{y}$, i.e. the distribution of $\mathbf{y}$ given $\mathbf{w}$, is the Dirichlet distribution with parameters $\mathbf{d} + \mathbf{w}$.

3. (Updating of gamma distributions.) Let $y_1, y_2, \ldots, y_n$ be a random sample from the Poisson distribution with mean τ, which is gamma-distributed with parameters a and b.

   Show that the posterior distribution of τ, i.e. the distribution of τ given $y_1, \ldots, y_n$, is the gamma distribution with parameters $a + \sum_{i=1}^{n} y_i$ and $b + n$.

4. Show that, in the basic HMM,

$$\Pr(\mathbf{X}_{t+1}^T, \mathbf{C}_{t+2}^T \mid \mathbf{X}^{(t)}, C_t, C_{t+1}) = \Pr(\mathbf{X}_{t+1}^T, \mathbf{C}_{t+2}^T \mid C_{t+1}).$$

   Hint: either use the methods of Appendix B or invoke d-separation, for which see, for example, Pearl (2000, pp. 16–18).

5. Consider the estimator $\widehat{I}$ as defined by equation (7.5):

$$\widehat{I} = \sum_j w_j\, p(\mathbf{x}^{(T)} \mid m, \boldsymbol{\theta}_m^{(j)}) \Big/ \sum_j w_j.$$

   (a) Show that the weight $w_j = p(\boldsymbol{\theta}_m^{(j)} \mid m)/p(\boldsymbol{\theta}_m^{(j)} \mid \mathbf{x}^{(T)}, m)$ can be written as $p(\mathbf{x}^{(T)} \mid m)/p(\mathbf{x}^{(T)} \mid m, \boldsymbol{\theta}_m^{(j)})$.

   (b) Deduce that the summand $w_j\, p(\mathbf{x}^{(T)} \mid m, \boldsymbol{\theta}_m^{(j)})$ is equal to $p(\mathbf{x}^{(T)} \mid m)$ (and is therefore independent of j).

   (c) Hence show that $\widehat{I}$ is the harmonic mean displayed in equation (7.6).

6. Let the observations $\mathbf{x}$ be distributed with parameter (vector) $\boldsymbol{\theta}$. Prove that

$$\frac{1}{p(\mathbf{x})} = \mathrm{E}\left(\frac{1}{p(\mathbf{x} \mid \boldsymbol{\theta})} \;\middle|\; \mathbf{x}\right),$$

   i.e. the integrated likelihood $p(\mathbf{x})$ is the harmonic mean of the likelihood $p(\mathbf{x} \mid \boldsymbol{\theta})$ computed under the posterior distribution $p(\boldsymbol{\theta} \mid \mathbf{x})$ for $\boldsymbol{\theta}$. (This is the 'harmonic mean identity', as in equation (1) of Raftery et al. (2007), and suggests the use of the harmonic mean of likelihoods sampled from the posterior as estimator of the integrated likelihood.)

CHAPTER 8

R packages

In this chapter, we describe three **R** packages that deal with HMMs, namely `depmixS4`, `HiddenMarkov` and `msm`. We discuss briefly how these packages can be used to fit basic HMMs to data, and to decode. We also give one example of the use of `R2OpenBUGS`, which provides an interface to `OpenBUGS`. For detailed descriptions of the various features available in these packages, the reader should consult the help files available on the Comprehensive R Archive Network (CRAN). The **R** package `HMM` deals only with categorical time series (see Section 9.3) and is not covered here.

In the sample code that follows, it is assumed that the **R** vector `quakes` contains the series of counts in Table 1.1. To download these data, use

```
quakes <- read.table(
  "http://www.hmms-for-time-series.de/second/data/earthquakes.txt")$V2
```

8.1 The package `depmixS4`

8.1.1 Model formulation and estimation

The package `depmixS4` (Visser and Speekenbrink, 2010) is a powerful **R** package that incorporates various types of dependent mixture model. In particular, it allows the user to fit HMMs and to decode.

To use `depmixS4`, the first thing one does is to specify a model via the function `depmix`. For example, after `depmixS4` has been installed and loaded to **R**, the following command creates an object that specifies the components of the model to be fitted (a two-state Poisson–HMM) and the data.

```
mod <- depmix(quakes~1,nstates=2,family=poisson(),ntimes=107)
```

In particular, several families of distributions can be specified for the state-dependent process, including binomial, gamma and normal distributions. Among other things, `depmixS4` also allows the user to incorporate covariates in both state process and state-dependent process, and to generate artificial data from an HMM.

Using the object `mod` as specified above, one can carry out inference, including model fitting and decoding. For example, the MLEs for the two-state Poisson–HMM for the earthquakes are accessed as follows.

```
fitted.mod.depmix <- fit(mod)
summary(fitted.mod.depmix)
```

The resulting estimates are

$$\Gamma = \begin{pmatrix} 0.9283 & 0.0717 \\ 0.1190 & 0.8810 \end{pmatrix},$$

$\delta = (1.000, 0.000)$, and $\lambda = (15.419, 26.015)$, with log-likelihood $l = -341.8787$. These estimates are very close to those we obtained in Section 4.3, where we fitted a non-stationary HMM using EM; they differ in the way one would expect from the stationary HMM given in Section 3.5. The depmixS4 package does not currently allow for the fitting of stationary HMMs. The default option in depmixS4 is to perform likelihood maximization by means of the EM algorithm.

The three-state Poisson–HMM fitted to the earthquakes data with depmixS4 is

$$\Gamma = \begin{pmatrix} 0.9393 & 0.0321 & 0.0286 \\ 0.0404 & 0.9064 & 0.0532 \\ 0.0000 & 0.1904 & 0.8096 \end{pmatrix},$$

$\delta = (1.000, 0.000, 0.000)$, and $\lambda = (13.134, 19.713, 29.711)$, with log-likelihood $l = -328.5275$. These estimates are the same as those we obtained in Section 4.3.

8.1.2 Decoding

For the depmix object fitted.mod.depmix as above, the state probabilities, as described in Section 5.4.1, can be obtained with the command

```
forwardbackward(fitted.mod.depmix)$gamma
```

The most likely state sequence, obtained using the Viterbi algorithm, is provided by

```
posterior(fitted.mod.depmix)$state
```

8.2 The package HiddenMarkov

8.2.1 Model formulation and estimation

Another useful package for HMMs is HiddenMarkov (Harte, 2014). It provides a variety of state-dependent distributions, including the Poisson, beta, gamma, normal, lognormal and binomial. As with depmixS4, one must first create an HMM object, which in the case of the two-state Poisson–HMM for the earthquakes data is achieved as follows.

```
Pi <- rbind(c(.9,.1),c(.1,.9))
mod <- dthmm(quakes,Pi=Pi,delta=c(.5,.5),"pois", list(lambda=c(10,30)))
```

The specified parameter values serve as initial values in the numerical search for the MLEs, though the HMM object specified in this way can also be used to simulate data from a model with those parameters. The default method of finding MLEs is again the EM algorithm. In our example, the MLEs are obtained as follows.

```
fitted.mod.HM <- BaumWelch(mod)
summary(fitted.mod.HM)
```

As with depmixS4, the default option in HiddenMarkov is that the initial distribution is treated as a parameter requiring estimation. The estimates obtained in the earthquake example are identical to those obtained in Section 4.3 and by using depmixS4 as in Section 8.1.

HiddenMarkov does offer the option of fitting stationary models, but as Harte (2014) writes, this is done 'in a slightly ad-hoc manner by effectively disregarding the first term' of the complete-data log-likelihood. The result is that the estimates of Γ and λ are precisely the same as in the non-stationary case. The stationary distribution implied by the estimate of Γ (i.e. the solution of $\delta\Gamma = \Gamma$) is then taken as the initial distribution of the Markov chain. Indeed the resulting model is a stationary model, but it is not the maximum likelihood estimate.

A stationary two-state model of this kind for the earthquake data can be obtained by specifying a model with nonstat=FALSE and running BaumWelch.

```
fitted.mod.HM.stat <- dthmm(quakes,Pi=Pi,delta=c(.5,.5),"pois",
 list(lambda=c(10,30)),nonstat=FALSE)
summary(fitted.mod.HM.stat)
BaumWelch(fitted.mod.HM.stat)
```

The estimates obtained are

$$\Gamma = \left(\begin{array}{cc} 0.9284 & 0.0716 \\ 0.1191 & 0.8809 \end{array} \right),$$

$\lambda = (15.420, 26.017)$, and $\delta = (0.6243, 0.3757)$, with log-likelihood $l = -342.3480$. These estimates are, as one would expect, different from the MLEs we obtained in Section 3.5.

The three-state (non-stationary) Poisson–HMM fitted to the earthquakes data with HiddenMarkov is

$$\mathbf{\Gamma} = \begin{pmatrix} 0.9392 & 0.0322 & 0.0286 \\ 0.0404 & 0.9063 & 0.0532 \\ 0.0000 & 0.1904 & 0.8096 \end{pmatrix},$$

with $\boldsymbol{\delta} = (1.000, 0.000, 0.000)$, $\boldsymbol{\lambda} = (13.134, 19.714, 29.711)$, and log-likelihood $l = -328.5275$. These estimates are essentially the same as those we obtained in Section 4.3.

8.2.2 Decoding

In HiddenMarkov, state probabilities are obtained as a by-product of the EM-based maximization of the likelihood. For the object fitted.mod.HM as above, the state probabilities can be accessed by

```
fitted.mod.HM$u
```

The paths as decoded by the Viterbi algorithm are obtained by

```
Viterbi(fitted.mod.HM)
```

8.2.3 Residuals

Pseudo-residuals can be obtained with HiddenMarkov as follows.

```
residuals(fitted.mod.HM)
```

For discrete data, the **residuals** function makes a 'continuity adjustment' (Harte, 2014). As a consequence, the residuals obtained in case of the Poisson–HMM considered here are points, not segments as described in Section 6.2.1.

8.3 The package msm

8.3.1 Model formulation and estimation

One of the many things that can be done by means of the **R** package msm (Jackson *et al.*, 2003) is to fit HMMs with a range of state-dependent distributions, including Poisson, normal, gamma and binomial distributions. With the following code, a two-state Poisson–HMM is fitted to the earthquakes data.

```
timeq <- 1:length(quakes)
fitted.mod.msm <- msm(quakes~timeq,qmatrix=rbind(c(-.1,.1),
```

```
c(.1,-.1)),hmodel=list(hmmPois(rate=10),hmmPois(rate=20)))
fitted.mod.msm
```

The msm package was developed specifically for the purpose of fitting *continuous-time* Markov chains and HMMs. The object fitted.mod.msm obtained using the above code is a continuous-time HMM, with transition intensities (rather than probabilities) governing the state process. (See also Section 3.3.2.) In the example, the transition intensity matrix obtained is

$$\mathbf{Q} = \begin{pmatrix} -0.0795 & 0.0795 \\ 0.1319 & -0.1319 \end{pmatrix}.$$

The transition probability matrix of the implied discrete-time model is provided by the following code.

```
pmatrix.msm(fitted.mod.msm)
```

This computes the matrix exponential of $\mathbf{Q}$, and here returns the estimate

$$\mathbf{\Gamma} = \begin{pmatrix} 0.9284 & 0.0716 \\ 0.1189 & 0.8811 \end{pmatrix},$$

which differs only very slightly from the estimates obtained by depmixS4 and HiddenMarkov. The default initial distribution used by msm assigns a probability of 1 to state 1; the resulting models are therefore directly comparable to the models obtained from depmixS4 and HiddenMarkov, and also to those which we have fitted by EM using our own code (Section 4.3). The two-state model fitted by msm corresponds closely to our model as given in Table 4.1 and to the models fitted using either depmixS4 or HiddenMarkov; its log-likelihood is $l = -341.8787$, and the state-dependent means are estimated as 15.420 and 26.015.

The three-state (discrete-time) model found in this way by msm does not correspond quite so closely to the models fitted by depmixS4, by HiddenMarkov, and by our EM code. The best three-state model we found by msm has $l = -328.5914$ (cf. $l = -328.5275$), state-dependent means 13.134, 19.719 and 29.714, and t.p.m.

$$\begin{pmatrix} 0.9393 & 0.0323 & 0.0285 \\ 0.0389 & 0.9083 & 0.0528 \\ 0.0040 & 0.1853 & 0.8106 \end{pmatrix}.$$

Examination of the four estimates of γ_{31} (i.e. from msm, depmixS4, HiddenMarkov and our code) suggests a reason. The estimate provided by msm is 0.0040, the others are zero to four decimal places. We know from Theorem 3.1 of Israel *et al.* (2001), reproduced in Exercise 11(b) of Chapter 3, that if a discrete-time Markov chain has the property that $\gamma_{31} = 0$ but $\gamma_{32}\gamma_{21} > 0$, that Markov chain is not embeddable in a continuous-time chain. That is, there does not exist a generator matrix $\mathbf{Q}$ such that $\exp(\mathbf{Q}) = \mathbf{\Gamma}$ (exactly). One cannot therefore expect msm to produce a discrete-time model with $\widehat{\gamma}_{31}$ equal to zero and $\widehat{\gamma}_{32}\widehat{\gamma}_{21} > 0$. But it comes close, and the log-likelihood achieved is only slightly lower than that of the three models fitted more directly, that is, not via a continuous-time Markov chain.

8.3.2 Decoding

The state probabilities and the most likely state sequence are obtained as follows from the object fitted.mod.msm.

```
viterbi.msm(fitted.mod.msm)
```

8.4 The package R2OpenBUGS

The software OpenBUGS and its predecessor WinBUGS are designed to produce samples from the posterior distributions of a wide range of statistical models. (BUGS stands for 'Bayesian analysis using Gibbs sampling'.) It is possible to access OpenBUGS from within **R** by using the **R** package R2OpenBUGS, and we demonstrate here how this can be done for the three-state Poisson–HMM for the earthquakes series as described in Section 7.3.

The crucial function in R2OpenBUGS is bugs, which takes data and starting values as input. It automatically writes an OpenBUGS script, calls the model, here specified by HMM.txt, and stores the output for easy access in **R**.

```
library(R2openBUGS)
x      <- quakes
n      <- length(x)
m      <- 3
data <- list("x", "n", "m" )

#Run Gibbs sampler
eq.sims  <- bugs(data,inits=NULL,model.file="HMM.txt",
        parameters=c("tau","lambda","Gamma"),n.iter=100000,n.chains=1)

eq.sims$summary
par(mfrow=c(3,3))
for(i in 1:9) hist(eq.sims$sims.matrix[,6+i],freq=F,ylim=c(0,15),
        xlim=c(0,1),xlab="",ylab="",main="")
```

The above code does one run of length 100 000 of the Gibbs sampler, and produces summaries similar to those in presented in Table 7.1 and Figure 7.3. The model HMM.txt used is:

```
model
{
        for(i in 1:m)
        {
                delta[i]<-1/m
                v[i]<-1
        }
        s[1] ~dcat(delta[])
        for (i in 2:100)
        {
                s[i] ~ dcat(Gamma[s[i-1],])
        }

        states[1] ~ dcat(Gamma[s[100],])
        x[1]~dpois(lambda[states[1]])
        for(i in 2:n)
        {
                states[i]~dcat(Gamma[states[i-1],])
                x[i]~dpois(lambda[states[i]])
        }

        for(i in 1:m)
        {
                tau[i]~dgamma(1,0.08)
                Gamma[i,1:m]~ddirch(v[])
        }
        lambda[1]<-tau[1]
        for(i in 2:m)
        {
        lambda[i]<-lambda[i-1]+tau[i]
        }
}
```

One feature of this code that may need explanation is this. Given a t.p.m. Γ, we need to be able to deduce the corresponding stationary distribution. It seems difficult or impossible to do this directly in OpenBUGS; instead we approximate the stationary distribution by running the Markov chain with t.p.m. Γ for 100 steps from a uniform initial distribution.

Table 8.1 presents a comparison of the means and medians of the posterior distributions of the state-dependent means λ_i as found here by R2OpenBUGS and as shown in Table 7.1. Agreement is fairly close but perhaps not as close as one might expect from 100 000 iterations.

8.5 Discussion

The packages described above provide a wide range of operations that can be carried out on HMMs, including many that we have not described. Little coding effort is required of the user if only standard operations such as those described above are to be carried out. Nevertheless, we believe that, if researchers wish to exploit the full flexibility (and extendability)

Table 8.1 *Earthquakes data: comparison of selected results from output of* R2OpenBUGS *and Table 7.1.*

	λ_1		λ_2		λ_3	
	Median	Mean	Median	Mean	Median	Mean
R2OpenBUGS	13.23	13.23	19.84	19.84	29.66	29.72
Table 7.1	13.15	13.12	19.74	19.71	29.59	29.64

of HMMs, they will find it advantageous to understand either our code or that of the packages. They will then be able to modify and extend such code in order to implement models that are tailor-made for particular applications of interest to them.

PART II

Extensions

## CHAPTER 9

HMMs with general
state-dependent distribution

9.1 Introduction

A notable advantage of HMMs is the ease with which the basic model can be modified or generalized, in many different ways, to provide models for a wide range of types of observations. We begin this chapter by describing the use of univariate state-dependent distributions other than the Poisson (Section 9.2), and then move on to discuss how to construct HMMs for multivariate observations, distinguishing the important special case of multinomial-like observations (Section 9.3) from other multivariate observations (Section 9.4).

9.2 General univariate state-dependent distribution

In Part I we focused on the (univariate) Poisson–HMM which we called the basic model. One may, however, use any distribution – discrete, continuous, or a mixture of the two – as the state-dependent distribution; in fact one can even use a different family of distributions for each state. One simply redefines the diagonal matrices containing the state-dependent probabilities, and in the estimation process accommodates whatever constraints the state-dependent parameters must observe, either by transforming the constraints away or by applying constrained optimization. We now outline a number of such HMMs.

9.2.1 HMMs for unbounded counts

The Poisson distribution is the canonical model for unbounded counts. However, a popular alternative, for overdispersed counts, is the negative binomial distribution. This is given, for all non-negative integers x, by the probability function

$$p_i(x) = \frac{\Gamma\left(x + \frac{1}{\eta_i}\right)}{\Gamma\left(\frac{1}{\eta_i}\right)\Gamma(x+1)} \left(\frac{1}{1+\eta_i\mu_i}\right)^{\frac{1}{\eta_i}} \left(\frac{\eta_i\mu_i}{1+\eta_i\mu_i}\right)^x,$$

where the parameters μ_i (the mean) and η_i are positive. (But note that this is only one of several possible parametrizations of the negative binomial.) A negative binomial–HMM may be useful if one wishes to allow for overdispersion within each state. Conceivable examples for the application of Poisson– or negative binomial–HMMs include series of counts of stoppages or breakdowns of technical equipment, earthquakes, sales, insurance claims, accidents reported, defective items and stock trades. The negative binomial distribution includes the geometric distribution as the special case $\eta_i = 1$, and the Poisson as the limit as $\eta_i \to 0$.

9.2.2 HMMs for binary data

The Bernoulli–HMM for binary time series is the simplest HMM. Its state-dependent probabilities for the two possible outcomes are, for some probabilities π_i, just

$$p_i(0) = \Pr(X_t = 0 \mid C_t = i) = 1 - \pi_i \qquad \text{(failure)},$$
$$p_i(1) = \Pr(X_t = 1 \mid C_t = i) = \pi_i \qquad \text{(success)}.$$

An example of a Bernoulli–HMM appears in Section 2.3.1. Possible applications of Bernoulli–HMMs are to daily rainfall occurrence (rain or no rain), daily trading of a share (traded or not traded) and quality control (a product meets design specifications or it does not).

9.2.3 HMMs for bounded counts

Binomial–HMMs may be used to model series of bounded counts. The state-dependent binomial probabilities are given by

$$_tp_i(x_t) = \binom{n_t}{x_t}\pi_i^{x_t}(1 - \pi_i)^{n_t - x_t}, \quad x_t = 0, 1, \ldots, n_t, \qquad (9.1)$$

where n_t is the number of trials at time t and x_t the number of successes. (The prefix t in $_tp_i()$ is used to indicate the time-dependence that arises when the n_t are not all equal.)

Possible examples for series of bounded counts that may be described by a binomial–HMM are series of:

- purchasing preferences, for example, n_t = number of purchases of all brands on day t, x_t = number of purchases of brand A on day t;

- sales of newspapers or magazines, for example, n_t = number available on day t, x_t = number purchased on day t.

Notice, however, that there is a complication when one computes the forecast distribution of a binomial–HMM. Either n_{T+h}, the number of trials at time $T + h$, must be known, or one has to fit a separate model to forecast n_{T+h}. Alternatively, by setting $n_{T+h} = 1$ one can simply compute the forecast distribution of the 'success proportion'.

9.2.4 HMMs for continuous-valued series

For continuous-valued series one replaces the state-dependent probability functions by probability density functions, for example, for the exponential, gamma, normal and t distributions. The state-dependent density functions can also be estimated nonparametrically, hence avoiding the need to make any distributional assumption (Langrock *et al.*, 2015).

Normal–HMMs have been used for modelling share returns series because the resulting mixture of normals can accommodate the large values of kurtosis that are a distinctive feature of such series. (See Section 20.1 for a multivariate model for returns on four shares.) HMMs with t state-dependent distributions are sometimes used instead for this purpose.

Note that the likelihood function of a normal–HMM is unbounded; it is possible to increase the likelihood without bound by fixing one of the state-dependent means μ_i at one of the observations and letting the corresponding variance σ_i approach zero. This can lead to problems in parameter estimation because here one usually wishes to find the best *finite* local maximum of the likelihood. In practice the infinite 'spikes' of the likelihood function are often missed by the maximization algorithm, but one can also avoid the problem by specifying positive lower bounds for the variances of the state-dependent distributions. Alternatively one can use the discrete likelihood; see Section 1.2.3.

9.2.5 HMMs for proportions

An obvious default choice of state-dependent model for time series of proportions (i.e. observations in $[0, 1]$) is the beta distribution. Its p.d.f. with parameters $\alpha, \beta > 0$ is given by

$$f(x) = \frac{1}{\mathrm{B}(\alpha, \beta)} x^{\alpha-1}(1 - x)^{\beta-1}, \quad 0 \leq x \leq 1, \tag{9.2}$$

where B denotes the beta function.

For some purposes (e.g. to specify initial values for iterative maximization of the likelihood function of a beta–HMM), it is convenient to parametrize this distribution by the mean $\mu = \alpha/(\alpha+\beta)$ and a precision parameter $\nu = \alpha + \beta$. The uniform distribution is the special case with $\mu = 1/2$ and $\nu = 2$. The variance decreases as ν increases.

9.2.6 HMMs for circular-valued series

Circular-valued series, such as observations of compass directions or times of day, require different models to those applicable to 'linear data'. There is no difficulty in constructing HMMs having circular-valued state-dependent distributions. Important cases here are the wrapped normal, the wrapped t, and the von Mises distributions, the last of which is often regarded as the natural first choice for unimodal continuous observations on the circle; see, for example, Fisher (1993, pp. 49–50, 55). The p.d.f. of the von Mises distribution with parameters $\mu \in (-\pi, \pi]$ (location) and $\kappa > 0$ (concentration) is

$$f(x) = (2\pi I_0(\kappa))^{-1} \exp(\kappa \cos(x - \mu)), \quad \text{for } x \in (-\pi, \pi], \qquad (9.3)$$

where I_0 denotes the modified Bessel function of the first kind of order 0. In some texts the interval $[0, 2\pi)$ is used instead of $(-\pi, \pi]$ in the p.d.f. (9.3). The von Mises–HMM is simply an HMM having von Mises state-dependent distributions.

9.3 Multinomial and categorical HMMs

Before describing more general multivariate HMMs we discuss the important special case of HMMs with multinomial state-dependent distribution. A special case thereof provides in turn a model for series of categorical observations such as biological sequence data.

9.3.1 Multinomial–HMM

A **multinomial–HMM** is the generalization of the binomial–HMM to cases in which there are $q \geq 2$, rather than two, mutually exclusive and exhaustive possible outcomes to each trial. The observations are then q series of counts, $\{x_{tj} : t = 1, 2, \ldots, T; j = 1, \ldots, q\}$ with $\sum_{j=1}^{q} x_{tj} = n_t$, where n_t is the (known) number of trials at time t. Thus, for example, $x_{28,3}$ represents the number of outcomes at time $t = 28$ that were of type 3. Let $\mathbf{x}_t = (x_{t1}, x_{t2}, \ldots, x_{tq})$.

We shall suppose that, conditional on the Markov chain $\mathbf{C}^{(T)}$, the T random vectors $\{\mathbf{X}_t = (X_{t1}, X_{t2}, \ldots, X_{tq}) : t = 1, \ldots, T\}$ are mutually independent.

The parameters of an m-state multinomial–HMM model are as follows. As in the basic model, the matrix $\mathbf{\Gamma}$ has $m(m-1)$ free parameters. The ith state of the Markov chain is associated with a multinomial distribution with parameters n_t (known) and q unknown probabilities $\pi_{i1}, \pi_{i2}, \ldots, \pi_{iq}$, constrained by $\sum_{j=1}^{q} \pi_{ij} = 1$, that is, $q-1$ free parameters. The model therefore has in all $m^2 - m + m(q-1) = m(m+q-2)$ parameters to be estimated.

The likelihood function for such a multinomial–HMM is given by

$$L_T = \boldsymbol{\delta}\,_1\mathbf{P}(\mathbf{x}_1)\boldsymbol{\Gamma}\,_2\mathbf{P}(\mathbf{x}_2)\cdots\boldsymbol{\Gamma}\,_T\mathbf{P}(\mathbf{x}_T)\mathbf{1}',$$

where $_t\mathbf{P}(\mathbf{x}_t) = \mathrm{diag}\,(_tp_1(\mathbf{x}_t),\ldots,\,_tp_m(\mathbf{x}_t))$. This has the same form as the likelihood for the binomial–HMM, but the binomial probabilities appearing in (9.1) are here replaced by multinomial probabilities

$$_tp_i(\mathbf{x}_t) = \Pr(\mathbf{X}_t = \mathbf{x}_t \mid C_t = i) = \binom{n_t}{x_{t1}, x_{t2}, \ldots, x_{tq}}\pi_{i1}^{x_{t1}}\pi_{i2}^{x_{t2}}\cdots\pi_{iq}^{x_{tq}}.$$

Note that these probabilities are indexed by the time t because the number of trials n_t is permitted to be time-dependent.

The parameters of the model can then be estimated by maximizing the likelihood as a function of $m(q-1)$ of the 'success probabilities', for example, π_{ij} for $j = 1, 2, \ldots, q-1$ and all i, and of the $m^2 - m$ off-diagonal transition probabilities. One must observe the usual 'generalized upper bound' constraints $\sum_{j\neq i}\gamma_{ij} \leq 1$ on the transition probabilities, and also the m similar constraints $\sum_{j=1}^{q-1}\pi_{ij} \leq 1$ on the probabilities π_{ij}, one constraint for each state i – as well as the lower bound of 0 on all these probabilities.

Once the parameters have been estimated, these can be used to apply local and global decoding. Note, however, that the forecast distributions, being mixtures of multinomial distributions, are multivariate for $q > 2$. Furthermore, they depend on future values of the number of trials. Thus, in order to compute the h-step-ahead forecast distribution, one needs to know n_{T+h}, the number of trials that will take place at time $T + h$. This will be known in some applications, but there are cases in which n_{T+h} is a random variable whose value remains unknown until time $T + h$. For the latter it is not possible to compute the forecast distribution of the counts at time $T + h$ but, as in the case of the binomial–HMM, it is possible to forecast the count proportions by setting $n_{T+h} = 1$. An alternative approach is to fit a separate model to the series $\{n_t\}$, to use that model to compute the forecast distribution of n_{T+h} and hence the required forecast distribution for the counts of the multinomial–HMM.

9.3.2 HMMs for categorical data

A simple but important special case of the multinomial–HMM is that in which $n_t = 1$ for all t. This provides a model for categorical series, such as DNA base sequences or amino-acid sequences, in which there is exactly one symbol at each position in the sequence: one of A, C, G, T in the former example, one of 20 amino acids in the latter. In this case the state-dependent probabilities $_tp_i(\mathbf{x})$ and the matrix expression for the likelihood simplify somewhat.

Because n_t is constant, the prefix t in $_tp_i(\mathbf{x})$ is no longer necessary, and because $\sum_{k=1}^{q} x_{tk} = 1$, the q-vector $\mathbf{x}_t$ has one entry equal to 1 and the others equal to zero. It follows that, if

$$\mathbf{x} = (\underbrace{0,\ldots,0}_{j-1}, 1, \underbrace{0,\ldots,0}_{q-j}),$$

then $p_i(\mathbf{x}) = \pi_{ij}$ and

$$\mathbf{P}(\mathbf{x}) = \mathrm{diag}\,(\pi_{1j}, \ldots, \pi_{mj}) = \mathbf{\Pi}(j), \tag{9.4}$$

say. For categorical series, the expression (5.5) for the h-step-ahead forecast distribution simplifies. Conditional on the observations, the probability that category j will occur at time $T + h$ is given by

$$\sum_{i=1}^{m} \xi_i(h)\pi_{ij}\,.$$

9.3.3 HMMs for compositional data

The Dirichlet distribution, introduced in Section 7.1.1, is a popular model for compositional data, that is, multivariate continuous-valued observations that are positive and sum to 1. The p.d.f. is of the form

$$f(x_1, x_2, \ldots, x_q) = \frac{\Gamma(\alpha_1)\Gamma(\alpha_2)\cdots\Gamma(\alpha_q)}{\Gamma(\alpha_1 + \alpha_2 + \cdots + \alpha_q)} x_1^{\alpha_1-1} x_2^{\alpha_2-1} \cdots x_q^{\alpha_q-1},$$

where Γ represents the gamma function, $x_j > 0$ is the jth entry of an observation vector such that $\sum_{j=1}^{q} x_j = 1$, and the parameters α_j (for $j = 1, 2, \ldots, q$) are positive. The Dirichlet–HMM generalizes the beta–HMM in much the same way as the multinomial–HMM generalizes the binomial–HMM. An application of the Dirichlet–HMM to data on sleep-disordered breathing is described by Langrock *et al.* (2013).

9.4 General multivariate state-dependent distribution

The series of multinomial-like counts and of compositional data discussed in the above sections are examples of multivariate series, but with a specific structure. In this section we illustrate how it is possible to develop HMMs for different and more complex types of multivariate series.

9.4.1 Longitudinal conditional independence

Consider q time series $\{\mathbf{X}_t = (X_{t1}, X_{t2}, \ldots, X_{tq}) : t = 1, \ldots, T\}$. As we did for the basic HMM, we assume that, conditional on $\mathbf{C}^{(T)} = \{C_t : t = 1, \ldots, T\}$, the random vectors $\mathbf{X}_1, \ldots, \mathbf{X}_T$ are mutually independent. We refer to this property as **longitudinal conditional independence** in order to distinguish it from contemporaneous conditional independence which is defined in Section 9.4.2.

To specify an HMM for such a series it is necessary to postulate a model for the distribution of the random vector $\mathbf{X}_t$ in each of the m states of the parameter process. That is, one requires the following probabilities to be specified for $t = 1, 2, \ldots, T$, $i = 1, 2, \ldots, m$, and all relevant $\mathbf{x} = (x_1, x_2, \ldots, x_q)$:

$$_tp_i(\mathbf{x}) = \Pr(\mathbf{X}_t = \mathbf{x} \mid C_t = i).$$

(For generality, we keep the time index t here, that is, we allow the state-dependent probabilities to change over time.) In the case of multinomial–HMMs these probabilities are supplied by m multinomial distributions.

We note that it is not required that each of the q component series represent observations of the same type. For example, in the bivariate model discussed in Section 18.3, the state-dependent distributions of X_{t1} are gamma distributions, and those of X_{t2} von Mises distributions; the first component is linear-valued and the second circular-valued.

Secondly, as in the univariate case, the various state-dependent distributions need not belong to the same family of distributions. In the univariate case one could, for example, use a gamma distribution in state 1 and an extreme-value distribution in state 2. Or one could use a Poisson distribution in state 1 and a degenerate distribution at 0 ($X = 0$ with probability 1) in state 2, in order to accommodate zero inflation; see Exercise 2 for such a model. In the multivariate case one could use a multivariate normal distribution in one state and a multivariate t in another.

What is necessary is to specify models for m joint distributions, a task that can be anything but trivial. For example, there is no single bivariate Poisson distribution; different versions are available and they have different properties. One has to select a version that is appropriate in the context of the particular application being investigated. In contrast, one can reasonably speak of *the* bivariate normal distribution because, for most practical purposes, there is only one.

Once the required joint distributions have been selected, (i.e. once one has specified the state-dependent probabilities $_tp_i(\mathbf{x}_t)$), the likelihood of a general multivariate HMM is easy to write down. It has the same form as that of the basic model, namely

$$L_T = \boldsymbol{\delta}\,_1\mathbf{P}(\mathbf{x}_1)\boldsymbol{\Gamma}\,_2\mathbf{P}(\mathbf{x}_2) \cdots \boldsymbol{\Gamma}\,_T\mathbf{P}(\mathbf{x}_T)\mathbf{1}',$$

where $\mathbf{x}_1, \ldots, \mathbf{x}_T$ are the observations and, as before,

$$_t\mathbf{P}(\mathbf{x}_t) = \operatorname{diag}\left(_tp_1(\mathbf{x}_t), \ldots, \,_tp_m(\mathbf{x}_t)\right).$$

The above expression for the likelihood also holds if some of the series are continuous-valued, provided that, where necessary, probabilities are replaced by densities.

9.4.2 Contemporaneous conditional independence

The task of finding suitable joint distributions is greatly simplified if one can in addition assume **contemporaneous conditional independence**. We illustrate the meaning of this term by means of the multisite precipitation model discussed by Zucchini and Guttorp (1991). In their work there are five binary time series representing the presence or absence of rain at each of five sites which are regarded as being linked by a common weather process $\{C_t\}$. There the random variables X_{tj} are binary. Let π_{ij} be defined as

$$\pi_{ij} = \Pr(X_{tj} = 1 \mid C_t = i) = 1 - \Pr(X_{tj} = 0 \mid C_t = i).$$

The assumption of contemporaneous conditional independence is that the state-dependent joint probability $p_i(\mathbf{x}_t)$ is just the product of the corresponding marginal probabilities:

$$p_i(\mathbf{x}_t) = \prod_{j=1}^{q} \pi_{ij}^{x_{tj}} (1 - \pi_{ij})^{1 - x_{tj}}. \tag{9.5}$$

Thus, for example, given weather state i, the probability that on day t it will rain at sites 1, 2, and 4, but not at sites 3 and 5, is the product of the (marginal) probabilities that these events occur: $\pi_{i1}\pi_{i2}(1 - \pi_{i3})\pi_{i4}(1 - \pi_{i5})$.

For general multivariate HMMs that are contemporaneously conditionally independent, the state-dependent probabilities are given by a product of the corresponding q marginal probabilities:

$$_t p_i(\mathbf{x}_t) = \prod_{j=1}^{q} \Pr(X_{tj} = x_{tj} \mid C_t = i).$$

Additional examples of multivariate HMMs with the contemporaneous conditional independence assumption are the bivariate models for animal movement discussed in Chapter 18, where X_{t1} is the distance travelled by an animal in period t and X_{t2} the corresponding 'turning angle'.

We wish to emphasize that the above two conditional independence assumptions, namely longitudinal and contemporaneous conditional independence, *do not imply* that

- the individual component series are serially independent; nor that
- the component series are mutually independent.

The parameter process, namely the Markov chain, induces both serial

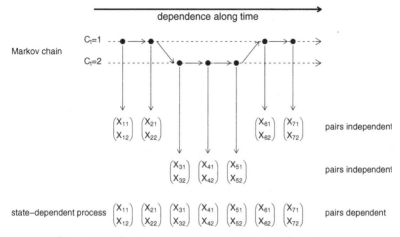

Figure 9.1 *Contemporaneous conditional independence.*

dependence and cross-dependence in the component series, even when the model is assumed to have both of these conditional independence properties. This is illustrated in Figure 9.1, and Exercise 4 provides a concrete example.

In summary, conditional independence does not imply unconditional independence.

9.4.3 Further remarks on multivariate HMMs

Details of the serial correlation and cross-correlation functions of some multivariate HMMs are given in Section 3.4 of MacDonald and Zucchini (1997), both for models assuming contemporaneous conditional independence and for those not making that assumption; see Exercise 7 in this chapter for the former case. Also given in the same section are several slightly different classes of model for multivariate HMMs, such as models with time lags and multivariate models in which some of the components are discrete and others continuous.

HMMs have been applied by a number of authors to model the returns on financial time series. In the simplest case of a two-state HMM, the states of the Markov chain can be interpreted as calm and turbulent phases of the stock market. HMMs appear to replicate most of what Rydén, Teräsvirta and Åsbrink (1998) – and others – term the 'stylized facts' associated with such series. The multivariate case can be modelled either by assuming contemporaneous conditional independence, or by fitting a (dependent) multivariate distribution in each of the states.

An example of the latter is given in Section 20.1, where a two-state multivariate normal–HMM is fitted to the daily returns on four shares.

Exercises

1. Give an expression for the likelihood function for each of the following HMMs. The list below refers to the state-dependent distributions.

   (a) Geometric with parameters $\theta_i \in (0, 1)$, $i = 1, 2, \ldots, m$.
   (b) Exponential with parameters $\lambda_i \geq 0$, $i = 1, 2, \ldots, m$.
   (c) Bivariate normal with parameters

   $$\boldsymbol{\mu}_i = \begin{pmatrix} \mu_{1i} \\ \mu_{2i} \end{pmatrix}, \qquad \boldsymbol{\Sigma}_i = \begin{pmatrix} \sigma_{1i}^2 & \sigma_{12i} \\ \sigma_{12i} & \sigma_{2i}^2 \end{pmatrix},$$

   for $i = 1, 2, \ldots, m$.

2. Consider a two-state stationary HMM in which the state-dependent distribution is Poisson with mean λ in state 1, and identically zero in state 2.

   (a) Give an expression for the likelihood function of the model.
   (b) Find the marginal mean and variance.
   (c) Find the autocorrelation function.

   Hint: see Exercise 1 in Chapter 1, and Exercise 4 in Chapter 2.

3. Find the autocorrelation functions for stationary HMMs with (a) normal and (b) binomial state-dependent distributions. (Assume in the latter case that the number of trials in each binomial distribution is some constant n.)

4. Consider the following bivariate normal–HMM on two states. The underlying stationary Markov chain has t.p.m.

   $$\boldsymbol{\Gamma} = \begin{pmatrix} 0.9 & 0.1 \\ 0.2 & 0.8 \end{pmatrix}.$$

   In state 1, X_{t1} and X_{t2} are independently distributed as standard normal, $N(0,1)$. In state 2, they are independently distributed as $N(5,1)$.

   (a) Generate a realization of length $10\,000$ (say) from such an HMM, and plot the values of X_{t1} against those of X_{t2}.
   (b) Compute the (sample) correlation of X_{t1} and X_{t2}.

5. (Runlengths in Bernoulli–HMMs.) Let $\{X_t\}$ be a Bernoulli–HMM in which the underlying stationary irreducible Markov chain has the states 1, 2, $\ldots$, m, transition probability matrix $\boldsymbol{\Gamma}$ and stationary distribution $\boldsymbol{\delta}$. The probability of an observation being 1 in state i is

denoted by p_i. Define a run of ones as follows: such a run is initiated by the sequence 01, and is said to be of length $k \in \mathbb{N}$ if that sequence is followed by a further $k - 1$ ones and a zero (in that order).

(a) Let K denote the length of a run of ones. Show that

$$\Pr(K = k) = \frac{\delta \mathbf{P}(0)(\mathbf{\Gamma P}(1))^k \mathbf{\Gamma P}(0) \mathbf{1}'}{\delta \mathbf{P}(0) \mathbf{\Gamma P}(1) \mathbf{1}'},$$

where $\mathbf{P}(1) = \mathrm{diag}(p_1, \ldots, p_m)$ and $\mathbf{P}(0) = \mathbf{I}_m - \mathbf{P}(1)$.

(b) Suppose that $\mathbf{B} = \mathbf{\Gamma P}(1)$ has distinct eigenvalues w_i. Show that the probability $\Pr(K = k)$ is, as a function of k, a linear combination of the kth powers of these eigenvalues, and hence of $w_i^{k-1}(1 - w_i)$, $i = 1, \ldots, m$.

(c) Does this imply that K is a mixture of geometric random variables? (Hint: will the eigenvalues w_i always lie between zero and one?)

(d) Assume for the rest of this exercise that there are only two states, with the t.p.m. given by

$$\mathbf{\Gamma} = \begin{pmatrix} 1 - \gamma_{12} & \gamma_{12} \\ \gamma_{21} & 1 - \gamma_{21} \end{pmatrix},$$

and that $p_1 = 0$ and $p_2 \in (0, 1)$.

Show that the distribution of K is as follows, for all $k \in \mathbb{N}$:

$$\Pr(K = k) = ((1 - \gamma_{21})p_2)^{k-1}(1 - (1 - \gamma_{21})p_2).$$

(So although not itself a Markov chain, this HMM has a geometric distribution for the length of a run of ones.)

(e) For such a model, will the length of a run of zeros also be geometrically distributed?

6. Consider the three two-state stationary Bernoulli–HMMs specified below. For instance, in model (a), $\Pr(X_t = 1 \mid C_t = 1) = 0.1$ and $\Pr(X_t = 1 \mid C_t = 2) = 1$, and X_t is either one or zero. The states are determined in accordance with a stationary Markov chain with t.p.m. $\mathbf{\Gamma}$. (Actually, model (c) is a Markov chain; there is in that case a one-to-one correspondence between states and observations.)

Let K denote the length of a run of ones. In each of the three cases, determine the following: $\Pr(K = k)$, $\Pr(K \leq 10)$, $\mathrm{E}(K)$, σ_K and $\mathrm{corr}(X_t, X_{t+k})$.

(a)

$$\mathbf{\Gamma} = \begin{pmatrix} 0.99 & 0.01 \\ 0.08 & 0.92 \end{pmatrix}, \qquad \mathbf{p} = (0.1, 1).$$

(b)

$$\Gamma = \begin{pmatrix} 0.98 & 0.02 \\ 0.07 & 0.93 \end{pmatrix}, \qquad \mathbf{p} = (0, 0.9).$$

(c)

$$\Gamma = \begin{pmatrix} 0.9 & 0.1 \\ 0.4 & 0.6 \end{pmatrix}, \qquad \mathbf{p} = (0, 1).$$

Notice that these three models are comparable in that (i) the unconditional probability of an observation being one is in all cases 0.2; and (ii) all autocorrelations are positive and decrease geometrically.

7. Consider a two-state bivariate HMM $\{(X_{t1}, X_{t2}) : t \in \mathbb{N}\}$, based on a stationary Markov chain with t.p.m. Γ and stationary distribution $\boldsymbol{\delta} = (\delta_1, \delta_2)$. Let μ_{i1} and μ_{i2} denote the means of X_{t1} and X_{t2} in state i, and similarly σ_{i1}^2 and σ_{i2}^2 the variances. Assume both contemporaneous conditional independence and longitudinal conditional independence.

(a) Show that, for all non-negative integers k,

$$\mathrm{Cov}(X_{t1}, X_{t+k,2}) = \delta_1 \delta_2 (\mu_{11} - \mu_{21})(\mu_{12} - \mu_{22})(1 - \gamma_{12} - \gamma_{21})^k.$$

(b) Hence find the cross-correlations $\mathrm{Corr}(X_{t1}, X_{t+k,2})$.

(c) Does contemporaneous conditional independence imply independence of X_{t1} and X_{t2}?

(d) State a sufficient condition for X_{t1} and X_{t2} to be uncorrelated.

(e) Generalize the results of (a) and (b) to any number (m) of states.

CHAPTER 10

Covariates and other extra dependencies

10.1 Introduction

This chapter outlines some of the ways in which covariate time series can be incorporated in HMMs. Trends and seasonal components, modelled as parametric functions of time, can then be handled as covariates. The chapter also describes a number of extensions to the standard HMMs covered in Part I. One such extension is achieved by allowing the latent process to be a second-order, or even higher-order, Markov chain rather than the first-order Markov chain that we have considered until now. One can also relax the assumption of conditional independence of the observations given the states, by allowing for additional dependencies between the latent and the observation processes, thereby substantially increasing the flexibility and thus extending the scope of HMMs. For example, the so-called Markov-switching models allow for serial dependence at the observation level in addition to that induced by the Markov chain. One can construct models that allow for 'feedback' in that the state of the process at a given time is influenced by preceding observations. Some of these extensions are illustrated in the applications covered in Part III.

10.2 HMMs with covariates

Covariates can be included in an HMM by allowing some of its parameters to depend on covariates. For this one can use parameters of the state-dependent distributions or those of the transition probability matrix (or even some of each).

For the purpose of model fitting, the values of the covariates are regarded as given, irrespective of whether they are deterministic (e.g. time) or realizations of some random process. However, in order to compute forecasts the relevant future values of the covariates are needed. If necessary, one can fit a separate model to the covariates in order to forecast the covariate values needed to provide forecasts of future values of the HMM. The situation is analogous to that discussed in Section 9.2 for the case of binomial–HMMs in which one needs the future values of the

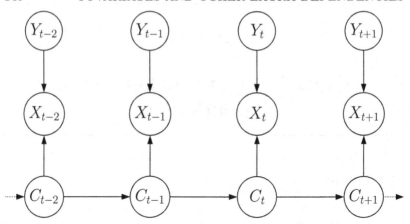

Figure 10.1 *Structure of HMM with covariates Y_t in the state-dependent probabilities.*

'number of trials' in order to compute the forecast distribution for the 'number of successes'.

10.2.1 Covariates in the state-dependent distributions

HMMs can be modified to allow for the influence of covariates by postulating dependence of the state-dependent probabilities $_t p_i(x_t)$ on those covariates, as demonstrated in Figure 10.1.

Here we take $\{C_t\}$ to be the homogeneous, or stationary, Markov chain, and suppose, in the case of a Poisson–HMM, that the state-dependent mean $_t\lambda_i = \mathrm{E}(X_t \mid C_t = i)$ depends on a (row) vector $\mathbf{y}_t$ of q covariates, for instance as follows:

$$\log \, {}_t\lambda_i = \boldsymbol{\beta}_i \mathbf{y}_t'.$$

In the case of a binomial–HMM with state-dependent 'success probabilities' $_t\pi_i$, the corresponding assumption is that

$$\mathrm{logit} \, {}_t\pi_i = \boldsymbol{\beta}_i \mathbf{y}_t'.$$

The elements of $\mathbf{y}_t$ could include deterministic functions of time (t) to model trend and seasonal variation. For example, a binomial–HMM with

$$\mathrm{logit} \, {}_t\pi_i = \beta_{i1} + \beta_{i2}t + \beta_{i3}\cos(2\pi t/12) + \beta_{i4}\sin(2\pi t/12) + \beta_{i5}z_t + \beta_{i6}w_t$$

allows for a (logit-linear) time trend, 12-period seasonality and the influence of covariates z_t and w_t, in the success probabilities $_t\pi_i$. Additional sine–cosine pairs can if necessary be included, to model more complex seasonal patterns. Similar models for the log of the state-dependent mean $_t\lambda_i$ are possible in the Poisson–HMM case. Clearly link functions other than the canonical ones used here could instead be used.

The expression for the likelihood of T consecutive observations x_1, $\ldots$, x_T from such a model involving covariates is similar to that of the basic model:

$$L_T = \delta_1\mathbf{P}(x_1,\mathbf{y}_1)\Gamma_2\mathbf{P}(x_2,\mathbf{y}_2)\cdots\Gamma_T\mathbf{P}(x_T,\mathbf{y}_T)\mathbf{1}',$$

the only difference being the allowance for covariates y_t in the state-dependent probabilities $_tp_i(x_t,\mathbf{y}_t)$ and in the corresponding matrices

$$_t\mathbf{P}(x_t,\mathbf{y}_t) = \mathrm{diag}(_tp_1(x_t,\mathbf{y}_t),\ldots,{}_tp_m(x_t,\mathbf{y}_t)).$$

Examples of models that incorporate a time trend can be found in Sections 21.2 and 22.2. In Section 21.3 there are models incorporating both a time trend and a seasonal component. Also illustrated in Chapter 21 are models with a change point, such as Poisson–HMMs with

$$\log{}_t\lambda_i = \left\{ \begin{array}{ll} \mu_{i1} & \text{for } t \leq t^*, \\ \mu_{i2} & \text{for } t > t^*, \end{array} \right.$$

where t^* is a prespecified time point.

It is worth noting that the binomial– and Poisson–HMMs which allow for covariates in this way provide important generalizations of logistic regression and Poisson regression respectively, generalizations that drop the independence assumption of such regression models and allow serial dependence. For such generalizations of Poisson regression, see also Wang and Puterman (1999a,b).

10.2.2 Covariates in the transition probabilities

An alternative way of incorporating covariate information in HMMs is to drop the assumption that the Markov chain is homogeneous, and assume instead that the transition probabilities are functions of time or other covariates, denoted as follows:

$$_t\Gamma = (_t\gamma_{ij}).$$

For models with three or more states it is generally less straightforward to incorporate covariates in the Markov chain than in the state-dependent distributions. One must ensure that the row-sum constraints for the transition probability matrix, $_t\Gamma$, are respected for all possible values of the covariates. One reason why it could nevertheless be worthwhile to model $_t\Gamma$ in terms of covariates is that the resulting Markov chain may have a useful substantive interpretation.

The two-state case is easy because only one element in each row of $_t\Gamma$ needs to be expressed in terms of the covariates, and so the row-sum constraints are automatically satisfied if one sees to it that the modelled entry lies in the interval $(0,1)$. One simple option for this case is to express the off-diagonal elements of $_t\Gamma$ as

$$\text{logit}\,_t\gamma_{i,3-i} = \boldsymbol{\beta}_i \mathbf{y}_t'.$$

where $\mathbf{y}_t$ represents a (row) vector of covariates. The resulting t.p.m. for transitions between times t and $t+1$ is given by

$$_t\Gamma = \begin{pmatrix} \dfrac{1}{1+\exp(\boldsymbol{\beta}_1\mathbf{y}_t')} & \dfrac{\exp(\boldsymbol{\beta}_1\mathbf{y}_t')}{1+\exp(\boldsymbol{\beta}_1\mathbf{y}_t')} \\ \dfrac{\exp(\boldsymbol{\beta}_2\mathbf{y}_t')}{1+\exp(\boldsymbol{\beta}_2\mathbf{y}_t')} & \dfrac{1}{1+\exp(\boldsymbol{\beta}_2\mathbf{y}_t')} \end{pmatrix}.$$

For example, a model incorporating r-period seasonality is that with

$$\text{logit}\,_t\gamma_{i,3-i} = \beta_{i1} + \beta_{i2}\cos(2\pi t/r) + \beta_{i3}\sin(2\pi t/r).$$

In this example the Markov chain is non-homogeneous and therefore does not have a stationary distribution. That makes it necessary to regard the initial distribution, that is, the distribution for C_1, as an additional vector of parameters to be estimated. The same will be the case in general unless the covariates $\mathbf{y}_t$ are stationary.

A broad class of models in which the Markov chain is non-homogeneous and which allows for the influence of covariates is that of Hughes (1993). Additional details of those models are discussed in Chapter 3 of MacDonald and Zucchini (1997).

10.3 HMMs based on a second-order Markov chain

One can enlarge the class of HMMs by replacing the underlying first-order Markov chain by a higher-order chain. Here we describe only the case of a stationary second-order chain. The chain is characterized by the transition probabilities

$$\gamma(i,j,k) = \Pr(C_t = k \mid C_{t-1} = j, C_{t-2} = i),$$

and has stationary bivariate distribution $u(j,k) = \Pr(C_{t-1} = j, C_t = k)$. (See Section 1.3.5 for a more detailed description of higher-order Markov chains.)

We mention here two important aspects of such a second-order HMM, which is depicted in Figure 10.2. The first is that it is possible to evaluate the likelihood of a second-order HMM in very similar fashion to that of the basic model; the computational effort is in this case cubic in m, the

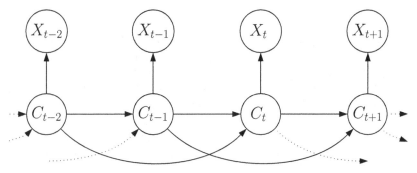

Figure 10.2 *Directed graph of second-order HMM.*

number of states, and, as before, linear in T, the number of observations. (See Exercise 2.) This then enables one to estimate the parameters by direct maximization of the likelihood, and also to compute the forecast distributions.

The second aspect is the number of free parameters of the (second-order) Markov chain component of the model. In general this number is $m^2(m-1)$, which rapidly becomes prohibitively large as m increases. Overparametrization can be circumvented by using some restricted subclass of second-order Markov chain models, for example those of Pegram (1980) or those of Raftery (1985a). Such models are necessarily less flexible than the general class of second-order Markov chains. They maintain the second-order structure of the chain, but trade some flexibility in return for a reduction in the number of parameters. We note in passing that the complications caused by the constraints on the parameters in a Raftery model have been discussed by Schimert (1992) and by Raftery and Tavaré (1994), who describe how one can reduce the number of constraints.

Having pointed out that it is possible to increase the order of the Markov chain in an HMM from 1 to 2 (or higher), in what follows we shall restrict our attention almost entirely to the simpler case of a first-order Markov chain. One exception is Section 17.3.2, where we present (among other models) an example of a second-order HMM for a binary time series. For more extensive discussions of higher-order HMMs, see Aston and Martin (2007) and Bartolucci (2011). In Chapter 12, we will discuss hidden semi-Markov models, which represent an alternative and much more popular way of relaxing the assumption that the latent process in an HMM has memory of one time lag only.

10.4 HMMs with other additional dependencies

We now describe a number of ways in which additional dependencies can be built into HMMs, apart from those covered in the previous two sections.

In the basic model, depicted in Figure 2.2 on p. 30, the only path to the observation X_t is from the current state C_t. In other words, the distribution of X_t is completely specified by the state C_t; there are no additional dependencies. Allowing the distribution of X_t to depend on some of the previous observations (as well as on C_t) leads to an important class of models, called Markov-switching models. Figure 10.3 depicts the Markov-switching AR(1) and AR(2). In the AR(1) case the distribution of X_t depends on X_{t-1} and C_t; in the AR(2) case it depends on X_{t-1}, X_{t-2} and C_t. For a discussion of models of this kind, see Hamilton (1994, Section 22.4).

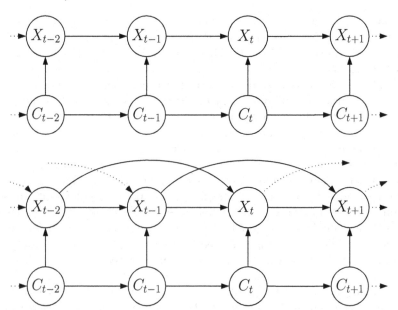

Figure 10.3 *Models with additional dependencies at observation level; the upper graph represents (for example) a Markov-switching AR(1) model, and the lower one a Markov-switching AR(2).*

Some examples of models with such dependencies at observation level are given in Section 21.3. Note that the conditional distribution of X_1 depends on the (unobserved) value x_0 in the AR(1) case, and on x_0 and x_{-1} in the AR(2) case. For the purpose of parameter estimation in the AR(p) case one can conveniently circumvent this complication by

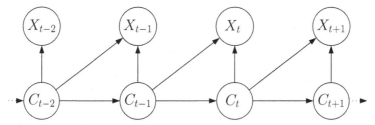

Figure 10.4 *Additional dependencies from latent process to observation level.*

conditioning on the first p observations. One maximizes the likelihood for the observations $x_{p+1}, x_{p+2}, \ldots, x_T$, given $x_1, x_2, \ldots, x_p$.

Apart from hidden Markov AR(k) models, that is, Markov-switching autoregressions, models with additional dependencies at observation level that have appeared in the literature include the double-chain Markov model of Berchtold (1999), the M1–Mk models of Nicolas *et al.* (2002) for DNA sequences, and those of Boys and Henderson (2004).

Another type of additional dependency allows the distribution of X_t to depend on previous states, as well as on C_t, the current state. We consider the simplest case, in which a single preceding state is used; see Figure 10.4. That is, we replace the usual assumption

$$\Pr(X_t \mid \mathbf{X}^{(t-1)}, \mathbf{C}^{(t)}) = \Pr(X_t \mid C_t), \quad t \in \mathbb{N},$$

by the weaker assumption that

$$\Pr(X_t \mid \mathbf{X}^{(t-1)}, \mathbf{C}^{(t)}) = \Pr(X_t \mid C_{t-1}, C_t).$$

If here we define the $m \times m$ matrix $\mathbf{Q}(x)$ to have as its (i, j) element the product $\gamma_{ij} \Pr(X_t = x \mid C_{t-1} = i, C_t = j)$, and denote by $\boldsymbol{\delta}$ the distribution of C_0 (not C_1), the likelihood is given by

$$L_T = \boldsymbol{\delta} \mathbf{Q}(x_1) \mathbf{Q}(x_2) \cdots \mathbf{Q}(x_T) \mathbf{1}'. \tag{10.1}$$

Notice that $\mathbf{Q}(x)$ is defined as the elementwise product of two matrices (also known as the 'Hadamard' or 'Schur' product), which is easily coded in **R**. We defer the proof of equation (10.1) to Exercise 3(a).

We now describe in detail the application of one model of this kind to the earthquakes series. Another is described more briefly in Exercise 4. Suppose that, given $C_{t-1} = i$ and $C_t = j$, the observation X_t has a Poisson distribution with mean $\alpha \lambda_i + (1 - \alpha) \lambda_j$. That is, we assume a Poisson distribution and mix the means corresponding to the current state and to the previous one. Equation (10.1) for the likelihood was used in order to fit a stationary model of this kind, for two and for three states, and the results were compared by AIC and BIC to the corresponding models without the extra dependency; see Table 10.1. By AIC, and *a*

Table 10.1 *Earthquakes data: comparison by AIC and BIC of stationary hidden Markov models without and with dependence on previous state.*

Model	k	$-\log L$	AIC	BIC
2-state	4	342.3183	692.6	703.3
2-state, extra dependency	5	341.8365	693.7	
3-state	9	329.4603	**676.9**	**701.0**
3-state, extra dependency	10	328.5283	677.1	

fortiori by BIC, the extra parameter α which allows for dependence of the mean on the previous state as well has not been worthwhile here. The parameter estimates for the three-state model fitted are as follows. The 'mixing parameter' α is 0.327, the state-dependent means are

$$\boldsymbol{\lambda} = (13.069, 19.699, 30.310),$$

and the t.p.m. is

$$\boldsymbol{\Gamma} = \begin{pmatrix} 0.955 & 0.023 & 0.022 \\ 0.049 & 0.906 & 0.044 \\ 0.000 & 0.190 & 0.810 \end{pmatrix},$$

the implied stationary distribution of which is

$$\boldsymbol{\delta} = (0.445, 0.407, 0.147).$$

These estimates do not differ much from their values in the basic model, which is the special case $\alpha = 0$ of the models we are considering here; see p. 55.

Extra dependencies of yet another kind appear in Section 20.2.3 and Chapter 23, which present respectively a discrete state-space stochastic volatility model with leverage (see Figure 20.3), and a model for animal behaviour which incorporates feedback from observation level to motivational state (see Figure 23.2). In both of these applications the extra dependencies are from observation level to the latent process.

Exercises

1. Let $\{X_t\}$ be a second-order HMM, based on a stationary second-order Markov chain $\{C_t\}$ on m states. How many parameters does the model have in the following cases? (Assume for simplicity that one parameter is needed to specify each of the m state-dependent distributions.) If necessary, see p. 22 for the relevant definitions.

   (a) $\{C_t\}$ is a general second-order Markov chain.

(b) $\{C_t\}$ is a Pegram model.

(c) $\{C_t\}$ is a Raftery model (i.e. an MTD model).

2. Let $\{X_t\}$ be a second-order HMM, based on a stationary second-order Markov chain $\{C_t\}$ on m states. For integers $t \geq 2$, and integers i and j from 1 to m, define

$$\nu_t(i, j; \mathbf{x}^{(t)}) = \Pr(\mathbf{X}^{(t)} = \mathbf{x}^{(t)}, C_{t-1} = i, C_t = j).$$

Note that these probabilities are just an extension to two dimensions of the forward probabilities $\alpha_t(i)$, which could more explicitly be denoted by $\alpha_t(i; \mathbf{x}^{(t)})$. For the case $t = 2$ we have

$$\nu_2(i, j; \mathbf{x}^{(2)}) = u(i, j) p_i(x_1) p_j(x_2).$$

(a) Show that, for integers $t \geq 3$,

$$\nu_t(j, k; \mathbf{x}^{(t)}) = \left(\sum_{i=1}^{m} \nu_{t-1}(i, j; \mathbf{x}^{(t-1)}) \gamma(i, j, k) \right) p_k(x_t). \quad (10.2)$$

(b) Show how the recursion (10.2) can be used to compute the likelihood of a series of T observations, $\Pr(\mathbf{X}^{(T)} = \mathbf{x}^{(T)})$.

(c) Show that the computational effort required to find the likelihood thus is $O(Tm^3)$.

3. Consider the generalization of the basic HMM which allows the distribution of an observation to depend on the current state of a Markov chain *and* the previous one; see Section 10.4.

(a) Show that the likelihood of such a model is given by equation (10.1). (Hint: use the multiple-sum result of Exercise 7 in Chapter 2.)

(b) Show that, if we denote by $\boldsymbol{\delta}$ the distribution of C_1 (not C_0), the likelihood is (also) given by

$$L_T = \boldsymbol{\delta} \mathbf{P}(x_1) \mathbf{Q}(x_2) \mathbf{Q}(x_3) \cdots \mathbf{Q}(x_T) \mathbf{1}'.$$

(Here, as usual, $\mathbf{P}(x_1)$ denotes the diagonal matrix of probabilities of observation x_1 conditional on the current state only.)

4. Consider again the earthquakes series. Let $\{C_t\}$ be a stationary Markov chain. Suppose that, instead of the usual assumption that

$$\Pr(X_t \mid \mathbf{X}^{(t-1)}, \mathbf{C}^{(t)}) = \Pr(X_t \mid C_t), \quad t \in \mathbb{N},$$

we assume that

$$\Pr(X_t \mid \mathbf{X}^{(t-1)}, \mathbf{C}^{(t)}) = \Pr(X_t \mid C_{t-1}, C_t).$$

Suppose in particular that, given that $C_{t-1} = i$ and $C_t = j$, the observation X_t has with probability α a Poisson distribution with mean λ_i, otherwise a Poisson distribution with mean λ_j. That is, we mix Poisson distributions indexed by the current state and the previous one.

Use equation (10.1) for the likelihood in order to fit a model of this kind, for two and for three states. Do your models (by AIC or BIC) improve on the models with $\alpha = 0$?

## CHAPTER 11

Continuous-valued state processes

The world is continuous, but the mind is discrete.

D.B. Mumford (International Congress of Mathematicians, 2002)

11.1 Introduction

The state architecture of HMMs is sometimes not well suited to the problem at hand. In some examples the choice of the number of states can be straightforward, possibly using model selection criteria and residual analyses, and the interpretation of the states can be intuitive. This need not be the case in general, however. In fact, the determination of the number of states often remains difficult in practice, and the assumption of a finite number of states is often not intuitive. Another possible problem with the HMM state architecture is that the number of parameters can become very large as the number of states increases. In these instances it may sometimes be advantageous to consider alternative model formulations, in which the state process is continuous-rather than discrete-valued and is specified by a model that is relatively parsimonious in terms of the number of parameters.

For example, consider again the time series of earthquake counts. There seem to be no obvious seismological grounds for assuming that there are three states, nor indeed any finite number of different rates of occurrence of major earthquakes. It would be more intuitive to assume that the rate of occurrence is continuous-valued, so that gradual change over the years is possible. For example, a simple model allowing for gradual change would be

$$X_t \sim \text{Poisson}\big(\beta \exp(C_t)\big), \qquad (11.1)$$

$$C_t = \phi C_{t-1} + \sigma \eta_t, \qquad (11.2)$$

with $|\phi| < 1$, $\beta, \sigma > 0$ and $\eta_t \overset{\text{iid}}{\sim} \text{N}(0,1)$. Here the state process, $\{C_t\}$, which determines the rate of occurrence of major earthquakes, is an autoregressive process of order 1, specified by only two parameters, as opposed to the six parameters needed for the Markov chain in a three-state Poisson–HMM. Since the values of $\{C_t\}$ fluctuate around zero, the mean of the state-dependent process $\{X_t\}$ fluctuates around β, with the parameters ϕ and σ controlling the strength of the mean-reverting effect and the variability of the occurrence rates, respectively. This model has previously been considered by, among others, Zeger (1988) and Chan and Ledolter (1995). We shall show below that, for the earthquake series, this model is by AIC superior to the Poisson–HMMs.

The model (11.1)–(11.2) for the earthquake series is an example of a state-space model (SSM) with continuous-valued state process. In their basic form, SSMs have precisely the same dependence structure as HMMs, comprising an observed state-dependent process (typically called the observation process in the SSM literature) and an unobserved state process satisfying the Markov property (typically called the system process). The state process of an SSM can be continuous-valued or discrete-valued; HMMs are the special case in which the state process is discrete-valued. (But note that some authors, e.g. Cappé *et al.*, 2005, do not distinguish between SSMs and HMMs.)

To calculate the likelihood of an SSM, one needs to integrate over all possible values of the state process at each time an observation is made; this is analogous to summing over all possible values in the case of an HMM. Thus, the likelihood is in general given by a high-order multiple integral and cannot be evaluated directly. Linear Gaussian SSMs can be fitted relatively easily via the use of the Kalman filter (see, for example, Harvey, 1989). Cases involving either nonlinearities or non-Gaussian conditional distributions, such as the model for earthquake counts given above, are usually more involved. However, it turns out that in many such cases a simple modification, namely a fine discretization of the state space of an SSM, renders the entire HMM methodology applicable to that SSM. (The discretization amounts to numerical integration and hence an approximation of the likelihood.) In particular, the HMM forward algorithm can be used to evaluate the approximate likelihood of the SSM, which makes parameter estimation via numerical maximization of the likelihood feasible in many cases. To the best of our knowledge, such discretization was first proposed by Kitagawa (1987). The description given here is based on those of Langrock (2011) and Langrock, MacDonald and Zucchini (2012).

11.2 Models with continuous-valued state process

We consider a basic SSM in which the state process is one-dimensional. Such an SSM is characterized by two processes: a continuous-valued Markov state process, $\{C_t\}$, and an observed process, $\{X_t\}$, whose realizations are assumed to be conditionally independent, given the states. In other words, it satisfies exactly the same conditional independence assumptions as does an HMM:

$$p(c_t \mid \mathbf{c}^{(t-1)}) = p(c_t \mid c_{t-1}), \quad t = 2, 3, \ldots,$$
$$p(x_t \mid \mathbf{x}^{(t-1)}, \mathbf{c}^{(t)}) = p(x_t \mid c_t), \quad t \in \mathbb{N},$$

where p is used as a general symbol for either a density or a probability. This model differs from an HMM in that the Markov process $\{C_t\}$ is continuous-valued rather than discrete-valued. After appropriately discretizing the state space into a sufficiently large but finite number of states, however, we can evaluate an approximation of the likelihood of any given SSM by using the HMM forward algorithm.

We first describe how the likelihood of any given SSM can be approximated arbitrarily accurately by discretizing the state space (Section 11.2.1). Then we describe how the HMM forward algorithm can be used to evaluate the approximate likelihood in a computationally feasible way (Section 11.2.2). Some technical issues relating to estimation are discussed in Section 11.2.3.

11.2.1 Numerical integration of the likelihood

The discretization (and hence numerical integration) procedure is as follows. For a given SSM, consider an essential range $[b_0, b_m]$ of possible values of C_t, and split this range into m subintervals $B_i = (b_{i-1}, b_i)$, $i = 1, \ldots, m$. These subintervals need not be of equal length, but for ease of exposition we present only that case; they are then all of length $h = (b_m - b_0)/m$. We denote by b_i^* a representative point in B_i (e.g. the midpoint). By making use of the SSM dependence structure, and repeatedly approximating integrals $\int_a^b f(c)\, dc$ by expressions of the form $(b - a)f(c^*)$, the likelihood of the observations $x_1, \ldots, x_T$ can be approximated as follows:

$$
\begin{aligned}
L_T &= \int \cdots \int p(x_1, \ldots, x_T, c_1, \ldots, c_T)\, dc_T \ldots dc_1 \\
&= \int \cdots \int p(x_1, \ldots, x_T \mid c_1, \ldots, c_T) p(c_1, \ldots, c_T)\, dc_T \ldots dc_1 \\
&= \int \cdots \int p(c_1) p(x_1 \mid c_1) \prod_{t=2}^{T} p(c_t \mid c_{t-1}) p(x_t \mid c_t)\, dc_T \ldots dc_1 \\
&\approx \int_{b_0}^{b_m} \cdots \int_{b_0}^{b_m} p(c_1) p(x_1 \mid c_1) \prod_{t=2}^{T} p(c_t \mid c_{t-1}) p(x_t \mid c_t)\, dc_T \ldots dc_1 \\
&\approx h^T \sum_{i_1=1}^{m} \cdots \sum_{i_T=1}^{m} p(b_{i_1}^*) p(x_1 \mid b_{i_1}^*) \prod_{t=2}^{T} p(b_{i_t}^* \mid b_{i_{t-1}}^*) p(x_t \mid b_{i_t}^*). \quad (11.3)
\end{aligned}
$$

In the final step, the innermost integral has been approximated as follows:

$$\int_{b_0}^{b_m} p(c_T \mid c_{T-1}) p(x_T \mid c_T) \, \mathrm{d}c_T \approx h \sum_{i_T=1}^{m} p(b_{i_T}^* \mid c_{T-1}) p(x_T \mid b_{i_T}^*).$$

This is a simple midpoint quadrature, and we note that this is not the only way in which the integral could be approximated; see, for example, Langrock, MacDonald and Zucchini (2012), and also Section 20.2, for an alternative. The other integrals are approximated analogously. The factor h^T is replaced by the appropriate product if one does not use the same (equal) subdivision of $[b_0, b_m]$ at each time point.

All the terms appearing in the approximation given in (11.3) are simple, yet the likelihood cannot be evaluated by using this expression as it stands, because of the extremely high number of summands (m^T). However, the numerical integration corresponds to a discretization of the state space, and with a discrete state space we are in the HMM framework and can apply the corresponding techniques. We show how to do so in the next subsection.

11.2.2 Evaluation of the approximate likelihood via forward recursion

The discretization of the state space into m intervals essentially corresponds to an approximation of the SSM by an m-state HMM. Furthermore, it is straightforward to specify the components of this approximating HMM. First, we consider the initial distribution of the state process. To obtain the exact expression given in (11.3), we define the ith component of the m-dimensional vector $\boldsymbol{\delta}$ to be $\delta_i = hp(b_i^*)$. Then δ_i gives the approximate probability of the state process falling in the interval B_i at time 1.

As an example, consider the AR(1) process given in (11.2), and suppose that the process is in its stationary distribution at the time of the first observation. Then $p(b_i^*)$ is the density of the normal distribution with mean zero and variance $\sigma^2/(1 - \phi^2)$ – the stationary distribution of the AR(1) process – evaluated at b_i^*.

Similarly, we define an $m \times m$ matrix $\boldsymbol{\Gamma} = (\gamma_{ij})$ by specifying $\gamma_{ij} = hp(b_j^* \mid b_i^*)$. Then γ_{ij} gives the approximate probability of the value of the state process falling into the interval B_j at time t, given that the process is in interval B_i at time $t - 1$. In the case of a Gaussian AR(1) state process, the corresponding value of γ_{ij} is h times the density of the normal distribution with mean ϕb_i^* and variance σ^2, evaluated at b_j^*. Finally, we define $\mathbf{P}(x_t)$ to be the $m \times m$ diagonal matrix with ith diagonal entry equal to $p(x_t \mid b_i^*)$. This corresponds to an approximation of the conditional density of x_t, given that the state process takes some value in the interval B_i at time t. In the observation process specified by (11.1), the ith diagonal entry of $\mathbf{P}(x_t)$ would be the probability mass function of the Poisson distribution with parameter $\beta \exp(b_i^*)$, evaluated at x_t.

Putting all the pieces together, we can rewrite the multiple-sum expression for the approximate likelihood, given in (11.3), in the form of a matrix product:

$$h^T \sum_{i_1=1}^{m} \cdots \sum_{i_T=1}^{m} p(b_{i_1}^*)p(x_1 \mid b_{i_1}^*) \prod_{t=2}^{T} p(b_{i_t}^* \mid b_{i_{t-1}}^*)p(x_t \mid b_{i_t}^*)$$

$$= \boldsymbol{\delta}\mathbf{P}(x_1)\boldsymbol{\Gamma}\mathbf{P}(x_2)\boldsymbol{\Gamma}\mathbf{P}(x_3) \cdots \boldsymbol{\Gamma}\mathbf{P}(x_{T-1})\boldsymbol{\Gamma}\mathbf{P}(x_T)\mathbf{1}'. \qquad (11.4)$$

Using this matrix product expression, the computational effort required to evaluate the approximate likelihood is linear in T, the number of observations, and quadratic in m, the number of intervals used in the discretization, exactly as for HMMs. The **R** function below evaluates the likelihood of the model specified by (11.1) and (11.2) for given m, b_m (with $b_0 = -b_m$), ϕ, σ and β.

```
# Function that calculates minus the approximate log-likelihood
mllk<-function(x,m,bm,phi,sigma,beta)
{
 gb<-seq(-bm,bm,length=m+1)
 h<-gb[2]-gb[1]
 g<-(gb[-1]+gb[-(m+1)])*0.5
 lambda<-exp(g)*beta
 Gamma<-matrix(0,m,m)
 for (i in 1:m){
  Gamma[i,]<-dnorm(g,phi*g[i],sigma)*h
 }
 delta<-dnorm(g,0,sigma/sqrt(1-phi^2))*h
 lscale<-0
 foo<-delta*dpois(x[1],lambda)
 for (t in 2:length(x)){
  foo<-foo%*%Gamma*dpois(x[t],lambda)
  sumfoo<-sum(foo)
  lscale<-lscale+log(sumfoo)
  foo<-foo/sumfoo
 }
 return(-lscale)
}
```

Although this code implements one particular example of an SSM, its structure usually remains virtually unchanged when modifications in the model formulation are considered. For example, only two lines of the above code are affected if a distribution other than the Poisson is specified for the state-dependent process.

It should perhaps be noted that, with the specifications of $\boldsymbol{\delta}$, $\boldsymbol{\Gamma}$ and $\mathbf{P}(x_t)$ given above, the row sums of $\boldsymbol{\Gamma}$ will only approximately equal 1, and the components of the vector $\boldsymbol{\delta}$ will only approximately total 1. This can be remedied by scaling the vector $\boldsymbol{\delta}$ and each row of $\boldsymbol{\Gamma}$ to total 1; this is not done in the code displayed.

11.2.3 Parameter estimation and related issues

Numerical maximization of the likelihood given in (11.4) is feasible in many cases, even for a large number of observations T and fairly large m (and hence a very close approximation to the likelihood). In practice one has to select the value of m and the range of possible C_t-values considered in the numerical integration. In our experience, estimates usually stabilize for values of m around 50 (cf. Langrock, MacDonald and Zucchini, 2012). Note that, although m needs to be large enough to provide a good approximation, the number of model parameters does not depend on m; the entries of the $m \times m$ matrix $\boldsymbol{\Gamma}$ depend only on the parameters determining the state process of the SSM. The range $[b_0, b_m]$ has to be chosen sufficiently large to cover the essential range of the state process, but not too large, in order to maintain sufficient fineness of the grid.

Whether or not the chosen range is indeed sufficiently large can easily be checked after fitting the model by examining the marginal distribution of the fitted state process. The other main technical issues that may arise in the numerical maximization of the likelihood are exactly the same as for HMMs, namely local maxima, parameter constraints and numerical underflow or overflow, and essentially the same strategies as for HMMs can be used to deal with these problems (cf. Chapter 3).

Quantification of uncertainty can be carried out by using either the parametric bootstrap or the Hessian of the log-likelihood for the estimated parameters. Furthermore, the approximate HMM representation can be used, *inter alia*, for forecasting, decoding or model checking of the SSM. Indeed, all standard HMM methods become applicable. In particular, the Viterbi algorithm can be used for decoding, pseudo-residuals can be used for model checking, and the closed-form expressions for forecasts in HMMs are applicable.

11.3 Fitting an SSM to the earthquake data

We now illustrate the approach using the earthquake series. To fit the model defined by (11.1) and (11.2) to the earthquake counts, we first discretize the state space of $\{C_t\}$ into $m = 200$ states, with $-b_0 = b_m = 1.5$ for the range of the C_t-values. Maximizing the resulting (approximate) likelihood numerically, and making use of the forward recursion described in Section 11.2.2, we find the parameter estimates

$$\widehat{\phi} = 0.89 \ (0.09), \quad \widehat{\sigma} = 0.14 \ (0.03), \quad \widehat{\beta} = 17.8 \ (2.07).$$

The values in brackets are the standard errors obtained from a parametric bootstrap using 500 samples. For the fitted model, the variance of the stationary distribution of $\{C_t\}$ is $\hat{\sigma}^2/(1 - \hat{\phi}^2)$, approximately 0.3^2. The essential range chosen is hence very conservative, with the stationary distribution of $\{C_t\}$ having mass less than 10^{-6} outside the interval $[-1.5, 1.5]$.

The application of the forward recursion renders this model-fitting exercise computationally inexpensive, with the numerical maximization of the approximate likelihood taking less than a second on an octa-core i7 CPU, at 2.7 GHz and with 4 GB RAM. The effect of the choice of m and the essential range $[b_0, b_m]$ is illustrated in Table 11.1. In this application, the likelihood approximation is virtually exact for $m = 40$, provided that the specified essential range is neither too small nor too large.

Table 11.1 *SSM fitted to earthquakes data: maximum log-likelihood values obtained for various values of m and b_{max}, where $-b_0 = b_m = b_{max}$. The approximation with $m = 20$ and $b_{max} = 4$ is too coarse to produce reasonable estimates.*

b_{max}	$m = 20$	$m = 40$	$m = 70$	$m = 100$	$m = 200$
0.5	−337.61050	−337.66812	−337.68095	−337.68412	−337.68640
1	−332.26895	−332.26918	−332.26924	−332.26925	−332.26926
2	−332.21761	−332.26789	−332.26789	−332.26789	−332.26789
4	−	−332.21761	−332.26789	−332.26789	−332.26789

The minus log-likelihood and AIC values for this model are obtained as 332.27 and 670.54, respectively. Therefore the AIC favours the model with continuous-valued state process over the standard Poisson–HMMs; recall that the smallest value of the AIC for Poisson–HMMs, 676.92, was obtained when using three states. Figure 11.1 displays the decoded most likely sequences of mean numbers of quakes, under the model with AR(1) state process and under the three-state Poisson–HMM; both were decoded by using the Viterbi algorithm. In the HMM the mean number of earthquakes changes among a finite number of levels, whereas in the SSM changes occur according to an autoregressive process and the mean number of earthquakes takes on values in $\mathbb{R}$, which seems more plausible as there is no *a priori* reason to suppose that there is a finite number of rates for the occurrence of major earthquakes.

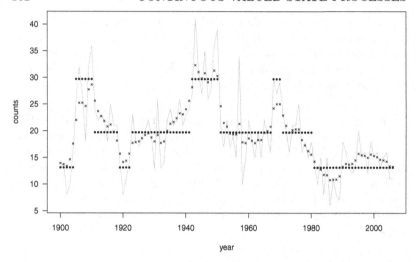

Figure 11.1 *Earthquake counts (continuous line) and decoded mean sequences of the three-state Poisson–HMM and the model with AR(1) state process (dots and crosses respectively).*

11.4 Discussion

We note here that various alternative methods can be used to fit nonlinear and non-Gaussian SSMs, including, among others, MCMC (see, for example, Carter and Kohn, 1994), simulation-based approaches (see, for example, Durbin and Koopman, 1997), the AD Model Builder (Fournier *et al.*, 2012) and integrated nested Laplace approximations (see, for example, Ruiz-Cárdenas, Krainski and Rue, 2012).

The discretization technique described in this chapter is easy to implement and has been successfully used in a variety of applications. Examples include stochastic volatility models for financial time series (Bartolucci and De Luca, 2003; Langrock, MacDonald and Zucchini, 2012), capture–recapture models (Langrock and King, 2013; see also Chapter 24) and models for animal movement (Pedersen *et al.*, 2011). The method is usually feasible for one-dimensional state spaces. However, it suffers from the 'curse of dimensionality' and is rather difficult to apply in the case of high-dimensional state spaces (Kitagawa, 1996).

An approximation technique similar to that described in this chapter can sometimes be used if the state space of the underlying Markov process is discrete but infinite. In such a case, the state process needs to be coarsened through binning in order to render the forward algorithm feasible. For example, in models for animal population dynamics, observations might correspond to the number of animals captured at different occasions, and the underlying process to the total number of animals (i.e. those captured plus those not captured). In that case, the state process takes values in $\{0, 1, 2, \ldots\}$, the non-negative integers. If it can reasonably be assumed that the state process will remain below some upper bound, say 5000, then an approximate likelihood can be built by binning the states into, say, $\{0, 1, \ldots, 49\}$, $\{50, 51, \ldots, 99\}$, $\ldots$, $\{4950, \ldots, 4999\}$, which leads to an approximation analogous to that described above for models with continuous-valued state process.

CHAPTER 12

Hidden semi-Markov models and their representation as HMMs

12.1 Introduction

The use of a Markov chain of order greater than 1 as the latent process of an HMM, outlined in Section 10.3, represents one way of relaxing the assumption that the latent process has memory of one time lag only. A disadvantage of that generalization is that the number of parameters increases rapidly as the order of the Markov chain increases. In this chapter, we discuss an alternative and probably more important relaxation of the assumption.

We start by noting that the number of consecutive time points that a first-order Markov chain spends in a given state, called the *dwell time* or *sojourn time*, follows a geometric distribution. For a Markov chain with transition probability matrix $\mathbf{\Gamma}$, the probability mass function (p.m.f.) of the dwell time in state i is given by

$$d_i(r) = (1 - \gamma_{ii})\gamma_{ii}^{r-1}, \quad r \in \mathbb{N}.$$

Hence the most likely dwell time for every state of an HMM with underlying first-order Markov chain is 1.

Hidden semi-Markov models (HSMMs) are designed to relax this restrictive, and in some applications unrealistic, condition. In such a model the latent process is a 'semi-Markov' process rather than a Markov process. We begin with a brief introduction to such processes. For a fuller discussion, see, for instance, Barbu and Limnios (2008, Chapter 3) or Kulkarni (2010, Section 8.9).

12.2 Semi-Markov processes, hidden semi-Markov models and approximating HMMs

The semi-Markov processes we shall employ, in discrete time and on m states, are defined as follows. Let $\{Y_t\}$ be a homogeneous Markov chain on $\{1, 2, \ldots, m\}$, with its t.p.m. $\mathbf{\Omega}$ having the special feature that all its diagonal elements are zero. So in a realization of $\{Y_t\}$, no two successive values Y_t, Y_{t+1} are equal. Now allow $\{Y_t\}$, plus a set of dwell-time distributions d_i on the positive integers, to generate a new

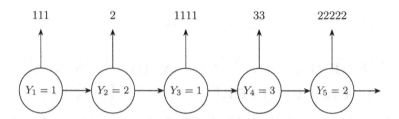

Figure 12.1 *Example of semi-Markov process with state switches driven by the Markov chain* $\{Y_t\}$.

process $\{C_t\}$, also on $\{1, 2, \ldots, m\}$, as follows. Each Y_t gives rise to a run of 'C-values' all equal to Y_t. The length of the run is a realization of the corresponding dwell-time distribution; that is, if $Y_t = i$, the distribution is d_i. All dwell times are independent of each other and of earlier values of Y_t.

For example, if $m = 3$ and $\{Y_t\}$ begins with 1, 2, 1, 3, 2, a possible realization of $\{C_t\}$ is as follows.

$$111 \mid 2 \mid 1111 \mid 33 \mid 22222 \mid \ldots$$

Here the sequences 111 and 1111 arise from two (independent) realizations of d_1, 2 and 2222 from two realizations of d_2, and 33 from d_3. This example is depicted in Figure 12.1.

The resulting process $\{C_t\}$, which is not in general a Markov process, is called a semi-Markov process. The probabilities of self-transition in the process $\{C_t\}$, which determine the times spent in the states, are now implied by the distributions d_i. In the special case in which all the distributions d_i are geometric, $\{C_t\}$ will be a Markov process; an HMM is therefore a special case of an HSMM.

A hidden semi-Markov model then consists of a latent semi-Markov process $\{C_t\}$ and an observed process $\{X_t\}$, the distribution of which is driven in the usual way by the latent process. The dwell time in each state of an HSMM can follow any distribution on the positive integers, for example, a Poisson shifted up by one, or a zero-truncated Poisson, or some distribution of finite support.

HSMMs and their applications are discussed by, among others, Ferguson (1980), Sansom and Thomson (2001), Guédon (2003), Bulla, Bulla and Nenadić (2009), Yu (2010), and Langrock and Zucchini (2011). The additional generality offered by HSMMs carries a computational cost; they are much more demanding to apply than are HMMs, particularly since the forward and backward recursions become more involved (Johnson, 2005, p. 409). Furthermore, although the inclusion of covariates,

trend, seasonality or random effects in the model has been broadly explored and is fairly standard for HMMs, the same cannot be said of
HSMMs.

We come now to the most important point of this chapter, discussed
also by Langrock and Zucchini (2011). In Sections 12.3 and 12.4 we
shall describe a way of constructing HMMs, with expanded state space,
whose properties closely approximate those of HSMMs. The corresponding general model formulation can be used to fit, at least approximately,
an HSMM with any desired dwell-time distributions. The idea is to use
so-called state aggregates, which in similar, but not identical, ways have
been used by, for example, Russell and Cook (1987) and Guédon (2005).
An overview of the various existing approaches can be found in Johnson (2005), who argues (in the context of speech recognition) that an
expanded-state HMM is 'almost certainly a much better practical choice
for duration modeling than development and implementation of more
complex and computationally expensive models with explicit modifications to handle duration probabilities'. For a contrary view, however, see
Guédon (2005).

By using the specially structured HMMs described below, the whole
standard HMM methodology, including state prediction, local and global
decoding, forecasting and model checking, becomes applicable to the
more flexible class of HSMMs. Furthermore, the incorporation of trend,
seasonality and covariates in the models becomes straightforward.

12.3 Examples of HSMMs represented as HMMs

To aid understanding of the general discussion in Section 12.4, we first
present several illustrative examples. We focus on the expansion of the
state space that enables one to represent the HSMM, at least approximately, as an HMM on an expanded state space.

12.3.1 A simple two-state Poisson–HSMM

Consider a two-state Poisson–HSMM, with state-dependent means $\lambda_1 = 5$ and $\lambda_2 = 10$, and with the p.m.f.s of the state dwell times given by

r :	1	2	3	4	≥ 5
$d_1(r)$:	$\frac{1}{4}$	$\frac{3}{8}$	$\frac{1}{4}$	$\frac{1}{8}$	0

and by $d_2(r) = 0.1 \times 0.9^{r-1}$ (for r a positive integer). Clearly, the dwell
time in state 1 is not geometrically distributed.

We now set out to display a five-state HMM that will replicate the
dwell-time distributions of the HSMM, by representing state 1 of the
HSMM by the state aggregate $\{1, 2, 3, 4\}$, a set of states within the

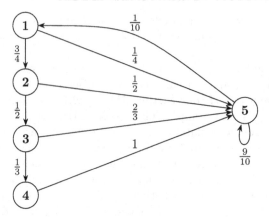

Figure 12.2 *Transition graph of five-state expanded Markov chain* $\{C_t^\star\}$ *for the example in Section 12.3.1.*

HMM. In Sections 12.3.3 and 12.4 we shall show more generally how to arrive at such a representation. Consider the five-state Poisson–HMM, with state-dependent means $\lambda_1 = \lambda_2 = \lambda_3 = \lambda_4 = 5$ and $\lambda_5 = 10$, and with the t.p.m. of the underlying Markov chain $\{C_t^\star\}$ given by

$$
\Gamma = \left(
\begin{array}{cccc|c}
0 & 3/4 & 0 & 0 & 1/4 \\
0 & 0 & 1/2 & 0 & 1/2 \\
0 & 0 & 0 & 1/3 & 2/3 \\
0 & 0 & 0 & 0 & 1 \\
\hline
1/10 & 0 & 0 & 0 & 9/10
\end{array}
\right).
$$

The transition graph of the five-state Markov chain is as shown in Figure 12.2.

At observation level there is no difference between state 1 of the HSMM and states 1, 2, 3, 4 of the HMM. In other words, as long as the Markov chain underlying the HMM is in the set of states $\{1, 2, 3, 4\}$ it gives rise to the same state-dependent distribution as does state 1 of the HSMM, namely a Poisson distribution with mean 5. Now let us examine the distribution of the length of a stay in the set of states $\{1, 2, 3, 4\}$. First of all, we observe that a switch away from state 5 in the HMM always results in state 1.

We shall in general denote by $d_i^\star$ the p.m.f. of the time spent by $\{C_t^\star\}$ in the ith state aggregate. The p.m.f. of the time spent in $\{1, 2, 3, 4\}$, denoted by $d_1^\star$, is given by

$$d_1^\star(1) = \Pr\big(C_{t+1}^\star \notin \{1,2,3,4\} \mid C_t^\star = 1\big) = \Pr(C_{t+1}^\star = 5 \mid C_t^\star = 1) = \frac{1}{4};$$

$$d_1^\star(2) = \Pr\big(C_{t+2}^\star \notin \{1,2,3,4\}, C_{t+1}^\star \in \{1,2,3,4\} \mid C_t^\star = 1\big)$$
$$= \Pr(C_{t+1}^\star = 2 \mid C_t^\star = 1)\Pr(C_{t+2}^\star = 5 \mid C_{t+1}^\star = 2)$$
$$= \Pr(1 \to 2 \to 5) = \frac{3}{4} \times \frac{1}{2} = \frac{3}{8};$$

$$d_1^\star(3) = \Pr(1 \to 2 \to 3 \to 5) = \frac{3}{4} \times \frac{1}{2} \times \frac{2}{3} = \frac{1}{4};$$

$$d_1^\star(4) = \Pr(1 \to 2 \to 3 \to 4 \to 5) = \frac{3}{4} \times \frac{1}{2} \times \frac{1}{3} \times 1 = \frac{1}{8};$$

$$d_1^\star(r) = 0 \text{ for } r \geq 5.$$

(In expressions such as $\Pr(1 \to 2 \to 5)$, the numbering of states refers to the states of the expanded Markov chain $\{C_t^\star\}$.)

Hence $d_1^\star(r) = d_1(r)$ for all positive integers r. Furthermore, state 5 of the HMM has the same (geometric) dwell-time distribution as does state 2 of the HSMM, characterized by the p.m.f. $d_2(r)$. We have thus represented the HSMM (exactly) as an HMM with an expanded state space. The trick is simply to structure the transitions within a set of states so that the desired dwell-time distribution is represented, or at least well approximated. The advantage of doing so is that we can now use the HMM representation to perform statistical inference for the given HSMM, which means that the entire set of HMM techniques becomes applicable also to the illustrative HSMM considered here, without any further amendments being needed.

In the two examples that follow, we shall slightly extend the generality of the HMM representation of HSMMs, before moving on to a general discussion in Section 12.4. We shall see that the t.p.m. of the HMM representing the HSMM always has the same structure, irrespective of what dwell-time distribution is to be fitted.

12.3.2 Example of HSMM with three states

Let $\mathbf{\Omega}$, the t.p.m. that drives the state switches in a three-state HSMM, be given by

$$\mathbf{\Omega} = \begin{pmatrix} 0 & 0.8 & 0.2 \\ 0.4 & 0 & 0.6 \\ 0.7 & 0.3 & 0 \end{pmatrix}.$$

Let the three dwell-time distributions be

$$d_1(r) = 0.1 \times 0.9^{r-1} \quad (r \in \mathbb{N});$$

$$d_2(r) = \begin{cases} 0.5 & (r = 1) \\ 0.3 & (r = 2) \\ 0.2 & (r = 3); \end{cases}$$

$$d_3(r) = \begin{cases} 0.5 & (r = 1) \\ 0.4 \times 0.2^{r-2} & (r \geq 2). \end{cases}$$

These three distributions are of different types: the first is geometric, the second has finite support, and the third has what we term 'unstructured start and geometric tail'; see Section 12.6.5.

We now give the t.p.m. $\mathbf{\Gamma}$ of a Markov chain on $\{1, 2, 3, 4, 5, 6\}$ that represents $\mathbf{\Omega}$ and the dwell-time distributions d_i: let

$$\mathbf{\Gamma} = \begin{pmatrix} 0.90 & 0.08 & 0 & 0 & 0.02 & 0 \\ \hline 0.20 & 0 & 0.50 & 0 & 0.30 & 0 \\ 0.24 & 0 & 0 & 0.40 & 0.36 & 0 \\ 0.40 & 0 & 0 & 0 & 0.60 & 0 \\ \hline 0.35 & 0.15 & 0 & 0 & 0 & 0.50 \\ 0.56 & 0.24 & 0 & 0 & 0 & 0.20 \end{pmatrix}.$$

See Figure 12.3 for the corresponding transition graph.

Although, as in the previous example, it will not yet be clear how one arrives at the matrix $\mathbf{\Gamma}$ (for which, see Section 12.4), it is possible to verify that the dwell-time distributions are replicated. For instance, the matrix $\mathbf{\Gamma}$ implies the following distribution for the dwell time in the third state aggregate, $\{5, 6\}$:

$$d_3^\star(1) = \Pr\big(C_{t+1}^\star \notin \{5,6\} \mid C_t^\star = 5\big) = \Pr(5 \to 1 \text{ or } 2) = 0.35 + 0.15;$$
$$d_3^\star(2) = \Pr(5 \to 6 \to 1 \text{ or } 2) = 0.5 \times (0.56 + 0.24) = 0.5 \times 0.8;$$
$$d_3^\star(3) = \Pr(5 \to 6 \to 6 \to 1 \text{ or } 2) = 0.5 \times 0.2 \times 0.8;$$

and more generally, for all $r \geq 2$,

$$d_3^\star(r) = \Pr(5 \to 6 \to \dots \to 6 \to 1 \text{ or } 2) = 0.5 \times 0.2^{r-2} \times 0.8 = 0.4 \times 0.2^{r-2}.$$

So $d_3^\star$ matches d_3 exactly. It is left to the reader as an exercise (Exercise 2) to verify that $d_1^\star$ and $d_2^\star$ match d_1 and d_2 exactly.

Suppose, for example, that the three state-dependent distributions of the HSMM are Poisson with means λ_1, λ_2, λ_3. Then the equivalent process is obtained by using an HMM with $\mathbf{\Gamma}$ as above and choosing the six state-dependent distributions to be Poisson with means λ_1, λ_2, λ_2, λ_2, λ_3, λ_3.

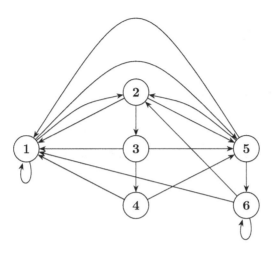

Figure 12.3 *Transition graph of six-state expanded Markov chain $\{C_t^*\}$ for the example in Section 12.3.2.*

12.3.3 A two-state HSMM with general dwell-time distribution in one state

Suppose now that a two-state HSMM is to be fitted, with state-dependent distributions p_1 and p_2. The p.m.f. d_1 is that of some non-geometric distribution with infinite support (e.g. a Poisson distribution shifted up by one), and d_2 is a geometric distribution. We construct an HMM that approximates this HSMM, by expanding that state which has a non-geometric dwell time, state 1, into a set of m_1 states (where m_1 is some large number) and by choosing the transition probabilities among those m_1 states as follows.

Consider an HMM with $m_1 + 1$ states and state-dependent distributions given by $p_i^* = p_1$ for $i = 1, \ldots, m_1$ and $p_{m_1+1}^* = p_2$. Each state in the state aggregate $I_1 = \{1, \ldots, m_1\}$ is associated with the same distribution of the state-dependent process, so that at the observation level there is nothing that distinguishes the states in I_1 from each other. We further define the $(m_1 + 1) \times (m_1 + 1)$ t.p.m. of the HMM as

$$\Gamma = \left(\begin{array}{ccccc|c} 0 & 1-c(1) & 0 & \ldots & 0 & c(1) \\ & 0 & 1-c(2) & & & c(2) \\ \vdots & & \ddots & \ddots & & \vdots \\ & & & 0 & 1-c(m_1-1) & c(m_1-1) \\ 0 & & \ldots & & 1-c(m_1) & c(m_1) \\ \hline d_2(1) & 0 & \ldots & & 0 & 1-d_2(1) \end{array} \right),$$

with $c(r) = d_1(r)/(1-F_1(r-1))$ and F_1 the cumulative distribution function associated with d_1. See Figure 12.4 for the corresponding transition graph.

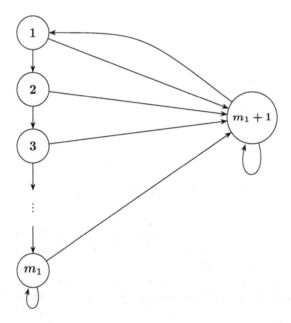

Figure 12.4 *Transition graph of expanded Markov chain $\{C_t^*\}$ for the example in Section 12.3.3.*

The structure of the t.p.m. is such that a switch out of state $m_1 + 1$ necessarily results in the Markov chain moving to state 1. Thereafter it visits the states $2, 3, \ldots, m_1$ (in that order), or it switches at some stage to state $m_1 + 1$. If the chain does reach state m_1, it can remain there for some time before switching to $m_1 + 1$. The probabilities of transitions within state aggregate I_1 are defined in such a way that the distribution of the length of a stay in I_1 matches, at least approximately, the distribution of a stay in state 1 of the HSMM to be represented. Specifically, it can be shown (see Exercise 4) that the p.m.f. $d_1^\star(r)$ of the dwell time in I_1 matches $d_1(r)$ for $r = 1, \ldots, m_1$. For $r > m_1$, $d_1^\star(r)$ decays geometrically, which in general is only an approximation to $d_1(r)$. The approximation can however be made arbitrarily close by increasing m_1.

We have thus represented the dwell time in state 1 of the HSMM as the dwell time in the state aggregate I_1 of the approximating HMM. Recall that at the observation level there is no difference between state 1 of the HSMM and states $1, \ldots, m_1$ of the HMM. This formulation does not entail an increase in the number of parameters, compared to the HSMM that is to be represented. In order to get an accurate approximation, this representation can involve large matrices, which makes these models computationally more demanding to fit than basic HMMs. The same is true of alternative strategies for estimating HSMMs which use the EM algorithm (Guédon, 2003). Crucially, however, the structure of the t.p.m. of the approximating HMM is the same irrespective of what dwell-time distribution is to be fitted.

12.4 General HSMM

We now consider a general m-state HSMM with observed process $\{X_t\}$, where the distribution of X_t is determined by the state of an unobserved m-state semi-Markov process $\{C_t\}$. Let d_i, as before, denote the p.m.f. of the dwell time in state i, $i = 1, \ldots, m$, and let F_i denote its distribution function. Let p_i be the state-dependent distribution in state i.

Consider the subsequence of $\{C_t\}$ consisting of the first occurrences of states in each run. (For example, the subsequence of 1, 1, 2, 2, 2, 1, 3, 3, 3, 3 consisting of first occurrences is 1, 2, 1, 3.) This subsequence is generated by a Markov chain (the 'embedded Markov chain') having t.p.m. $\boldsymbol{\Omega}$, where

$$\omega_{ij} = \Pr\big(C_{t+1} = j \mid C_t = i, C_{t+1} \neq i\big), \quad i, j = 1, \ldots, m,$$

$\sum_{j=1}^m \omega_{ij} = 1$, and $\omega_{ii} = 0$ for $i = 1, \ldots, m$. (If $m = 2$, then $\omega_{12} = \omega_{21} = 1$.) Note that these are conditional transition probabilities, given that the current state is left; the probabilities of self-transitions, and thus the dwell-time distributions, are determined by the p.m.f.s d_i, $i = 1, \ldots, m$.

Let $m_1, m_2, \ldots, m_m$ be positive integers, $m_0 = 0$, and consider an HMM with state-dependent process $\{X_t^\star\}$ (observable) and Markov chain (MC) $\{C_t^\star\}$ (unobservable) on states $1, 2, \ldots, \sum_{i=1}^m m_i$. We consider the

Table 12.1 *Notation used for HSMM and for approximating HMM. For the construction of the t.p.m. $\boldsymbol{\Gamma}$, see equations (12.2)–(12.5).*

	Observed process	Latent process	Parameters driving latent process	Initial distribution of MC
HSMM	X_t	C_t	$\boldsymbol{\Omega}$, distributions d_i	
HMM	$X_t^\star$	$C_t^\star$	$\boldsymbol{\Gamma}$	$\boldsymbol{\delta}$

state aggregates

$$I_k = \left\{ n : \sum_{i=0}^{k-1} m_i < n \le \sum_{i=0}^{k} m_i \right\}, \quad k = 1, \ldots, m,$$

defining $I_k^- = \min(I_k)$ and $I_k^+ = \max(I_k)$. We assume that all the states of an aggregate I_i are associated with the same state-dependent distribution, p_i. In other words, the distribution of $X_t^\star$ given state $C_t^\star = l$ is defined to be the same for all $l \in I_i$, and the same as that of X_t given $C_t = i$:

$$\Pr\big(X_t^\star \mid C_t^\star \in I_i\big) = \Pr\big(X_t \mid C_t = i\big), \quad t = 1, \ldots, T; \ i = 1, \ldots, m \,. \tag{12.1}$$

We denote the t.p.m. of $\{C_t^\star\}$ by $\boldsymbol{\Gamma}$, where $\gamma_{ij} = \Pr\big(C_{t+1}^\star = j \mid C_t^\star = i\big)$, $i, j = 1, \ldots, \sum_{i=1}^{m} m_i$. (It may help the reader to refer to the summary of notation given in Table 12.1.)

We partition $\boldsymbol{\Gamma}$ as

$$\boldsymbol{\Gamma} = \begin{pmatrix} \boldsymbol{\Gamma}_{11} & \cdots & \boldsymbol{\Gamma}_{1m} \\ \vdots & \ddots & \vdots \\ \boldsymbol{\Gamma}_{m1} & \cdots & \boldsymbol{\Gamma}_{mm} \end{pmatrix}. \tag{12.2}$$

The $m_i \times m_i$ diagonal block $\boldsymbol{\Gamma}_{ii}$, $i = 1, \ldots, m$, is defined, for $m_i \ge 2$, as

$$\boldsymbol{\Gamma}_{ii} = \begin{pmatrix} 0 & 1 - c_i(1) & 0 & \cdots & & 0 \\ \vdots & 0 & \ddots & & & \vdots \\ \vdots & & & & & 0 \\ 0 & 0 & \cdots & 0 & 1 - c_i(m_i - 1) \\ 0 & 0 & \cdots & 0 & 1 - c_i(m_i) \end{pmatrix}, \tag{12.3}$$

where $c_i(r)$ is defined, for $r = 1, 2, 3, \ldots, m_i$, by

$$c_i(r) = \frac{d_i(r)}{1 - F_i(r-1)}. \qquad (12.4)$$

For $m_i = 1$, we define $\mathbf{\Gamma}_{ii} = 1 - c_i(1)$. Notice the simple structure of $\mathbf{\Gamma}_{ii}$: non-zero entries can appear only on the 'superdiagonal' and in the last diagonal position. The $m_i \times m_j$ off-diagonal block $\mathbf{\Gamma}_{ij}$ is defined (for $i \neq j$) as

$$\mathbf{\Gamma}_{ij} = \begin{pmatrix} \omega_{ij} c_i(1) & 0 & \cdots & 0 \\ \omega_{ij} c_i(2) & 0 & \cdots & 0 \\ \vdots & & & \\ \omega_{ij} c_i(m_i) & 0 & \cdots & 0 \end{pmatrix}. \qquad (12.5)$$

In the case $m_j = 1$, the columns of zeros disappear from $\mathbf{\Gamma}_{ij}$.

Note that the functions c_i play the key role in our HMM formulation, since they are used to render the desired dwell-time distributions.* Different dwell-time distributions lead to different c_is, while the transition probabilities ω_{ij} and the overall structure of the t.p.m. $\mathbf{\Gamma}$ remain unaffected. The functions c_i are generated solely from the parameters of the dwell-time distributions; no additional parameters or constraints arise in comparison to the HSMM.

Note also that the matrix $\mathbf{\Gamma}$ indeed constitutes a t.p.m. since the entries all lie in the interval $[0, 1]$ and the row sums are equal to 1. Although $\mathbf{\Gamma}$ may appear somewhat complicated, its structure is not difficult to interpret. All transitions within state aggregate I_i are governed by the diagonal block $\mathbf{\Gamma}_{ii}$, which thus determines the dwell-time distribution of that state aggregate. The off-diagonal blocks determine the probabilities of transitions between different state aggregates. For example, for $i \neq j$, the matrix $\mathbf{\Gamma}_{ij}$ contains the probabilities of all possible transitions between the state aggregates I_i and I_j. Note that in this construction a transition from I_i to I_j must enter I_j in the lowest state, I_j^-. We will now show that this choice of $\mathbf{\Gamma}$ yields an HMM that is an approximate representation of the given HSMM.

We first consider the probability of a transition (in the MC with t.p.m. $\mathbf{\Gamma}$) from state aggregate I_i to I_j, $\Pr(C_{t+1}^\star \in I_j \mid C_t^\star \in I_i, C_{t+1}^\star \notin I_i)$. This is analogous to the transition probability ω_{ij} in the MC underlying the HSMM, and it can be shown that, for $i \neq j$, $1 \leq i, j \leq m$,

$$\Pr(C_{t+1}^\star \in I_j \mid C_t^\star \in I_i, C_{t+1}^\star \notin I_i) = \omega_{ij}; \qquad (12.6)$$

---

* If the support of d_i is finite, one chooses m_i to be (at most) the upper limit of that support, and a zero denominator will not arise in (12.4). If the support is infinite, $F_i(r-1)$ may sometimes be very close to 1, which can cause numerical instability. Computing the upper-tail probability $1 - F_i(r-1)$ directly (e.g. via `lower.tail=FALSE` where available) can prevent this.

see Exercise 3.

We focus now on the accuracy of the representation of the dwell-time distributions d_i in the HMM formulation. We denote the p.m.f. of the dwell-time distribution in state aggregate I_i by $d_i^\star$ $(i = 1, \ldots, m)$. With the possible exception of the state aggregate that is active at time $t = 1$, the stay in a given aggregate I_i begins in state I_i^-. By $d_i^\star$ we refer to the distribution of those dwell times that do start in state I_i^-. Then for $i = 1, \ldots, m$ we have the key result that

$$d_i^\star(r) = \begin{cases} d_i(r) & \text{for } r \le m_i, \\ d_i(m_i)(1 - c_i(m_i))^{r-m_i} & \text{for } r > m_i; \end{cases} \qquad (12.7)$$

see Exercise 4. The two p.m.f.s can therefore differ only for $r > m_i$ (i.e. in the right tail), where $d_i^\star(r)$ exhibits a geometric tail. Clearly, the difference between d_i and $d_i^\star$ can be made arbitrarily small by choosing m_i sufficiently large. It also follows that, for any dwell-time distribution with finite support, we can ensure that $d_i^\star(r) = d_i(r)$ for all r by choosing m_i to be the maximum possible dwell time in state i.

In summary, (12.1), (12.6) and (12.7) imply that our HMM formulation is capable of representing, at least approximately, any given dwell-time distribution. In other words, the distribution of the m-state HSMM $\{X_t\}$ can be approximated by that of $\{X_t^\star\}$, and the approximation can be made arbitrarily close by choosing the m_is sufficiently large.

12.5 R code

It is far from obvious how one writes general code to produce the matrix $\mathbf{\Gamma}$ (the t.p.m. of the expanded Markov chain, defined by (12.2)–(12.4)) from the inputs ($\mathbf{\Omega}$ and the dwell-time distributions d_i). We therefore give below an **R** function that does so. In this code we have taken care to ensure that probabilities that are identically zero (e.g. γ_{44} in the example in Section 12.3.2) do not (because of rounding) emerge as very small negative quantities; that can happen in the case of dwell-time distributions of finite support.

```
hsmm2hmm<-function(omega,dm,eps=1e-10){
  mv<-sapply(dm,length)
  m<-length(mv)
  G<-matrix(0,0,sum(mv))
  for (i in 1:m){
    mi<-mv[[i]]
    F<-cumsum(c(0,dm[[i]][-mi]))
    ci<-ifelse(abs(1-F)>eps,dm[[i]]/(1-F),1)
    cim<-ifelse(1-ci>0,1-ci,0)
    Gi<-matrix(0,mi,0)
    for (j in 1:m){
      if(i==j) { if(mi==1)
      { Gi<-cbind(Gi,c(rep(0,mv[[j]]-1),cim))} else
```

```
    { Gi<-cbind(Gi,rbind(cbind(rep(0,mi-1),diag(cim[-mi],mi-1,mi-1)),
                         c(rep(0,mi-1),cim[[mi]])))}
    } else   { if(mi==1)
    { Gi<-cbind(Gi,matrix(c(omega[[i,j]]*ci,rep(0,mv[[j]]-1)),1))} else
    { Gi<-cbind(Gi,cbind(omega[[i,j]]*ci,matrix(0,mv[[i]],mv[[j]]-1)))}
    }
  }
  G<-rbind(G,Gi)
 }
 G
}
```

The first argument omega of hsmm2hmm specifies the $m \times m$ matrix Ω. The list dm specifies the dwell-time distributions as follows: the ith component of dm is a vector of m_i probabilities $d_i(r)$, $r = 1, \ldots, m_i$, that sum either to 1 or to less than 1. The former case is applicable to distributions with finite support, in which case $d_i(r) = 0$ for $r > m_i$. The latter is applicable to distributions with unbounded support.

The output of the function is the matrix Γ. For a dwell-time distribution with finite support, Γ implies an exact representation of that distribution. For a dwell-time distribution $d_i(r)$ with infinite support, Γ implies a dwell-time distribution $d_i^*(r)$ which matches $d_i(r)$ exactly for $r = 1, \ldots, m_i$ and represents the values of $d_i(r)$ in the right tail (for $r > m_i$) by

$$d_i^*(r) = d_{m_i}(r) \left(\frac{1 - \sum_{s=1}^{m_i} d_i(s)}{1 - \sum_{s=1}^{m_i - 1} d_i(s)} \right)^{r - m_i}. \tag{12.8}$$

If the right tail of $d_i(r)$ is of that form, then the representation is exact, otherwise an approximation.

The R code below illustrates the use of hsmm2hmm in the examples discussed in Sections 12.3.1 and 12.3.2.

```
# For example in 12.3.1:
Omega<-matrix(c(0,1,1,0),2,2)
dm<-list(c(2,3,2,1)/8, 0.1)
(Gamma<-hsmm2hmm(Omega,dm))
```

```
# For example in 12.3.2:
Omega<-matrix(c(0,8,2, 4,0,6, 7,3,0)/10,3,3,byrow=T)
dm<-list(0.1, c(5,3,2)/10, c(5,4)/10)
(Gamma<-hsmm2hmm(Omega,dm))
```

For the example in Section 12.3.3, the following code takes the dwell-time distribution in state 1 of the HSMM to be a Poisson with mean 3 shifted up by one, and produces the matrix Γ for $m_1 = 10$.

```
# For example in 12.3.3:
Omega<-matrix(c(0,1,1,0),2,2)
```

```
dm<-list(dpois(0:9,3), 0.1)
(Gamma<-hsmm2hmm(Omega,dm))
```

In this last example, the HMM representation is not exact. For example, with $m_1 = 10$ we have $d_i^\star(11) = 0.00078 \neq 0.00081 = d_i(11)$. If higher accuracy is desired, then this can be achieved by using a larger m_1. In this case, using $m_1 = 20$ renders the approximation virtually exact.

12.6 Some examples of dwell-time distributions

We now discuss five examples of particular dwell-time distributions. In all cases we know from the first line of (12.7) that $d_i^\star(r) = d_i(r)$ at least for $r \leq m_i$, and we shall see in some of these examples that equality holds for all positive integers r.

12.6.1 Geometric distribution

Let $d_i(r)$ be the p.m.f. of the geometric distribution:

$$d_i(r) = \pi_i(1 - \pi_i)^{r-1}, \quad r = 1, 2, \ldots.$$

Then, for $r \geq 1$,

$$c_i(r) = \frac{d_i(r)}{1 - \sum_{s=1}^{r-1} d_i(s)} = \frac{\pi_i(1 - \pi_i)^{r-1}}{\sum_{s=r}^{\infty} \pi_i(1 - \pi_i)^{s-1}}$$

$$= \frac{\pi_i}{\sum_{s=0}^{\infty} \pi_i(1 - \pi_i)^s} = \pi_i.$$

Thus, for $r \leq m_i$, we have $d_i^\star(r) = d_i(r)$, and for $r > m_i$, we have

$$d_i^\star(r) = d_i(m_i)\big(1 - c_i(m_i)\big)^{r-m_i} = \pi_i(1 - \pi_i)^{m_i-1}(1 - \pi_i)^{r-m_i}$$

$$= \pi_i(1 - \pi_i)^{r-1} = d_i(r).$$

In this example the value chosen for m_i does not play any role in $d_i^\star$: the choice $m_i = 1$ is simplest. If the dwell time is geometric for all states of an HSMM, the HSMM reduces to an HMM.

12.6.2 Shifted Poisson distribution

Let $d_i(r)$ be the p.m.f. of a shifted Poisson distribution:

$$d_i(r) = \exp(-\lambda_i)\frac{\lambda_i^{r-1}}{(r - 1)!}, \quad r = 1, 2, \ldots.$$

Then

$$d_i^\star(r) = \begin{cases} d_i(r) & \text{for } r \leq m_i, \\ d_i(m_i)z^{r-m_i} & \text{for } r > m_i, \end{cases}$$

where

$$z = 1 - \frac{d_i(m_i)}{1 - \sum_{j=1}^{m_i-1} d_i(j)},$$

which is independent of r. Although the functions $d_i^\star(r)$ and $d_i(r)$ differ for $r > m_i$, the discrepancy between them becomes small as m_i increases. This is illustrated in Figure 12.5, which displays a shifted Poisson distribution with parameter $\lambda_i = 5$ and the corresponding $d_i^\star(r)$ for $m_i = 4, 6, 8$.

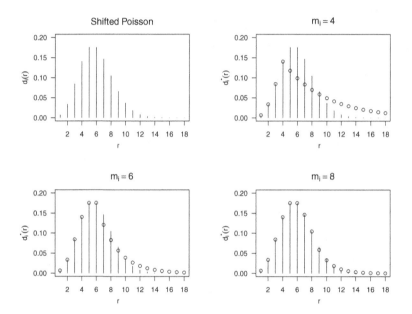

Figure 12.5 *The functions $d_i(r)$ (vertical lines) and $d_i^\star(r)$ (circles) for $m_i = 4, 6, 8$ in the shifted Poisson example of Section 12.6.2.*

12.6.3 Shifted negative binomial distribution

Let $d_i(r)$ be the p.m.f. of a shifted negative binomial distribution:

$$d_i(r) = \frac{\Gamma(r + k_i - 1)}{(r-1)!\Gamma(k_i)} \pi_i^{k_i}(1 - \pi_i)^{r-1}, \quad r = 1, 2, \ldots, \tag{12.9}$$

with parameters $k_i > 0$ and $\pi_i \in [0, 1]$. Then

$$d_i^\star(r) = \begin{cases} d_i(r) & \text{for } r \le m_i, \\ d_i(m_i)z^{r-m_i} & \text{for } r > m_i, \end{cases}$$

where again

$$z = 1 - \frac{d_i(m_i)}{1 - \sum_{j=1}^{m_i-1} d_i(j)}.$$

As in the previous example, the functions $d_i^*(r)$ and $d_i(r)$ differ only for $r > m_i$, with the discrepancy decreasing as m_i increases. The shifted negative binomial is an important example for a dwell-time distribution, since it includes the geometric distribution as the special case $k_i = 1$, and the corresponding HSMMs thus constitute a natural extension of conventional HMMs. This has made it a natural choice in applications of HSMMs.

12.6.4 Shifted binomial distribution

Consider a shifted binomial distribution with p.m.f.

$$d_i(r) = \binom{n_i}{r-1} \pi_i^{r-1} (1 - \pi_i)^{n_i - (r-1)}, \quad r = 1, \ldots, n_i + 1.$$

Since $\sum_{k=1}^{n_i+1} d_i(k) = 1$, we have, for $m_i = n_i + 1$,

$$c_i(m_i) = \frac{d_i(n_i + 1)}{1 - \sum_{k=1}^{n_i} d_i(k)} = 1.$$

Thus, for $r > m_i$, we have

$$d_i^*(r) = d_i(m_i)\big(1 - c_i(m_i)\big)^{r - m_i} = 0 = d_i(r).$$

Furthermore, we also have, as usual, $d_i^*(r) = d_i(r)$ for $r \leq m_i$. Choosing $m_i = n_i + 1$ ensures that $d_i^*(r)$ equals $d_i(r)$ for all r; the HMM representation of a corresponding HSMM is therefore exact.

12.6.5 A distribution with unstructured start and geometric tail

Another possible distribution on the positive integers involves an unstructured start and a geometric tail, with p.m.f. of the following form for some integer $q \geq 2$:

$$d_i(r) = \begin{cases} \theta_{ir} & \text{for } r \leq q, \\ \theta_{iq}\left(\frac{1 - F_i(q)}{1 - F_i(q-1)}\right)^{r-q} & \text{for } r > q, \end{cases}$$

where $F_i(r) = \sum_{k=1}^{r} \theta_{ik}$ and $F_i(q) = \sum_{k=1}^{q} \theta_{ik} < 1$. We say that the start is of order (or length) $q-1$ because, in comparison to the geometric distribution, $q-1$ additional parameters are required. Note that the case $q = 1$ (i.e. order 0) collapses as follows to a geometric distribution:

$$d_i(r) = \theta_{i1}(1 - \theta_{i1})^{r-1}, \quad r \geq 1.$$

An HSMM having state i of this kind (with general q) can be formulated as an HMM which has $d_i^\star(r) = d_i(r)$ (exactly) for all r if we choose q to be the corresponding m_i.[†] This we establish as follows. From equation (12.7) we know that, for $r \geq m_i$,

$$d_i^\star(r) = d_i(m_i)(1 - c_i(m_i))^{r-m_i}.$$

But

$$1 - c_i(m_i) = 1 - \frac{d_i(m_i)}{1 - F_i(m_i - 1)} = \frac{1 - F_i(m_i)}{1 - F_i(m_i - 1)}.$$

Hence, for $r \geq m_i$,

$$d_i^\star(r) = \theta_{im_i} \left(\frac{1 - F_i(m_i)}{1 - F_i(m_i - 1)} \right)^{r-m_i} = d_i(r).$$

12.7 Fitting HSMMs via the HMM representation

An important benefit of using the HMM formulation is that this enables us to apply all the well-established methods available for HMMs. In particular, since the likelihood function of the HMM representing or approximating the HSMM of interest is available in the standard HMM form, the parameters can be estimated by direct numerical maximization of likelihood. For an m-state HSMM $\{X_t\}$ with state-dependent distributions $p_1, \ldots, p_m$, which is represented or approximated by an HMM of the structure described in Section 12.4, the (approximate) likelihood of a sequence of observations $x_1, \ldots, x_T$ is given by

$$L_T = \delta \mathbf{P}(x_1)\mathbf{\Gamma P}(x_2)\mathbf{\Gamma} \cdots \mathbf{P}(x_{T-1})\mathbf{\Gamma P}(x_T)\mathbf{1}',$$

where

$$\mathbf{P}(x) = \mathrm{diag}\big(\underbrace{p_1(x), \ldots, p_1(x)}_{m_1 \text{ times}}, \ldots, \underbrace{p_m(x), \ldots, p_m(x)}_{m_m \text{ times}}\big).$$

A question of interest here concerns the choice of the initial distribution, δ, of the Markov chain in the HMM used to represent or approximate the HSMM of interest. Suppose that the HSMM to be fitted is in state i at time $t = 1$. Now unless the state process *entered* state i at $t = 1$, the distribution of the first dwell time will, in general, differ from d_i. It is to circumvent the difficulties of taking this difference into account that Sansom and Thomson (2001), as well as Guédon (2003) and Bulla *et al.* (2009), assume in HSMMs that the time instant $t = 1$ is a state boundary. The initial distribution of the HSMM then is explicitly modelled and can feasibly be estimated (see Guédon, 2003). This

[†] Note that the definition given by Langrock and Zucchini (2011) for a p.m.f. with unstructured start and geometric tail is incorrect and should be replaced by the definition given here.

assumption is unlikely to have much effect on parameter estimation except for short series. However, despite being natural in many applications (see, for example, Sansom and Thomson, 2001), it can be unrealistic in others. More seriously, the enforced state switch at the start of the series impedes stationarity of the HSMM in general.

If one uses the HMM representation, the assumption of a state switch at the start of the series is avoidable, by allowing the initial state probabilities δ_i to be non-zero for $i \notin \{I_1^-, I_2^-, \ldots, I_m^-\}$. Moreover, as we have seen in Chapter 3, it is straightforward to fit a stationary HMM, so that the model formulation considered here allows one to fit stationary HSMMs, by taking the initial distribution δ to be the solution to the linear system $\delta\Gamma = \delta$. On the other hand, the HMM-based representation can also deal with the case where the user wishes to assume that there is a state boundary at time $t = 1$.

12.8 Example: earthquakes

For illustrative purposes, we consider now a stationary three-state HSMM for the earthquake series, assuming Poisson state-dependent distributions and shifted Poisson dwell-time distributions. This model has the same number of parameters (nine) as does the three-state HMM presented in Section 3.5; the latter necessarily has geometric dwell-time distributions.

The HSMM that was fitted (via a stationary Poisson–HMM with 40 states in each state aggregate) has state-dependent means $\lambda = (11.714, 18.736, 29.278)$. The means of the shifted Poisson dwell-time distributions are 5.839, 11.678 and 5.924 (before shifting) for states 1, 2 and 3, respectively. The conditional transition probabilities are

$$\Omega = \begin{pmatrix} 0 & \omega_{12} & \omega_{13} \\ \omega_{21} & 0 & \omega_{23} \\ \omega_{31} & \omega_{32} & 0 \end{pmatrix} = \begin{pmatrix} 0.000 & 0.684 & 0.316 \\ 0.662 & 0.000 & 0.338 \\ 0.000 & 1.000 & 0.000 \end{pmatrix}.$$

Recall that within the HSMM framework the transition probabilities ω_{ij} are conditional on the state being left:

$$\omega_{ij} = \Pr\big(C_{t+1} = j \mid C_t = i, C_{t+1} \neq i\big), \quad i,j = 1,2,3; \ i \neq j.$$

The log-likelihood for this model is $l = -331.5757$, somewhat inferior to that of the fitted basic three-state Poisson–HMM ($l = -329.4603$).

Theorem 8.34 of Kulkarni (2010) tells us that the limiting distribution of $\{C_t\}$, the semi-Markov process, is proportional to $(\pi_1\tau_1, \pi_2\tau_2, \pi_3\tau_3)$, where τ_i is the mean of the ith dwell-time distribution and π is the stationary distribution implied by the t.p.m. Ω. Here $\pi = (0.299, 0.453,$

0.248), $\tau = (6.839, 12.678, 6.924)$, and the resulting limiting distribution for $\{C_t\}$ is $(0.216, 0.604, 0.181)$. This distribution can then be compared with that obtained by summing δ, the stationary distribution of the 120-state Markov chain, within each state aggregate. Agreement ought to be close, and indeed it is extremely close: the first 12 decimal places are identical.

We also fitted a stationary three-state Poisson–HSMM with one shifted Poisson and two geometric dwell-time distributions. (In the terminology of Guédon (2005), this is a hybrid HMM/HSMM.) We used 40 states for the state aggregate with shifted Poisson dwell time (and one state for each of the other two). It is not necessary to decide *a priori* which of the three states – with small, medium or large state-dependent mean – involves the Poisson dwell-time distribution, since the numerical maximization will choose the most appropriate state for the Poisson dwell-time distribution. In the earthquake example, this is state 2, the state with the medium state-dependent mean. The fitted model has

$$\Omega = \begin{pmatrix} 0.000 & 0.600 & 0.400 \\ 0.532 & 0.000 & 0.468 \\ 0.000 & 1.000 & 0.000 \end{pmatrix}$$

and state-dependent means $\lambda = (13.141, 19.671, 29.609)$. The dwell-time distributions in states 1 and 3 are geometric with parameters $\pi_1 = 0.065$ and $\pi_3 = 0.209$ respectively, and the distribution in state 2 is a shifted Poisson with mean 7.131 (before shifting). The log-likelihood for this model is $l = -327.884$, hence better than that of the basic three-state Poisson–HMM ($l = -329.4603$). Note that the two models being compared again have the same number of parameters.

In this case the stationary distribution of the 42-state Markov chain, suitably summed, is $(0.418, 0.416, 0.166)$, which also agrees very closely with the limiting distribution of the fitted HSMM.

Figure 12.6 displays the paths obtained, by the Viterbi algorithm, for the fitted three-state Poisson–HSMM with one shifted Poisson and two geometric dwell-time distributions. Although the state-dependent means of this HSMM are very similar to those of the three-state Poisson–HMM, there are three differences in the corresponding state sequences. The observations made in 1932, 1933 and 1957 are respectively allocated to states 1, 1 and 3 by the HSMM, but all three are allocated to state 2 by the HMM; see the top panel of Figure 5.6. This shows that the Viterbi path can be quite sensitive to model specification. Decoded paths should therefore be interpreted with caution.

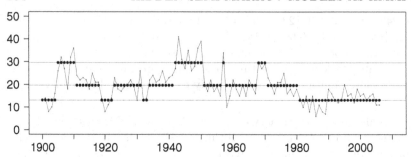

Figure 12.6 *Earthquakes: globally decoded state sequence according to the three-state Poisson–HSMM with one shifted Poisson and two geometric dwell-time distributions. The horizontal lines indicate the state-dependent means.*

12.9 Discussion

The proposed strategy for fitting HSMMs is based on the idea of using specially structured HMMs to represent the HSMM of interest. A strength of this approach is the ease with which it can be implemented, and that it is usually straightforward to modify existing code according to changes in the model formulation. By using an HMM to represent the HSMM, many of the non-standard model formulations that nowadays are routinely implemented in the class of HMMs – such as models which incorporate covariates, seasonality or random effects – can be transferred to the class of HSMMs. The HMM representation of HSMMs is particularly useful when the numbers of states in the state aggregates are not large and the modes of the dwell-time distributions are low.

The method becomes less efficient as the sizes of the state aggregates increase, since the dimensions of the matrices needing to be multiplied increase. In extreme cases, with dwell-time distributions having very large support, the method can be infeasible because of insufficient computer memory. In such cases, alternative estimation approaches based on the EM algorithm appear to be preferable. The R package hsmm (Bulla and Bulla, 2013) can be used to fit various types of HSMM by using the EM algorithm. For non-standard model formulations, EM-based approaches require substantial algorithmic development (Guédon, 2003).

Exercises

1. Give an HMM that (exactly) represents the two-state HSMM with state-dependent distributions N(0,1) (state 1) and N(1,1) (state 2),

and state dwell-time distributions given by

$$
d_1(r) = \begin{cases} 0.2 & \text{if } r = 1, \\ 0.6 & \text{if } r = 2, \\ 0.1 & \text{if } r = 3, \\ 0.1 & \text{if } r = 4, \end{cases}
$$

and

$$
d_2(r) = \begin{cases} 0.2 & \text{if } r = 1, \\ 0.4 & \text{if } r = 2, \\ 0.4 & \text{if } r = 3. \end{cases}
$$

2. In the example in Section 12.3.2, we showed that $d_3^\star(r) = d_3(r)$ (exactly) for all positive integers r. Verify that $d_1^\star$ and $d_2^\star$ match d_1 and d_2 exactly.

3. Prove equation (12.6). Hint: start from

$$
\Pr\left(C_{t+1}^\star \in I_j \mid C_t^\star \in I_i, C_{t+1}^\star \notin I_i\right) = \frac{\Pr\left(C_{t+1}^\star \in I_j, C_t^\star \in I_i\right)}{\Pr\left(C_{t+1}^\star \notin I_i, C_t^\star \in I_i\right)},
$$

for $i \neq j$.

4. Prove equation (12.7) as follows.

   (a) For $r \leq m_i$,

   $$
   d_i^\star(r) = \left(\prod_{j=1}^{r-1} (1 - c_i(j)) \right) c_i(r).
   $$

   (b) For $r > m_i$,

   $$
   d_i^\star(r) = \left(\prod_{j=1}^{m_i} (1 - c_i(j)) \right) (1 - c_i(m_i))^{r - m_i - 1} \times c_i(m_i).
   $$

## CHAPTER 13

HMMs for longitudinal data

13.1 Introduction

This chapter outlines how HMMs can be used to model longitudinal data, known in the econometric literature as panel data. Such data consist of K time series of the same type of observation on each of K subjects. Examples of application include time series of

- observed behaviour (e.g. speed of movement, or feeding events) for each of K animals;
- disease status for each of K individuals;
- recordings of brain activity on each of K sleeping subjects;
- precipitation amounts at each of K sites.

We will suppose, as is usually done in this context, that the same type of model is used to describe each of the K series (referred to as the 'component series') but the parameters of the model may differ across subjects.

In some applications the component series are assumed to be driven by the same sequence of states. An application of this type is the HMM for daily precipitation occurrence at multiple sites described by Zucchini and Guttorp (1991). The days are classified as wet or dry, and the probability of a wet day at each site is assumed to be independent of the wet/dry status of the other sites, given the state of the Markov chain. The state on a given day (interpreted as the 'weather state') is assumed to apply to all sites. A more sophisticated model for daily precipitation occurrence is the non-homogeneous HMM proposed by Hughes, Guttorp and Charles (1999), in which the parameters of the Markov chain are functions of covariates derived from atmospheric circulation information.

Another application in which the component series are driven by the same sequence of states is the modelling of returns on several shares on a stock market; see Section 20.1. The state on a given day, interpreted as 'market sentiment', is assumed to affect all the shares on that day.

However, in many applications of HMMs to longitudinal data it is not assumed that the component series are synchronized in this way. (Indeed, HMMs in which the component series share the same sequence of states could be regarded simply as multivariate HMMs; see Section 9.4.) As an example, consider the three-state HMM used to analyse the effect of sonar exposure on the diving behaviour of blue whales, described by DeRuiter *et al.* (2016). In that application the same types of observation are made during series of dives performed by 37 blue whales, tagged at different times and in different locations. It is reasonable to assume that each whale's behaviour is driven by its own state sequence.

In such a case one assumes, as we do for the remainder of this chapter, that the component series are independent, each with its own underlying sequence of states (i.e. there are K independent sequences of states). The (joint) likelihood is then just the product of the K individual likelihoods.

In some applications the component series are not long enough to fit a model to each series separately. In the blue whale example, the number of dives for the 37 whales ranged from 6 to 93. It is not feasible to fit a three-state HMM to a series of length 6 without making simplifying assumptions. One possibility, discussed in Section 13.2, is to assume that some of the parameters are the same for all subjects. Such 'pooling' of information across subjects can be advantageous even when the component series are long enough to be fitted individually.

Covariate information, when available, can be used to account for some of the variability between subjects. For example, in applications relating to animal behaviour, the sex, weight or age of the animals can be important in this respect. Covariates can be subject-specific (e.g. sex) or time-varying (e.g. time of day). In either case they can be incorporated into HMMs for longitudinal data in the same way as is done for single time series (cf. Chapter 10). Typically one expresses the parameters of the transition probability matrix, or those of the state-dependent distributions, by simple functions of the covariates. It is not always straightforward to decide which parameters are to be expressed as functions of the covariates. If a covariate is assumed to affect the rates of switching between states then it is the parameters of the t.p.m. that should be expressed in terms of the covariate, whereas if it assumed that the covariate does not affect those rates then it is (one or more of) the parameters of the state-dependent distribution that should be expressed in terms of the covariate.

Another way of reducing the total number of parameters needed to model longitudinal data is to regard some of the parameters of the t.p.m., or of the state-dependent distributions, as realizations of random variables. Models with random effects, also known as mixed HMMs, are the subject of Section 13.3. Two cases are described: continuous-valued and discrete-valued random effects.

To illustrate the main ideas we shall use the following simple running example throughout this chapter. Let $\{x_{tk} : t = 1, \ldots, T; k = 1, \ldots, K\}$ represent counts, and suppose that we wish to fit a stationary two-state

Poisson–HMM to each of the K component series. To simplify the notation we assume that the component series are all of length T; the formulae can be modified in the obvious way for cases in which the lengths differ. Let $\lambda_1^{(k)}$, $\lambda_2^{(k)}$ and $\boldsymbol{\Gamma}^{(k)} = (\gamma_{ij}^{(k)})$ represent the parameters for the kth subject. The four quantities $\lambda_1^{(k)}$, $\lambda_2^{(k)}$, $\gamma_{12}^{(k)}$ and $\gamma_{21}^{(k)}$ specify the model for that subject.

13.2 Models that assume some parameters to be constant across component series

Common features of the subjects can be modelled by pooling information across the component series, that is, by assuming that certain parameters are equal for all K subjects. Pooling has the advantage of reducing the total number of (free) parameters to be estimated, and hence leads to a reduction in the standard errors of the estimators.

'Complete pooling' refers to the extreme case in which it is assumed that all K component series have the same parameter values, that is, the model ignores possible heterogeneity across subjects. Clearly, estimates based on such a model are likely to be misleading if substantial differences between subjects do exist. Failing to account for inter-subject variability can lead to inaccurate estimates of the quantities of interest derived from the fitted model, including the standard errors of estimators. On the other hand, complete pooling provides a baseline model which can be useful in assessing the benefit of applying alternative models with partial pooling.

The opposite extreme to complete pooling is no pooling, where the K component series are modelled separately, that is, it is not assumed that any of the parameter values of the K subjects are equal. This option involves the largest number of parameters, in some cases more parameters than can be estimated with acceptable accuracy, or estimated at all. The parameter estimates obtained with a model with no pooling can provide a useful starting point in the search for a more suitable model. They can be used to distinguish the parameters of the K component models that are approximately constant across subjects from those that are clearly different. Such information can guide the choice of a suitable model with partial pooling, i.e. one in which some, but not all, of the parameters are assumed to be constant across component series.

Referring to the running example introduced at the end of the previous section, we now illustrate the difference between pooled and non-pooled parameters.

- *No pooling.* To fit a model with *no pooling* we need to estimate four parameters for each component series, i.e. $4K$ parameters in total.

- *Complete pooling.* Here we assume that $\gamma_{12}^{(1)} = \cdots = \gamma_{12}^{(K)} = \gamma_{12}$, say, and make analogous assumptions for $\gamma_{21}^{(k)}$, $\lambda_1^{(k)}$ and $\lambda_2^{(k)}$. So we need to estimate only four parameters.

- *Partial pooling (1).* If we pool the parameters of the t.p.m.s, but not those of the state-dependent distributions, we need to estimate the $2 + 2K$ parameters γ_{12}, γ_{21}, $\lambda_1^{(k)}$, $\lambda_2^{(k)}$, for $k = 1, 2, \ldots, K$.

- *Partial pooling (2).* If we pool the parameters in the state-dependent process, but not those of the t.p.m.s, we need to estimate the $2K + 2$ parameters $\gamma_{12}^{(k)}$ and $\gamma_{21}^{(k)}$ (for $k = 1, 2, \ldots, K$), λ_1 and λ_2.

Irrespective of the pooling strategy chosen, each component series is modelled by a two-state Poisson–HMM. As we have assumed that the component series are independent, the likelihood is given by the product of likelihoods for the component series,

$$L = \prod_{k=1}^{K} \boldsymbol{\delta}^{(k)} \mathbf{P}^{(k)}(x_{1k}) \boldsymbol{\Gamma}^{(k)} \mathbf{P}^{(k)}(x_{2k}) \boldsymbol{\Gamma}^{(k)} \cdots \boldsymbol{\Gamma}^{(k)} \mathbf{P}^{(k)}(x_{Tk}) \mathbf{1}'. \quad (13.1)$$

Here $\mathbf{P}^{(k)}(x)$ is a diagonal matrix whose ith diagonal element is the probability of observing x under a Poisson($\lambda_i^{(k)}$) distribution, and $\boldsymbol{\delta}^{(k)}$ denotes the initial distribution of the Markov chain for the kth component series. The latter can be estimated, or alternatively assumed to be the stationary distribution determined by the component-specific t.p.m. of the Markov chain. Expression (13.1) is applicable to models regardless of whether the parameters are pooled, non-pooled or partially pooled. One builds in the relevant assumptions by removing the superscript '(k)' from the pooled parameters. For example, in the case of *partial pooling (1)* the matrices $\boldsymbol{\Gamma}^{(k)}$ are all replaced by a single matrix $\boldsymbol{\Gamma}$.

The number of potentially suitable partially pooled models for a given data set can be large, especially if covariates are involved. The model-building process and the subsequent interpretation of the model are simplified if only one of the two components of the HMM (the Markov chain or the state-dependent process) has pooled parameters.

Up to this point we have regarded non-pooled parameters as constants. The next section describes an alternative way of modelling subject-specific parameters. Instead of regarding them as constants (usually referred to as 'fixed effects'), we model them as realizations of a probability distribution, in which case they are called 'random effects'.

13.3 Models with random effects

HMMs for longitudinal data with fixed effects can involve a large number of parameters, especially when the number of subjects, K, is large.

An alternative approach that allows for variability across subjects, but requires fewer parameters, is to use random effects. Following Altman (2007), we refer to models having random effects as 'mixed HMMs'.

Recent years have seen an increasing interest in mixed HMMs. We do not attempt to cover all the recent developments on this topic, but merely sketch the main ideas. A useful account of the subject is given by Altman (2007); Maruotti (2011), Schliehe-Diecks, Kappeler and Langrock (2012), and Bartolucci *et al.* (2013) provide overviews of mixed HMMs and related models.

We distinguish two types of random effect, those that are continuous-valued (Section 13.3.1) and those that are discrete-valued (Section 13.3.2).

13.3.1 HMMs with continuous-valued random effects

In longitudinal settings, a common way of allowing for heterogeneity is to assume that some model parameters are random effects, that is, are independently drawn from a distribution that is common to all component series, with one realization for each series. The random effects are usually assumed to be continuous-valued; the normal distribution is often used.

Consider the case of *partial pooling (1)*. The parameters of the t.p.m.s are pooled but those of the state-dependent distributions are not. Thus, in addition to γ_{12} and γ_{21}, one needs to estimate $2K$ constants if $\lambda_1^{(k)}$ and $\lambda_2^{(k)}$ (for $k = 1, 2, \ldots, K$) are taken to be fixed effects.

If instead $\lambda_1^{(k)}$, $\lambda_2^{(k)}$ (for $k = 1, 2, \ldots, K$) are regarded as random effects, then the number of quantities to be estimated can be reduced. To fit such a mixed HMM one first needs to specify distributions for the parameters of the state-dependent process, for example, $\lambda_1^{(k)} \sim$ Gamma(μ_1, σ_1) and (independently) $\lambda_2^{(k)} \sim$ Gamma(μ_2, σ_2). We regard $\lambda_1^{(1)}, \ldots, \lambda_1^{(K)}$ as independent realizations of a gamma distribution with unknown parameters μ_1 (mean) and σ_1 (standard deviation), and similarly for the $\lambda_2^{(k)}$. In order to fit this mixed HMM one needs to estimate only the six quantities γ_{12}, γ_{21}, μ_1, σ_1, μ_2, σ_2.

However, except for special cases, the likelihood of this model is much more demanding to compute than expression (13.1), the likelihood of the model with fixed effects. That is because for each subject we need to integrate over the possible values of the random effects. In the simple example given, which has two independent gamma-distributed random effects, the likelihood is given by

$$L = \prod_{k=1}^{K} \int_0^\infty \int_0^\infty \delta \mathbf{P}(x_{1k}) \mathbf{\Gamma} \mathbf{P}(x_{2k}) \mathbf{\Gamma} \cdots \mathbf{\Gamma} \mathbf{P}(x_{Tk}) \mathbf{1}'$$

$$\times f_1(\lambda_1; \mu_1, \sigma_1) f_2(\lambda_2; \mu_2, \sigma_2) \, d\lambda_1 \, d\lambda_2. \qquad (13.2)$$

Here $\mathbf{P}(x)$ is the diagonal matrix in which the ith diagonal element is the probability of observing x under a Poisson distribution with mean λ_i, $f_1(\lambda_1; \mu_1, \sigma_1)$ is the p.d.f. of the gamma distribution for the Poisson mean in state 1, and $f_2(\lambda_2; \mu_2, \sigma_2)$ for that in that state 2. The model is fitted by maximizing (13.2) with respect to γ_{12}, γ_{21}, μ_1, σ_1, μ_2 and σ_2. The estimates of σ_1 and σ_2 are measures of the variability between subjects. The integrals in (13.2) have to be computed numerically, by using the **R** function `integrate`, for example. In general, the number of integrals that appear in the likelihood of a mixed HMM equals the number of random effects in the model.

The case of *partial pooling (2)* also involves the estimation of $2 + 2K$ parameters if fixed effects are used. This can be reduced to six by regarding the entries of the t.p.m. as random effects and writing

$$\mathbf{\Gamma}^{(k)} = \begin{pmatrix} 1 - \varepsilon_1^{(k)} & \varepsilon_1^{(k)} \\ \varepsilon_2^{(k)} & 1 - \varepsilon_2^{(k)} \end{pmatrix},$$

where, for example,

$$\varepsilon_i^{(k)} \stackrel{\text{iid}}{\sim} \text{Beta}(\alpha_i, \beta_i), \quad \text{for } k = 1, \ldots, K; \ i = 1, 2.$$

(See equation (9.2) for the parametrization of the beta distribution.) In this case the parameters requiring estimation are α_1, β_1, α_2, β_2, λ_1 and λ_2, and the likelihood is

$$L = \prod_{k=1}^{K} \int_0^1 \int_0^1 \delta \mathbf{P}(x_{1k}) \mathbf{\Gamma} \mathbf{P}(x_{2k}) \mathbf{\Gamma} \cdots \mathbf{\Gamma} \mathbf{P}(x_{Tk}) \mathbf{1}'$$

$$\times f_1(\varepsilon_1; \alpha_1, \beta_1) f_2(\varepsilon_2; \alpha_2, \beta_2) \, d\varepsilon_1 \, d\varepsilon_2.$$

Here $\mathbf{P}(x)$ is as in (13.2), the off-diagonal elements of $\mathbf{\Gamma}$ are ε_1 and ε_2, $f_1(\varepsilon_1; \alpha_1, \beta_1)$ is the p.d.f. of the beta distribution for ε_1, and $f_2(\varepsilon_2; \alpha_2, \beta_2)$ that for ε_2.

A popular alternative to the beta distribution is to assume that

$$\text{logit}^{-1}\left(\varepsilon_i^{(k)}\right) \stackrel{\text{iid}}{\sim} \text{N}(\mu_i, \sigma_i^2), \quad \text{for } k = 1, \ldots, K,$$

that is, that the transition probabilities are transformed normal random variables. In this case, the parameters of the state process requiring estimation are μ_1, μ_2, σ_1 and σ_2.

In the above examples univariate distributions were formulated for each of two random effects. Instead one can specify bivariate distributions, thereby allowing for possible dependence between random effects. For instance, in the last example a bivariate normal distribution could be specified for the vector $\left(\text{logit}^{-1}\left(\varepsilon_1^{(k)}\right), \text{logit}^{-1}\left(\varepsilon_2^{(k)}\right)\right)$. This would involve one additional parameter, a correlation coefficient.

In general, each continuous-valued random effect adds one integral to the likelihood. The numerical evaluation of the likelihood can therefore be computationally expensive. A numerical or EM-based maximization of the likelihood is usually only feasible when there are few random effects, since the computational burden increases exponentially with the number of random effects. Simulation-based methods, such as Monte Carlo EM or MCMC, seem preferable when there are more than two random effects (Altman, 2007). A computationally less intensive approach uses discrete distributions for the random effects. This we discuss in the next subsection.

Examples of application of continuous-valued random effects appear in Sections 18.6 and 23.9.4. In the first of these, two independent beta-distributed random effects are used, in the second a truncated-normal random effect.

13.3.2 HMMs with discrete-valued random effects

Maruotti and Rydén (2009) propose the use of discrete-valued random effects. The main advantage of this approach over the use of continuous-valued random effects is that it avoids integration and often reduces the computational effort. Aitkin (1996) gives a detailed discussion of the use of discrete-valued random effects in the context of generalized linear models.

Consider the case of *partial pooling (1)*, for which in the previous subsection $\lambda_1^{(k)}$ and $\lambda_2^{(k)}$ were assumed to be gamma-distributed random effects. We now assume that they are discrete-valued random variables taking on finitely many values. Specifically, suppose that, for $k = 1, \ldots, K$ and $i = 1, 2$,

$$
\lambda_i^{(k)} = \begin{cases} \lambda_{1,i} & \text{with probability } \pi_{1,i}, \\ \lambda_{2,i} & \text{with probability } \pi_{2,i}, \\ \vdots \\ \lambda_{q_i,i} & \text{with probability } \pi_{q_i,i}. \end{cases}
$$

The quantities to be estimated are the possible values of the random effects $(\lambda_{1,1}, \ldots, \lambda_{q_1,1}, \lambda_{1,2}, \ldots, \lambda_{q_2,2})$, the corresponding probabilities $(\pi_{1,1}, \ldots, \pi_{q_1,1}, \pi_{1,2}, \ldots, \pi_{q_2,2})$, and the transition probabilities γ_{12} and γ_{21}. The total number of parameters is $2q_1 + 2q_2 + 2$. The likelihood of this model involves summations, rather than integrations, over the possible values of the random effects:

$$L = \prod_{k=1}^{K} \sum_{j_1=1}^{q_1} \sum_{j_2=1}^{q_2} \boldsymbol{\delta}\mathbf{P}(x_{1k})\boldsymbol{\Gamma}\mathbf{P}(x_{2k})\boldsymbol{\Gamma}\cdots\boldsymbol{\Gamma}\mathbf{P}(x_{Tk})\mathbf{1}'\pi_{j_1,1}\pi_{j_2,2}. \quad (13.3)$$

As before, the matrix $\mathbf{P}(x)$ is a diagonal matrix giving the probabilities of observing x in state 1 and in state 2. The Poisson mean in state 1 is $\lambda_{j_1,1}$ and that in state 2 is $\lambda_{j_2,2}$.

We now consider the case of *partial pooling (2)*. In the previous subsection we assumed the off-diagonal entries of $\boldsymbol{\Gamma}$ to be continuous-valued random effects (beta-distributed or transformed normals). One can assume instead that the random effects take on a finite number of values; that is, for $k = 1, \ldots, K$ and $i = 1, 2,$

$$\varepsilon_i^{(k)} = \begin{cases} \varepsilon_{1,i} & \text{with probability } \pi_{1,i}, \\ \varepsilon_{2,i} & \text{with probability } \pi_{2,i}, \\ \vdots \\ \varepsilon_{q_i,i} & \text{with probability } \pi_{q_i,i}. \end{cases}$$

The likelihood of this model is

$$L = \prod_{k=1}^{K} \sum_{j_1=1}^{q_1} \sum_{j_2=1}^{q_2} \boldsymbol{\delta}\mathbf{P}(x_{1k})\boldsymbol{\Gamma}\mathbf{P}(x_{2k})\boldsymbol{\Gamma}\cdots\boldsymbol{\Gamma}\mathbf{P}(x_{Tk})\mathbf{1}'\pi_{j_1,1}\pi_{j_2,2}. \quad (13.4)$$

Expressions (13.3) and (13.4) are identical. However, in (13.3) we sum over the possible values of the parameters of the state-dependent process whereas in (13.4) we sum over the possible values of the state transition probabilities. In (13.4) the off-diagonal elements of $\boldsymbol{\Gamma}$ are $\varepsilon_{j_1,1}$ and $\varepsilon_{j_2,2}$. The computational effort involved in evaluating this kind of likelihood is usually substantially less than in the case of continuous-valued random effects.

The use of discrete random effects also allows for a simple way to model dependence between multiple random effects. In the immediately preceding example, rather than specifying two independent univariate distributions for the two random effects, one could instead specify a single bivariate distribution:

$$\left(\varepsilon_1^{(k)}, \varepsilon_2^{(k)}\right) = \begin{cases} \left(\varepsilon_{11}^{(k)}, \varepsilon_{12}^{(k)}\right) & \text{with probability } \pi_1, \\ \left(\varepsilon_{21}^{(k)}, \varepsilon_{22}^{(k)}\right) & \text{with probability } \pi_2, \\ \vdots \\ \left(\varepsilon_{q1}^{(k)}, \varepsilon_{q2}^{(k)}\right) & \text{with probability } \pi_q. \end{cases} \quad (13.5)$$

In this case the likelihood simplifies to

$$L = \prod_{k=1}^{K} \sum_{j=1}^{q} \delta \mathbf{P}(x_{1k}) \mathbf{\Gamma} \mathbf{P}(x_{2k}) \mathbf{\Gamma} \cdots \mathbf{\Gamma} \mathbf{P}(x_{Tk}) \mathbf{1}' \pi_j.$$

This approach implies that each subject follows one of q possible state-switching patterns. In effect this partitions the K subjects into q homogeneous groups. In some applications it may be possible to find an interpretation for the grouping. For instance, in an analysis of series of daily counts of epileptic seizures in a group of patients, the state-switching patterns might well differ between patients who complied with medication and those who did not. But in the absence of further information such an interpretation would be speculative.

Examples of application of discrete-valued random effects appear in Section 18.6. There we describe models which use a random effect for each of the two off-diagonal transition probabilities in a two-state HMM. Both independent random effects and dependent are explored.

13.4 Discussion

A defining property of longitudinal data is that observations *of the same type* are available for each of the subjects. Such data offer the opportunity to identify those respects in which subjects differ and those in which they are similar. HMMs offer the means to accommodate such distinctions and similarities.

Pooling the parameters of the state-dependent distributions allows for variability in the state-switching dynamics of subjects, including the times spent in the different states. Alternatively, pooling the parameters of the t.p.m. allows for subject-specific variability within each state. Combinations of these two options are of course possible, but the resulting models are more difficult to interpret and, as pointed out by Altman (2007), '[i]t is easy to create complex models using latent variables. However, caution must be exercised in order to avoid unwanted implications of one's modeling choices.'

Subject-specific covariates, if available, can explain some of the differences between subjects. Covariate information is easily incorporated into HMMs. However, it is not always clear which parameters should be modelled as functions of the covariate and what functions to use. (For want of information, very simple functions are generally used.) Of course the deployment of covariates should be guided by expert knowledge of the subject matter. Where this is unavailable one must resort to trying different options and applying model selection criteria.

In some applications it is the effect of the covariate that is of primary interest. But unless unexplained variability between subjects is taken into account (e.g. by using a random effect), the effect of the covariate may be masked. For example, in the blue whale application mentioned above, it is the effect of a covariate indicating the presence or absence of sonar signals that is of primary interest. However, whale behaviour is affected by many other, unrecorded factors such as the abundance of food, site-specific conditions or individual characteristics. Such factors, which lead to enormous variability in the behaviour of the whales, need to be accounted for. In DeRuiter *et al.* (2016), this is accomplished by incorporating both the sonar covariate and a random effect in the state transition probabilities.

In general, the use of random effects is a convenient way to reduce the number of parameters that need to be estimated, especially when the number of subjects is large. There is a disadvantage to incorporating random effects in HMMs, which is that their implementation is computationally very demanding. The computational effort can be reduced by using discrete-valued rather than continuous-valued random effects.

Exercises

1. Prove (13.2) and (13.3).

2. Show that the first discrete random-effects approach discussed in Section 13.3.2 (independent random effects) is in fact a nested special case of the second approach (dependent random effects).

3. The discrete random-effects approach specified by (13.5) essentially means that each component series adopts one of q possible state-switching patterns. Suppose that the model has been fitted to data. Derive a formula for

$$\Pr\left(\left(\varepsilon_1^{(k)}, \varepsilon_2^{(k)}\right) = \left(\varepsilon_{j1}^{(k)}, \varepsilon_{j2}^{(k)}\right) \mid \mathbf{X}^{(T,k)} = \mathbf{x}^{(T,k)}\right), \quad j = 1, \ldots, q,$$

under the fitted model, where $\mathbf{X}^{(T,k)}$ denotes the history of the kth component series.

PART III

Applications

CHAPTER 14

Introduction to applications

In order to give the reader an idea of the versatility of HMMs, some of the areas of application of HMMs are listed in Table 14.1. This list is by no means exhaustive; in fact, several hundred papers on applications of HMMs had already been published by 2001 (Cappé, 2001). It is worth noting that, in some of these applications, the states of the HMM have a clear interpretation (e.g. for capture–recapture data, in which case the state indicates whether or not an animal is alive); see Chapter 24. In the context of supervised learning as discussed in Section 5.6, the states are defined *a priori* and observed in a training sample.

But in other applications the states may merely be convenient proxies, or have a plausible interpretation. In the case of share returns, the states can be regarded as proxies for 'market sentiment'. In the case of daily series of precipitation occurrence, the states can be regarded as weather conditions. (See Section 9.4.2.) In such cases the states of an HMM, like the components of an independent mixture model, are artefacts used to accommodate unobserved heterogeneity in the data. Identifying the source of the heterogeneity is tempting but essentially speculative.

The applications that we describe in detail in the following chapters were chosen from our own research work, and we have attempted to cover a wide variety of types of data. Several ecological and environmental applications are discussed, but other important areas of application, such as biological sequence analysis and psychology and other social sciences, are not represented in these chapters. For the former, see Durbin *et al.* (1998). For the latter, see especially the account of Visser (2010, Section 2).

Table 14.1 *Overview of some of the areas of applications of HMMs. Only one example reference is given for each type of observation listed.*

Observations	Interpretation of states	Literature
Fourier transforms of recorded speech	sequence of phonemes	Juang and Rabiner (1991)
features of bitmap image of handwriting	letters of the alphabet	Kundu, He and Bahl (1989)
EEG measurements during sleep	REM and non-REM sleep states	Langrock *et al.* (2013)
multiple sclerosis lesion count data	disease states	Altman and Petkau (2005)
times between a neuron's firing events	states of neuron	Camproux *et al.* (1996)
DNA sequence of bases	homogeneous segments of DNA sequence	Churchill (1989)
share returns	market sentiment	Rydén *et al.* (1998)
retailer transaction records	customer's propensity to buy	Mark *et al.* (2013)
traffic accidents	different weather conditions	Laverty, Miket and Kelly (2002)
volcanic eruptions	activity level of the volcano	Bebbington (2007)
responses in a learning experiment	guessing state vs. learned state	Visser *et al.* (2002)
daily rainfall occurrence	weather state	Zucchini and Guttorp (1991)
wave and wind measures	sea regime	Bulla *et al.* (2012)
animal movement metrics	behavioural states	Langrock, King *et al.* (2012)
recaptures of tagged animals	survival state	Laake (2013)

CHAPTER 15

Epileptic seizures

15.1 Introduction

Albert (1991) and Le, Leroux and Puterman (1992) describe the fitting of two-state Poisson–HMMs to series of daily counts of epileptic seizures in one patient. Such models appear to be a promising tool for the analysis of seizure counts, the more so as there are suggestions in the neurology literature that the susceptibility of a patient to seizures may vary in a fashion that can reasonably be represented by a Markov chain; see Hopkins, Davies and Dobson (1985). Another promising approach, not pursued here, is to use an AR(1) analogue based on thinning; see Franke and Seligmann (1993).

15.2 Models fitted

We analyse here a series of counts of myoclonic seizures suffered by one patient on 204 consecutive days*. The observations are given in Table 15.1 and displayed in Figure 15.1.

Table 15.1 *Counts of epileptic seizures in one patient on 204 consecutive days (to be read across rows).*

0	3	0	0	0	0	1	1	0	2	1	1	2	0	0	1	2	1	3	1	3
0	4	2	0	1	1	2	1	2	1	1	1	0	1	0	2	2	1	2	1	0
0	0	2	1	2	0	1	0	1	0	1	0	0	0	0	0	0	0	0	1	0
0	0	0	0	1	0	0	0	1	0	0	0	1	0	0	0	1	0	0	1	0
0	2	1	0	1	1	0	0	0	2	2	0	1	1	3	1	1	2	1	0	3
6	1	3	1	2	2	1	0	1	2	1	0	1	2	0	0	2	2	1	0	1
0	0	2	0	1	0	0	0	1	0	0	1	0	0	0	0	0	0	1	3	
0	0	0	0	0	1	0	1	1	1	0	0	0	0	1	0	1	2	1	0	
0	0	0	0	0	1	4	0	0	0	0	0	0	0	0	0	0	0	0		
0	0	0	0	0	0	0	0	0	0	0	0	0	0							

---

* The 225-day series published by Le *et al.* (1992) contained an accidental repeat of the observations for a 21-day period; see MacDonald and Zucchini (1997, p. 208). Table 15.1 gives the corrected series.

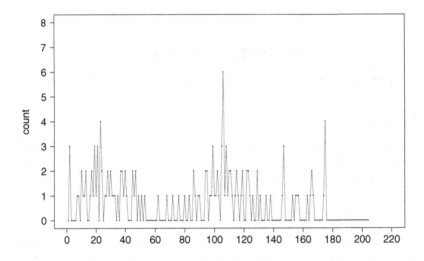

Figure 15.1 *Epileptic seizure counts on 204 days.*

Le *et al.* use an HMM of the type described by Leroux and Puterman (1992). Their model does not assume that the underlying Markov chain is stationary. It is fitted by maximizing the likelihood conditional on the Markov chain starting in a given state with probability 1, and then maximizing over the possible initial states.

We consider a similar HMM, but based on a *stationary* Markov chain and fitted by maximization of the unconditional likelihood of the observations. We investigate models with $m = 1$, 2, 3 and 4 states. (The one-state model is just the model which assumes that the observations are realizations of independent Poisson random variables with a common mean. That mean is the only parameter.)

Table 15.2 gives the AIC and BIC values for the models. From the table we see that, of the four models considered, the three-state model is chosen by AIC, but the two-state model is chosen, by a large margin, by BIC. We concentrate on the two-state model, the details of which are as follows. The Markov chain has transition probability matrix

$$
\begin{pmatrix}
0.965 & 0.035 \\
0.027 & 0.973
\end{pmatrix},
$$

and starts from the stationary distribution $(0.433, 0.567)$. The seizure rates in states 1 and 2 are 1.167 and 0.262, respectively.

The ACF of the model can be computed by the results of Exercise 4

Table 15.2 *Epileptic seizure counts: comparison of several stationary Poisson–HMMs by means of AIC and BIC.*

No. of states	k	$-l$	AIC	BIC
1	1	232.15	466.31	469.63
2	4	211.68	431.36	**444.64**
3	9	205.55	**429.10**	458.97
4	16	201.68	435.36	488.45

Table 15.3 *Sample ACF for the epileptic seizure counts.*

k	1	2	3	4	5	6	7	8
$\widehat{\rho}(k)$	0.236	0.201	0.199	0.250	0.157	0.181	0.230	0.242

of Chapter 2. It is given, for all positive integers k, by

$$
\begin{aligned}
\rho(k) &= \left(1 + \frac{\delta\boldsymbol{\lambda}'}{(\lambda_2 - \lambda_1)^2 \delta_1 \delta_2}\right)^{-1} (1 - \gamma_{12} - \gamma_{21})^k \\
&= 0.235 \times 0.939^k.
\end{aligned}
$$

Table 15.3 gives the corresponding sample ACF. Figure 15.2, which displays both, shows that the agreement between sample and theoretical ACF is reasonably close.

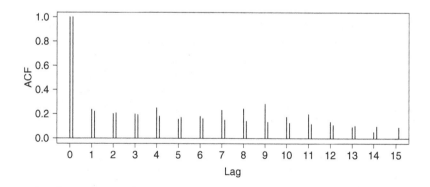

Figure 15.2 *Sample and theoretical ACF for the epileptic seizures data. At each lag the left bar represents the sample ACF, and the right bar the ACF of a stationary two-state Poisson–HMM.*

Table 15.4 *Observed and expected numbers of days with* $r = 0, 1, 2, \ldots$ *epileptic seizures.*

r	Observed no.	Expected no.
0	117	116.5
1	54	55.4
2	23	21.8
3	7	7.5
4	2	2.1
5	0	0.5
≥ 6	1	0.1
	204	203.9

The marginal properties of the model can be assessed from Table 15.4, which gives the observed and expected numbers of days on which there were 0, 1, 2, ..., 6 or more seizures. Agreement is excellent.

15.3 Model checking by pseudo-residuals

We now use the techniques of Section 6.2.2 to check for outliers under the two-state model we have chosen. Figure 15.3 is a plot of ordinary normal pseudo-residual segments. From it we see that three observations of the 204 stand out as extreme, namely those for days 106, 147 and 175. They all yield pseudo-residual segments lying entirely within the top $\frac{1}{2}\%$ of their respective distributions.

It is interesting to note that observations 23 and 175, both of which represent four seizures in one day, yield rather different pseudo-residual segments. The reason for this is clear when one notes that most of the near neighbours (in time) of observation 175 are zero, which is not true of observation 23 and its neighbours. Observation 23 is much less extreme relative to its neighbours than is 175, and this is reflected in the pseudo-residual. Similarly, observation 106 (six seizures in a day) is less extreme relative to its neighbours than is observation 175 (four seizures).

However, a more interesting exercise is to see whether, if a model had been fitted to (say) the first 100 observations only, day-by-day monitoring thereafter by means of forecast pseudo-residuals would have identified any outliers. The two-state model fitted from the first 100 observations has transition probability matrix

$$\begin{pmatrix} 0.983 & 0.017 \\ 0.042 & 0.958 \end{pmatrix},$$

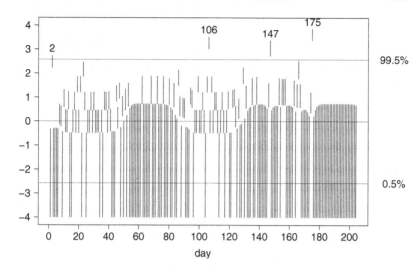

Figure 15.3 *Epileptic seizures data: ordinary (normal) pseudo-residual segments, relative to stationary two-state Poisson–HMM fitted to all 204 observations.*

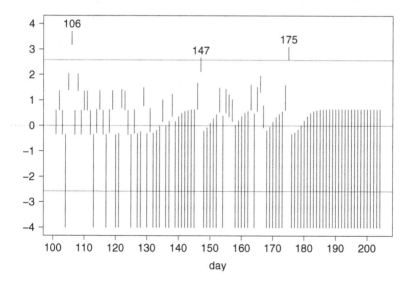

Figure 15.4 *Epileptic seizures data: forecast pseudo-residual segments, relative to stationary two-state Poisson–HMM fitted to data for days 1–100 only.*

and seizure rates 1.049 and 0.258.

From a plot of forecast pseudo-residuals (Figure 15.4), we see that the same three observations stand out: days 106, 147 and 175. Observation 106 emerges from such a monitoring procedure as the clearest outlier relative to its predecessors, then 175, then 147.

Exercises

1. Consider the two-state model for the epileptic seizures.

   (a) Compute the probabilities $\Pr(C_t = i \mid \mathbf{X}^{(T)})$ for this model, for $i = 1, 2$ and all t.

   (b) Perform both local and global decoding to estimate the most likely states. Do the results differ?

   (c) Perform state prediction for the next three time points; i.e. find the probabilities $\Pr(C_{T+h} = i \mid \mathbf{X}^{(T)})$ for $h = 1$, 2, 3.

   (d) Compute the forecast distribution $\Pr(X_{T+h} = x \mid \mathbf{X}^{(T)})$ for $h = 1$, ..., 10.

2. (a) Fit a stationary *three*-state Poisson–HMM to the epileptic seizures.

   (b) Find the general expression for the ACF of this model.

   (c) For lags $1, \ldots, 8$, compare this model ACF with the sample ACF given in Table 15.3.

3. Suppose that, instead of being given as exact counts, the seizures data are coded as 0, 1, and 'more than 1'.

   (a) Fit a stationary two-state Poisson–HMM to this series, and compare the resulting estimates with those given on p. 202.

   (b) Compare the expected and observed frequencies.

   (c) Compute the h-step-ahead forecast distribution for $h = 1, 2, 3$.

4. Suppose that the data are coded as 0 and 'at least 1'. Fit a stationary two-state Poisson–HMM to this series, and compare the resulting estimates with those given on p. 202 and those found in Exercise 3(a).

CHAPTER 16

Daily rainfall occurrence

16.1 Introduction

There exists a substantial literature on the stochastic modelling of daily precipitation; see, for example, Woolhiser (1992), Hughes *et al.* (1999), Srikanthan and McMahon (2001) and Neykov *et al.* (2012). Many of the proposed models are constructed from two submodels: the first describes rainfall occurrence (whether a particular day is wet or dry) and the second the rainfall amount on wet days. Here we restrict our attention to rainfall occurrence and fit Bernoulli–HMMs and Bernoulli–HSMMs to a binary sequence of dry and wet days. This application has previously been described by Langrock and Zucchini (2011). Our main objective here is to illustrate the application of HSMMs via the HMM representation presented in Chapter 12. We also illustrate the incorporation of seasonal effects into HMMs and HSMMs, as described in Section 10.2.

The data considered here are given by a binary series of dry and wet days over a period of about 47 years at a site in Bulgaria, namely Zlatograd, a town in the Rhodope mountains. As is usually done with hydrological series, we discarded observations for 29 February in order to avoid the complication of having 366 days in leap years. As can be expected, the series exhibits strong variation across seasons (see Table 16.1).

Table 16.1 *Zlatograd series: sample frequency of a rainy day by month.*

Jan	Feb	Mar	Apr	May	Jun	Jul	Aug	Sep	Oct	Nov	Dec
0.20	0.18	0.21	0.29	0.35	0.34	0.31	0.31	0.33	0.39	0.41	0.27

16.2 Models fitted

We begin by considering a simple model for the daily rainfall series from Zlatograd, namely a two-state Markov chain in which the transition probabilities are allowed to vary seasonally (called Model 1). Let $\{C_t\}$ $(t = 1, 2, \ldots, 16\,951)$ be a Markov chain representing rainfall occurrence, where $C_t = 1$ if day t is dry, and $C_t = 2$ if day t is wet. The transition probability matrix for Model 1 was assumed to depend on t:

$$_t\Gamma = \begin{pmatrix} \Pr(C_{t+1} = 1 \mid C_t = 1) & \Pr(C_{t+1} = 2 \mid C_t = 1) \\ \Pr(C_{t+1} = 1 \mid C_t = 2) & \Pr(C_{t+1} = 2 \mid C_t = 2) \end{pmatrix}.$$

To incorporate seasonality into the model, the logit transforms of the diagonal elements of the t.p.m. were modelled as linear combinations of trigonometric functions of the form

$$\alpha_0 + \sum_{k=1}^{q} \left(\alpha_k \sin\left(\frac{2\pi kt}{365}\right) + \beta_k \cos\left(\frac{2\pi kt}{365}\right) \right), \tag{16.1}$$

where the order q, controlling the flexibility of the model, was chosen by a model selection criterion (see below).

Next we fitted a two-state Bernoulli–HMM incorporating seasonality in the transition probabilities to the series. In this case the Markov chain $\{C_t\}$ is no longer taken to be an observation; it represents the unobserved state on day t. The observation on day t is regarded as a realization of a Bernoulli random variable whose parameter is determined by the state: the probability that day t is wet is π_1 if on day t the process is in state 1 (i.e. $C_t = 1$), and is π_2 if $C_t = 2$ (Model 2). Model 1 is the special case of Model 2 with $\pi_1 = 0$ and $\pi_2 = 1$. In both of these models, the initial distribution of states was estimated from the data, which adds one parameter.

An alternative way of introducing seasonality in the HMM is to assume that the entries of the t.p.m. are constant (i.e. do not depend on t), but that the Bernoulli parameters depend on t instead. Specifically, we considered a model where the logit transforms of the Bernoulli parameters, $\pi_1(t)$ and $\pi_2(t)$, have the form given in (16.1) (Model 3). In this model we assumed that the Markov chain is stationary.

Table 16.2 *Minus log-likelihood (−l) and AIC for Models 1–3 fitted to the Zlatograd series (q = 1).*

Model	−l	AIC
1	9 468.22	18 950.45
2	9 445.65	18 909.31
3	9 423.38	**18 862.75**

Table 16.2 gives minus the log-likelihood and the AIC for Models 1–3 with $q = 1$. By the AIC, the HMM with seasonality in the Bernoulli

parameters provides the best fit by a substantial margin, so in what follows we focus on that way of incorporating seasonality.

As the next step, we considered a Bernoulli–HSMM, extending Model 3 by allowing the dwell-time distribution in each state to have an un-structured start and a geometric tail, as described in Section 12.6.5. For the lengths of the unstructured starts we considered the values 0 (Marko-vian), 1, 2 and 3 for both the state associated with the dry periods and that of the wet periods. We also investigated different orders of season-ality in the Bernoulli parameters, namely $q = 1, 2, 3$, and all possible combinations of the lengths of unstructured starts and seasonality; one of the combinations is Model 3 above.

From all these models the AIC selected the (stationary) Bernoulli–HSMM which has dwell-time distribution with unstructured start of length 2 for the dry periods, unstructured start of length 3 for the wet periods, and order of seasonality $q = 2$ in the Bernoulli parameters. (We refer to this as Model 4.) The AIC value for Model 4 was obtained as 18 806.54, and a further increase in the order of the dwell-time distri-bution associated with the wet periods did not improve the AIC. The t.p.m. of the (exact) HMM representation of Model 4 is given by

$$\left(\begin{array}{ccc|cccc}
0 & 0.93 & 0 & 0.07 & 0 & 0 & 0 \\
0 & 0 & 0.74 & 0.26 & 0 & 0 & 0 \\
0 & 0 & 0.85 & 0.15 & 0 & 0 & 0 \\
\hline
0.15 & 0 & 0 & 0 & 0.85 & 0 & 0 \\
0.37 & 0 & 0 & 0 & 0 & 0.63 & 0 \\
0.34 & 0 & 0 & 0 & 0 & 0 & 0.66 \\
0.27 & 0 & 0 & 0 & 0 & 0 & 0.73
\end{array}\right). \qquad (16.2)$$

Note that all the zeros in this t.p.m. are structural zeros.

Here the state aggregate $I_1 = \{1, 2, 3\}$ is associated with low probab-ility of precipitation (dry periods), and $I_2 = \{4, 5, 6, 7\}$ with high prob-ability of precipitation (wet periods). The upper left block of the t.p.m. determines the dwell-time distribution in the dry periods, the lower right block that in the wet periods. The estimated p.m.f.s of the dwell-time distributions in the two state aggregates are displayed in Figure 16.1. The deviation from the p.m.f. of a geometric distribution is evident, and in particular, the modal dwell time is greater than 1 in both cases.

The estimated Bernoulli parameter functions $\pi_1(t)$ or $\pi_2(t)$, which have period 365, are displayed in Figure 16.2. State aggregate I_1 can be regarded as the 'dry' state of the HSMM; the probability of rain is generally low but peaks slightly in early November. In state aggregate I_2, the 'wet' state, the probability of rain is generally high but ranges between 0.5 and 0.8. The stationary probabilities for state aggregates I_1

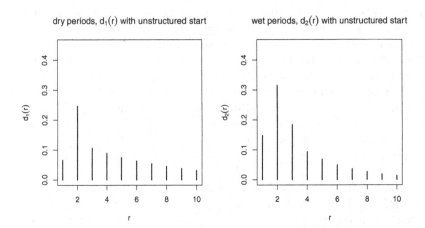

Figure 16.1 *Estimated p.m.f.s of the dwell-time distributions, with unstructured start, associated with dry and wet periods (Model 4).*

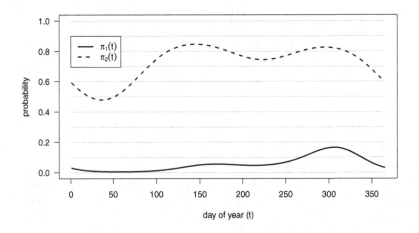

Figure 16.2 *Estimated Bernoulli parameter functions for the Zlatograd series (Model 4).*

and I_2 are 0.63 and 0.37, respectively. In other words, the system is in the 'dry' state 63% of the time, and in the 'wet' state 37% of the time.

An alternative dwell-time distribution worth investigating is the shifted negative binomial distribution, which is a natural generalization of the geometric and has two parameters. An advantage of using a dwell-time distribution with a fixed number of parameters is that the model need be fitted only once. In contrast, when one uses a distribution with un-structured start one needs to fit the model each time one changes the length of the unstructured start in order to determine the appropriate length.

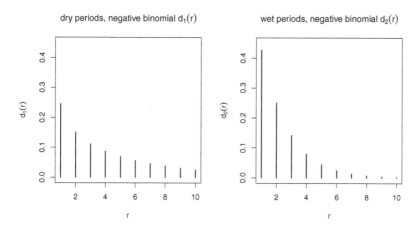

Figure 16.3 *Estimated p.m.f.s of (shifted) negative binomial dwell-time distri-butions associated with dry and wet periods (Model 5).*

Using the HMM approximation to fit a stationary HSMM with shifted negative binomial dwell-time distributions (and $q = 2$, Model 5) to the Zlatograd series yielded an AIC value of 18 815.53, which is higher than that of Model 4, the HSMM with dwell-time distributions having un-structured start (AIC = 18 806.54). For Model 5, the estimated p.m.f.s of the dwell-time distributions are displayed in Figure 16.3. The striking difference between the p.m.f.s shown in Figure 16.3 and those in Figure 16.1 is that the modes differ. The shifted negative binomial distribu-tion, having fewer parameters, is of course less flexible. Both types of HSMM led to substantially lower AIC values than those of the basic HMM with the same order of seasonality as the HSMM counterparts ($q = 2$, AIC = 18 832.49). The additional flexibility provided by allow-ing for non-geometric dwell-time distributions thus led to a substantial improvement in the fit.

CHAPTER 17

Eruptions of the Old Faithful geyser

17.1 Introduction

There are many published analyses, from various points of view, of data
relating to eruptions of the Old Faithful geyser in the Yellowstone National Park in the USA: for instance, Cook and Weisberg (1982, pp. 40–
42), Weisberg (1985, pp. 230–235), Silverman (1985; 1986, p. 7), Scott
(1992, p. 278) and Aston and Martin (2007). Some of these accounts
ignore the strong serial dependence in the behaviour of the geyser; see
the comments of Diggle (1993).

In this chapter we present:

- an analysis of a series of long and short eruption durations of the
  geyser (this series is a dichotomized version of one of the two series
  provided by Azzalini and Bowman, 1990);

- univariate models for the series of durations and waiting times, in
  their original, non-dichotomized, form; and

- a bivariate model for the durations and waiting times.

The models we describe in this chapter are mostly HMMs, but in Section
17.3 we also fit Markov chains of first and second order, and compare
them with the HMMs.

17.2 The data

The data of Azzalini and Bowman consist of 299 pairs of observations,
collected continuously from 1 August to 15 August 1985. The pairs are
(w_t, d_t), with w_t being the time between the starts of successive eruptions, and d_t being the duration of the subsequent eruption.

The durations are given in minutes and seconds, except that some
are identified only as short (S), medium (M) or long (L); the accurately
recorded durations range from 50 seconds to 5 minutes 27 seconds. There
were in all 20 short eruptions, 2 medium and 47 long. For certain purposes Azzalini and Bowman represented the codes S, M, L by durations
of 2, 3, and 4 minutes' length, and that is the form in which the data
have been made available in the **R** package MASS. The waiting times are
all given in minutes.

17.3 The binary time series of short and long eruptions

In both series, most of the observations can be described as either long
or short, with very few observations intermediate in length, and with
relatively low variation within the low and high groups. It is therefore
natural to treat these series as binary time series. Azzalini and Bowman
do so by dichotomizing the waiting times w_t at 68 minutes and the
durations d_t at 3 minutes, denoting short by 0 and long by 1. We do the
same. Some convention is needed here for the 'mediums', and we use the
convention that the (two) mediums are treated as long.

Table 17.1 *Short and long eruption durations of Old Faithful geyser (299 observations, to be read across rows).*

1 0 1 1 1 0 1 1 0 1 0 1 0 1 1 0 1 0 1 1 0 1 0 1 0 1 0 1 1 1
1 1 0 1 0 1 0 1 0 1 0 1 0 1 0 1 0 1 0 1 0 1 0 1 0 1 1 1 1 1
0 1 0 1 0 1 0 1 1 0 1 0 1 1 1 0 1 1 1 1 0 1 1 1 0 1 0 1 0
1 0 1 0 1 0 1 0 1 0 1 0 1 0 1 0 1 0 1 0 1 1 0 1 0 1 0 1 0 1
0 1 1 1 0 1 1 1 1 1 1 0 1 1 1 1 0 1 1 1 1 1 1 0 1 0 1
0 1 0 1 0 1 0 1 1 1 1 1 0 1 0 1 0 1 0 1 1 1 0 1 0 1 0 1 1
0 1 0 1 1 1 1 0 1 0 1 0 1 0 1 1 1 0 1 0 1 0 1 1 0 1 1 0 1 1
1 0 1 0 1 0 1 0 1 1 0 1 1 1 1 1 1 0 1 0 1 0 1 1 1 1 0 1 1
0 1 1 1 0 1 1 0 1 0 1 1 1 0 1 0 1 1 1 1 1 0 1 1 1 0 1 0 1 0
1 1 0 1 0 1 1 1 1 1 1 1 1 0 1 0 1 0 1 0 1 0 1 0 1 0 1 1 0

It emerges that $\{W_t\}$ and $\{D_t\}$, the dichotomized versions of the series
$\{w_t\}$ and $\{d_t\}$, are very similar – almost identical, in fact – and Azzalini
and Bowman therefore concentrate on the series $\{D_t\}$ as representing
most of the information relevant to the state of the system. Table 17.1
presents the series $\{D_t\}$. On examination of this series one notices that
0 is always followed by 1, and 1 by either 0 or 1. A summary of the data
is displayed in the 'observed no.' column of Table 17.3.

17.3.1 Markov chain models

Azzalini and Bowman first fitted a (first-order) Markov chain model.
This model seemed quite plausible from a geophysical point of view, but
did not match the sample ACF at all well. They then fitted a second-
order Markov chain model, which matched the ACF much better, but
did not attempt a geophysical interpretation for this second model. They
estimated the ACF of $\{D_t\}$ as displayed in Table 17.2. Since the sample
ACF is not even approximately of the form α^k, a Markov chain is not

Table 17.2 *Old Faithful eruptions: sample autocorrelation function, $\widehat{\rho}(k)$, of the series $\{D_t\}$ of short and long eruptions.*

k	1	2	3	4	5	6	7	8
$\widehat{\rho}(k)$	−0.538	0.478	−0.346	0.318	−0.256	0.208	−0.161	0.136

a satisfactory model; see Section 1.3.3. They therefore fitted a second-order Markov chain, which did not degenerate to a first-order model.

An estimate of the transition probability matrix of the first-order Markov chain, based on maximizing the likelihood conditional on the first observation as described in Section 1.3.4, is

$$
\begin{pmatrix} 0 & 1 \\ \frac{105}{194} & \frac{89}{194} \end{pmatrix} = \begin{pmatrix} 0 & 1 \\ 0.5412 & 0.4588 \end{pmatrix}. \tag{17.1}
$$

Although it is not central to this discussion, it is worth noting that unconditional maximum likelihood estimation is very easy in this case. Because there are no transitions from 0 to 0, the explicit result of Bisgaard and Travis (1991) applies; see our equation (1.6) on p. 20. The result is that the transition probability matrix is estimated as

$$
\begin{pmatrix} 0 & 1 \\ 0.5404 & 0.4596 \end{pmatrix}. \tag{17.2}
$$

This serves to confirm as reasonable the expectation that, for a series of length 299, estimation by conditional maximum likelihood differs very little from unconditional.

Since the sequence $(0,0)$ does not occur, the three states needed to express the second-order Markov chain as a first-order Markov chain are, in order: $(0,1)$, $(1,0)$, $(1,1)$. The corresponding t.p.m. is

$$
\begin{pmatrix} 0 & \frac{69}{104} & \frac{35}{104} \\ 1 & 0 & 0 \\ 0 & \frac{35}{89} & \frac{54}{89} \end{pmatrix} = \begin{pmatrix} 0 & 0.6635 & 0.3365 \\ 1 & 0 & 0 \\ 0 & 0.3933 & 0.6067 \end{pmatrix}. \tag{17.3}
$$

Model (17.3) has stationary distribution $\frac{1}{297}(104, 104, 89)$, and the ACF can be computed from

$$
\rho(k) = \frac{297^2 \Pr(D_t = D_{t+k} = 1) - 193^2}{193 \times 104}.
$$

The resulting figures for $\{\rho(k)\}$ are given in Table 17.4, and match the sample ACF well.

Table 17.3 *Old Faithful: observed numbers of short and long eruptions and various transitions, compared with those expected under the two-state HMM.*

	Observed no.	Expected no.
short eruptions (0)	105	105.0
long eruptions (1)	194	194.0
Transitions:		
from 0 to 0	0	0.0
from 0 to 1	104	104.0
from 1 to 0	105	104.9
from 1 to 1	89	89.1
from $(0, 1)$ to 0	69	66.7
from $(0, 1)$ to 1	35	37.3
from $(1, 0)$ to 1	104	104.0
from $(1, 1)$ to 0	35	37.6
from $(1, 1)$ to 1	54	51.4

17.3.2 Hidden Markov models

We now discuss the use of HMMs for the series $\{D_t\}$. Bernoulli–HMMs with $m = 1, 2, 3$ and 4 were fitted to this series.

We describe the two-state model in some detail. This model has log-likelihood -127.31, $\mathbf{\Gamma} = \begin{pmatrix} 0.000 & 1.000 \\ 0.827 & 0.173 \end{pmatrix}$ and state-dependent probabilities of a long eruption given by $(0.225, 1.000)$. That is, there are two (unobserved) states, state 1 always being followed by state 2, and state 2 by state 1 with probability 0.827. In state 1 a long eruption has probability 0.225; in state 2, probability 1. A convenient interpretation of this model is that it is a stationary two-state Markov chain, with some noise added in the first state; if the probability 0.225 were instead zero, the model would be exactly the Markov chain. In the usual notation, a long eruption has unconditional probability $\Pr(X_t = 1) = \boldsymbol{\delta}\mathbf{P}(1)\mathbf{1}' = 0.649$, and a long eruption is followed by a short one with probability $\boldsymbol{\delta}\mathbf{P}(1)\mathbf{\Gamma}\mathbf{P}(0)\mathbf{1}'/\boldsymbol{\delta}\mathbf{P}(1)\mathbf{1}' = 0.541$. A short is always followed by a long. A comparison of observed numbers of zeros, ones and transitions, with the numbers expected under this model, is presented in Table 17.3.

The ACF is given for all $k \in \mathbb{N}$ by $\rho(k) = (1 + \alpha)^{-1}w^k$, where $w = -0.827$ and $\alpha = 0.529$. Hence $\rho(k) = 0.654 \times (-0.827)^k$. In Figure 17.1 and Table 17.4 the resulting figures are compared with the sample ACF and with the theoretical ACF of the second-order Markov chain model (17.3). The HMM fits the sample ACF well; not quite as well as the second-

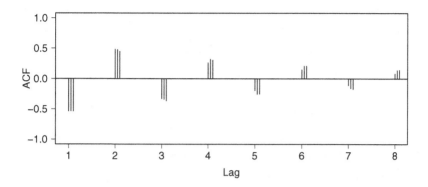

Figure 17.1 *Old Faithful, short and long eruptions: sample autocorrelation function, and ACF of two models. At each lag the centre bar represents the sample ACF, the left bar the ACF of the second-order Markov chain (i.e. model (17.3)), and the right bar that of the two-state HMM.*

Table 17.4 *Old Faithful, short and long eruptions: sample ACF compared with the ACF of the second-order Markov chain and the HMM.*

k	1	2	3	4	5	6	7	8
$\rho(k)$ for model (17.3)	−0.539	0.482	−0.335	0.262	−0.194	0.147	−0.110	0.083
sample ACF, $\hat{\rho}(k)$	−0.538	0.478	−0.346	0.318	−0.256	0.208	−0.161	0.136
$\rho(k)$ for HM model	−0.541	0.447	−0.370	0.306	−0.253	0.209	−0.173	0.143

order Markov chain model as regards the first three autocorrelations, but better for longer lags.

The parametric bootstrap, with a sample size of 100, was used to estimate the means and covariances of the maximum likelihood estimators of the four parameters γ_{12}, γ_{21}, p_1 and p_2. That is, 100 series of length 299 were generated from the two-state HMM described above, and a model of the same type fitted in the usual way to each of these series. The sample mean vector for the four parameters is (1.000, 0.819, 0.215, 1.000), and the sample covariance matrix is

$$\begin{pmatrix} 0 & 0 & 0 & 0 \\ 0 & 0.003303 & 0.001540 & 0 \\ 0 & 0.001540 & 0.002065 & 0 \\ 0 & 0 & 0 & 0 \end{pmatrix}.$$

The estimated standard deviations of the estimators are therefore (0.000, 0.057, 0.045, 0.000). (The zero standard errors are a consequence of the

Table 17.5 *Old Faithful, short and long eruptions: percentiles of bootstrap sample of estimators of parameters of two-state HMM.*

Percentile:	5th	25th	Median	75th	95th
$\widehat{\gamma}_{21}$	0.709	0.793	0.828	0.856	0.886
$\widehat{p}_1$	0.139	0.191	0.218	0.244	0.273

rather special nature of the model from which we are generating the series.)

As a further indication of the behaviour of the estimators we present in Table 17.5 selected percentiles of the bootstrap sample of values of $\widehat{\gamma}_{21}$ and $\widehat{p}_1$. From these bootstrap results it appears that, for this application, the maximum likelihood estimators have fairly small standard deviations and are not markedly asymmetric.

There is a further class of models that generalizes both the two-state second-order Markov chain and the two-state HMM as described above. This is the class of two-state second-order HMMs, described in Section 10.3. By using the recursion (10.2) for the probability $\nu_t(j, k; \mathbf{x}^{(t)})$, with the appropriate scaling, it is almost as straightforward to compute the likelihood of a second-order model as a first-order one and to fit models by maximum likelihood. In the present example the resulting probabilities of a long eruption are 0.0721 (state 1) and 1.0000 (state 2). The parameter process is a two-state second-order Markov chain with associated first-order Markov chain having t.p.m.

$$\begin{pmatrix} 1-a & a & 0 & 0 \\ 0 & 0 & 0.7167 & 0.2833 \\ 0 & 1.0000 & 0 & 0 \\ 0 & 0 & 0.4414 & 0.5586 \end{pmatrix}. \qquad (17.4)$$

Here a may be taken to be any real number between 0 and 1, and the four states used for this purpose are, in order: $(1, 1)$, $(1, 2)$, $(2, 1)$, $(2, 2)$. The log-likelihood is -126.9002. (Clearly the state $(1, 1)$ can be disregarded above without loss of information, in which case the first row and first column are deleted from the matrix (17.4).)

It should be noted that the second-order Markov chain used here as the underlying process is the general four-parameter model, not the Pegram–Raftery submodel, which has three parameters. From the comparison which follows it will be seen that an HMM based on a Pegram–Raftery second-order chain is in this case not worth pursuing, because with a total of five parameters it cannot produce a log-likelihood value better than -126.90. (The two four-parameter models fitted produce values of

Table 17.6 *Old Faithful, short and long eruptions: comparison of models on the basis of AIC and BIC.*

Model	k	$-l$	AIC	BIC
1-state HM (i.e. independence)	1	193.80	389.60	393.31
Markov chain	2	134.24	272.48	279.88
second-order Markov chain	4	127.12	**262.24**	**277.04**
2-state HM	4	127.31	262.62	277.42
3-state HM	9	126.85	271.70	305.00
4-state HM	16	126.59	285.18	344.39
2-state second-order HM	6	126.90	265.80	288.00

−127.31 and −127.12, which by AIC and BIC would be preferable to a log-likelihood of −126.90 for a five-parameter model.)

17.3.3 Comparison of models

From Table 17.6 it emerges that, on the basis of AIC and BIC, only the second-order Markov chain and the two-state (first-order) HMM are worth considering. In the comparison, both of these models are taken to have four parameters, because, although the observations suggest that the sequence (short, short) cannot occur, there is no *a priori* reason to make such a restriction.

Although it is true that the second-order Markov chain seems a slightly better model on the basis of the model selection exercise described above, and possibly on the basis of the ACF, both are reasonable models capable of describing the principal features of the data without using an excessive number of parameters. The HMM perhaps has the advantage of relative simplicity, given its nature as a Markov chain with some noise in one of the states. Azzalini and Bowman note that their second-order Markov chain model would require a more sophisticated interpretation than does their first-order model. Either a longer series of observations or a convincing geophysical interpretation for one model rather than the other would be needed to take the discussion further.

17.3.4 Forecast distributions

The ratio of likelihoods, as described in Section 5.3, can be used to provide the forecasts implied by the fitted two-state HMM. As it happens, the last observation in the series, D_{299}, is 0, so that under the model $\Pr(D_{300} = 1) = 1$. The conditional distribution of the next h values,

Table 17.7 *Old Faithful, short and long eruptions: the probabilities* $\Pr(D_{300} = 1, D_{301} = i, D_{302} = j \mid \mathbf{D}^{(299)})$ *for the two-state HMM (left) and the second-order Markov chain model (right).*

	$j = 0$	1		$j = 0$	1
$i = 0$	0.000	0.641	$i = 0$	0.000	0.663
1	0.111	0.248	1	0.132	0.204

given the history $\mathbf{D}^{(299)}$ (i.e. the joint h-step-ahead forecast), is easily computed. For $h = 3$ this is given in Table 17.7. The corresponding probabilities for the second-order Markov chain model are also given in the table.

17.4 Univariate normal–HMMs for durations and waiting times

Bearing in mind the dictum of van Belle (2002, p. 99) that one should not dichotomize unless absolutely necessary, we now describe normal–HMMs for the durations and waiting-times series in their original, non-dichotomized, form.

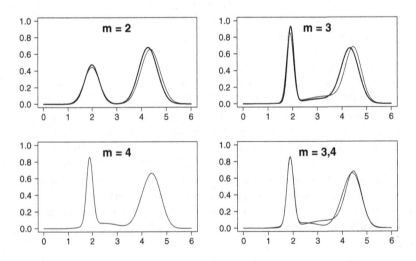

Figure 17.2 *Old Faithful durations, p.d.f.s of normal–HMMs. Thick lines (m = 2 and 3 only): models based on continuous likelihood and the* MASS *form of the data. Thin lines (all panels): models based on discrete likelihood.*

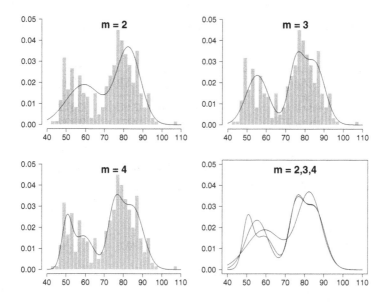

Figure 17.3 *Old Faithful waiting times, p.d.f.s of normal–HMMs. Models based on continuous likelihood and models based on discrete likelihood are essentially the same. Notice that the model for $m = 3$ is identical, or almost identical, to the three-state model of Robert and Titterington (1998); see their Figure 7.*

We use here the discrete likelihood, which accepts observations in the form of upper and lower bounds. (See Section 1.2.3 for the meaning of the term 'discrete likelihood'.) The recording of some of the durations only as S, M or L might appear to present a problem in finding the relevant intervals for the durations, but we treat durations S and L as lying in the intervals $(0, 3)$ and $(3, 20)$ minutes, respectively. We treat the two durations of medium length, somewhat arbitrarily, as lying in $(2.5, 3.5)$. Otherwise the interval for a duration is the observation $\pm \frac{1}{2}$ second. The waiting times are all given in minutes, so if (for instance) $w_1 = 80$, we use $w_1 \in (79.5, 80.5)$. In addition to the avoidance of unbounded (continuous) likelihood, the use of the discrete likelihood has the advantage that it can cope smoothly with interval censoring. It is noticeable that at least one of the published analyses of this data set avoids the durations series because of the interval censoring, and concentrates on the waiting-times series.

First, we present the likelihood and AIC and BIC values of the normal–HMMs; see Tables 17.8 and 17.9. We fitted univariate normal–HMMs with two to four states to durations and waiting times, and compared

Table 17.8 *Old Faithful durations: comparison of normal–HMMs and independent mixture models by AIC and BIC. All models were based on discrete likelihood.*

Model	k	$-\log L$	AIC	BIC
2-state HM	6	1168.955	2349.9	2372.1
3-state HM	12	1127.185	2278.4	**2322.8**
4-state HM	20	1109.147	**2258.3**	2332.3
indep. mixture (2)	5	1230.920	2471.8	2490.3
indep. mixture (3)	8	1203.872	2423.7	2453.3
indep. mixture (4)	11	1203.636	2429.3	2470.0

Table 17.9 *Old Faithful waiting times: comparison of normal–HMMs based on discrete likelihood.*

Model	k	$-\log L$	AIC	BIC
2-state HM	6	1092.794	2197.6	2219.8
3-state HM	12	1051.138	2126.3	**2170.7**
4-state HM	20	1038.600	**2117.2**	2191.2

Table 17.10 *Old Faithful: three-state (univariate) normal–HMMs, based on discrete likelihood.*

Durations

Γ			i	1	2	3
0.000	0.000	1.000	δ_i	0.291	0.195	0.514
0.053	0.113	0.834	μ_i	1.894	3.400	4.459
0.546	0.337	0.117	σ_i	0.139	0.841	0.320

Waiting times

Γ			i	1	2	3
0.000	0.000	1.000	δ_i	0.342	0.259	0.399
0.298	0.575	0.127	μ_i	55.30	75.30	84.93
0.662	0.276	0.062	σ_i	5.809	3.808	5.433

these on the basis of AIC and BIC. For both durations and waiting times, the four-state model is chosen by AIC and the three-state by BIC. (In the case of the durations, we show also the comparable figures relating to independent mixture models with two to four states.) We concentrate now on the three-state models, which are given in Table 17.10. In all cases the model quoted is that based on maximizing the discrete likelihood, and the states have been ordered in increasing order of mean. The t.p.m. and stationary distribution are, as usual, Γ and δ, and the state-dependent means and standard deviations are μ_i and σ_i, for $i = 1, 2, 3$. The marginal densities of the HMMs for the durations are presented in Figure 17.2, and for the waiting times in Figure 17.3.

One feature of these models that is noticeable is that, although the matrices Γ are by no means identical, in both cases the first row is $(0, 0, 1)$ and the largest element of the third row is γ_{31}, the probability of transition from state 3 to state 1.

17.5 Bivariate normal–HMM for durations and waiting times

Finally, we give here a stationary three-state bivariate model for durations d_t and waiting times w_{t+1}, for t running from 1 to 298. The pairing (d_t, w_t) would be another possibility; in Exercise 2 the reader is invited to fit a similar model to that bivariate series and compare the two models. The three state-dependent distributions are general bivariate normal distributions. Hence there are in all $21 = 15 + 6$ parameters: 5 for each bivariate normal distribution, and 6 for the transition probabilities. The model was fitted by maximizing the discrete likelihood of the bivariate observations (d_t, w_{t+1}); that is, by maximizing

$$L_T = \delta \mathbf{P}(d_1, w_2)\Gamma\mathbf{P}(d_2, w_3) \cdots \Gamma\mathbf{P}(d_{T-1}, w_T)\mathbf{1}',$$

where $\mathbf{P}(d_t, w_{t+1})$ is the diagonal matrix with each diagonal element not a bivariate normal density but a probability: the probability that the tth pair (duration, waiting time) falls in the rectangle $(d_t^-, d_t^+) \times (w_{t+1}^-, w_{t+1}^+)$. Here d_t^- and d_t^+ represent the lower and upper bounds available for the tth duration, and similarly for waiting times. Figure 17.4 displays perspective and contour plots of the marginal p.d.f. of this model, the parameters of which are given in Table 17.11. Note that here Γ, the t.p.m., is of the same form as the matrices displayed in Table 17.10.

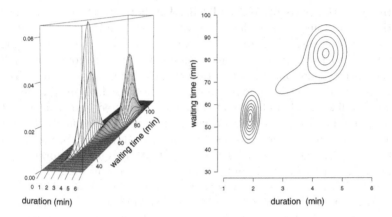

Figure 17.4 *Old Faithful durations and waiting times: perspective and contour plots of the p.d.f. of the bivariate normal–HMM. (Model fitted by discrete likelihood.)*

Table 17.11 *Old Faithful durations and waiting times: three-state bivariate normal–HMM, based on discrete likelihood.*

Γ			i (state)	1	2	3
0.000	0.000	1.000	δ_i	0.283	0.229	0.488
0.037	0.241	0.722	mean duration	1.898	3.507	4.460
0.564	0.356	0.080	mean waiting time	54.10	71.59	83.18
			s.d. duration	0.142	0.916	0.322
			s.d. waiting time	4.999	8.289	6.092
			correlation	0.178	0.721	0.044

Exercises

1. (a) Fit a bivariate normal–HMM with *two* states to the observations (d_t, w_{t+1}).

   (b) Compare the resulting marginal distributions for durations and waiting times to those implied by the three-state model reported in Table 17.11.

2. Fit a bivariate normal–HMM with three states to the observations (d_t, w_t), where t runs from 1 to 299. How much does this model differ from that reported in Table 17.11?

3. (a) Write an **R** function to generate observations from a bivariate

normal–HMM. (Hint: see the code in Section A.1.5, and use the package mvtnorm.)

(b) Write a function that will find maximum likelihood estimates of the parameters of a bivariate normal–HMM when the observations are assumed to be known exactly; use the 'continuous likelihood', i.e. use densities, not probabilities, in the likelihood.

(c) Use the function in (a) to generate a series of 1000 observations, and use the function in (b) to estimate the parameters.

(d) Apply varying degrees of interval censoring to your generated series and estimate the parameters. To what extent are the parameter estimates affected by interval censoring?

4. Consider a three-state model for (d_t, w_{t+1}) of the kind described in Section 17.5, except that contemporaneous conditional independence is assumed. That, is use bivariate normal distributions consisting of two *independent* normal random variables as the state-dependent distributions. This model requires three parameters fewer, as the three correlations are assumed to be zero.

Fit such a model, and compare your parameter estimates with those appearing in Table 17.11.

## CHAPTER 18

HMMs for animal movement

18.1 Introduction

Relating animal movement to environmental factors, such as habitat quality, contributes to the understanding of the effects of climate change or of human activity, such as land use, on the dispersal of animals. Individual movement potentially plays an important role in population dynamics (Morales *et al.*, 2010). These considerations, together with significant advances in tracking technology, have motivated a growing interest among ecologists in models for animal movement.

Over the past decade, much of the research on the topic has focused on partitioning movement patterns into categories, or states, that can plausibly be interpreted as the behavioural states of the animal. Corresponding models have been formulated and fitted primarily in a discrete-time framework, typically by using HMMs (Morales *et al.*, 2004; Holzmann *et al.*, 2006; Patterson *et al.*, 2009; Langrock, King *et al.*, 2012) or more general state-space models (Jonsen *et al.*, 2005; Patterson *et al.*, 2008). But a similar approach has also been proposed in continuous time (Blackwell, 2003). State-space models and associated Bayesian estimation approaches are commonly used in cases where there is considerable measurement error in the recorded locations, with the error being explicitly represented in the corresponding models. The approach described here, based on HMMs, deals only with cases in which the measurement error can be assumed negligible.

In this chapter we describe how HMMs can be used to model movement data (Section 18.3), and demonstrate the implementation thereof in three case studies, related to the movement of larvae of the fruit fly *Drosophila melanogaster* (Section 18.4), to the movement of a bison (Section 18.5) and to the movement of several groups of red-cockaded woodpeckers (Section 18.6). In the latter two examples, non-standard models are considered, namely HSMMs as described in Chapter 12, and mixed HMMs as described in Chapter 13.

18.2 Directional data

Since the modelling of animal movement often involves directional (or, more generally, circular) data, we begin by discussing some of the tools available for the analysis of directional data.

18.2.1 Directional means

One obvious difference between directional and 'linear' data is what is meant by a mean. Consider, for instance, the mean of the directions $355°$ and $5°$. Clearly it should be 0, not $180°$.

The usual approach is to convert any angle to corresponding x- and y-values on the unit circle, compute the averages of these Cartesian coordinates, and convert the resulting averages back to an angle. So the directional mean of a circular random variable X is simply an angle μ such that

$$\tan \mu = \frac{E \sin X}{E \cos X}.$$

But this does not define the angle uniquely. We use the convention used by **R** in the function of two arguments `atan2`: $\arctan(y, x) \in (-\pi, \pi]$ is the angle between the positive horizontal axis and the vector from the origin to (x, y). With this convention, we define the directional mean of a circular random variable X to be

$$\mu = \arctan(E \sin X, E \cos X).$$

If for instance an angle is ϕ_1 with probability η, otherwise ϕ_2, its directional mean is

$$\phi \quad = \quad \arctan\left(\eta \sin \phi_1 + (1 - \eta) \sin \phi_2, \eta \cos \phi_1 + (1 - \eta) \cos \phi_2\right).$$

Equivalently, the directional mean could be defined via the argument of a complex number, as implemented in **R** by the function `Arg`.

18.2.2 The von Mises distribution

We now discuss an important family of distributions designed for circular data, the von Mises distributions, which have properties that make them in some respects a natural first choice as a model for unimodal continuous observations on the circle; see, for example, Fisher (1993, pp. 49–50, 55). The probability density function of the von Mises distribution with parameters $\mu \in (-\pi, \pi]$ (location) and $\kappa > 0$ (concentration) is

$$f(x) = c \exp(\kappa \cos(x - \mu)) \quad \text{for } x \in (-\pi, \pi]. \tag{18.1}$$

One can in this p.d.f. replace the interval $(-\pi, \pi]$ by $[0, 2\pi)$, or by any other interval of length 2π. The normalizing constant c can be written in terms of a modified Bessel function of the first kind as $(2\pi I_0(\kappa))^{-1}$. The location parameter μ is not in the usual sense the mean of a random variable X having the above density; instead (see Exercise 3) it is the directional mean of X. In modelling time series of directional data one can consider using an HMM with von Mises distributions as the state-dependent distributions, although any other circular distribution is also possible, for example, the wrapped Cauchy distribution.

18.3 HMMs for movement data

18.3.1 Movement data

Data sets on animal movement typically give the locations in the horizontal plane, observed by using GPS telemetry technology, for instance. The sampling of locations is usually made at regular time intervals, and we restrict ourselves to the consideration of such data, since the HMMs that we will be considering here are suited only to observations that are sampled on a regular grid. Sampling intervals vary considerably across studies, ranging from fractions of seconds up to days. The time difference between observations affects what types of inference can be made, so that care should be taken when choosing the sampling interval. For example, if the goal of an analysis is to infer the behavioural states of an animal, or proxies thereof, then observations should be made at a temporal scale which is meaningful with regard to the behavioural dynamics of the animal. In any case, HMMs will typically provide only rough classifications of an animal's behavioural dynamics, as illustrated in the case studies below.

Various different movement metrics can be considered when modelling telemetry data, including

- the bivariate positions themselves,

- the increments in the two directions,

- the (Euclidean) distance between successively observed positions (usually referred to as the *step length*) and

- the change of direction between successive relocations (usually referred to as the *turning angle*).

We focus on bivariate time series comprising step lengths and turning angles. These are often considered together when analysing and interpreting movement data, in particular since they lead to intuitive interpretations (Marsh and Jones, 1988). We note that, conditional on the position and heading of an animal at the initial observation, the bivariate series of step lengths and turning angles completely determines the entire subsequent movement path, and hence all metrics listed above.

18.3.2 HMMs as multi-state random walks

For flexible modelling of data on step lengths and turning angles using HMMs, a class of building-blocks can be considered that comprises correlated and biased random walks (CRWs and BRWs) as well as walks that are both correlated and biased (BCRWs) (Codling, Plank and Benhamou, 2008). The most important of these are CRWs, which involve directional persistence and can be expressed by a turning-angle distribution with mass centred either on 0 (for positive correlation in direction) or on π (for negative correlation in direction). In fact, most animal movement models of HMM type involve only CRWs (see further comments below).

Bias in random walks can refer either to a general preference for some direction (e.g. east) or to a tendency to move towards a particular location. For example, a bias towards the location (x_0, y_0) is obtained by assuming that the turning angles are such that the expected movement direction at time $t + 1$ is the direction of the vector $(x_0, y_0) - (x_t, y_t)$, where (x_t, y_t) is the animal's position at time t. The location of a centre of attraction (x_0, y_0), or direction of bias, in a BRW component can, in principle, be estimated along with the other model parameters.

Random walks that involve both correlation and bias require a quantification of the animal's trade-off between maintaining directional persistence and possible turning towards a centre of attraction. Weighted averages of the corresponding directions must of course be directional means. If, for instance ϕ_p, denotes the previous direction, ϕ_b the direction of the bias and η the associated weight of the former in the average, the weighted average ϕ is

$$\phi = \arctan\left(\eta \sin \phi_p + (1 - \eta) \sin \phi_b, \eta \cos \phi_p + (1 - \eta) \cos \phi_b\right).$$

A typical HMM for animal movement will involve some combination of CRWs, BRWs and BCRWs, each of these being allocated to a different state of the underlying Markov chain. Figure 18.1 displays four movement trajectories simulated from HMMs with different components, illustrating the considerable flexibility of these models. Although the incorporation of BRWs and BCRWs is conceptually straightforward, most of the movement models of HMM type that have been considered in the literature involve only CRW components; an exception is McClintock *et al.* (2012). Most of the models that have been considered to date assume two distinct CRW patterns, associated with the different states of an underlying two-state Markov chain. The two states of the corresponding models are often associated with the animal being either 'encamped'/'resting' (with mostly short step lengths and many turns) or 'exploring'/'migrating' (with, on average, longer step lengths

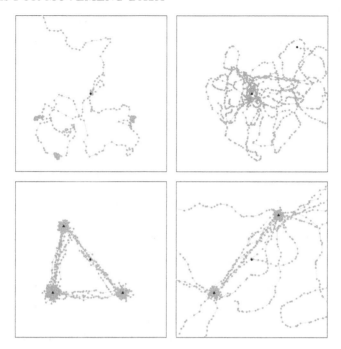

Figure 18.1 *Simulated animal movement trajectories. Clockwise from top left: two-state HMM comprising two CRWs, two-state HMM comprising CRW & BCRW, three-state HMM comprising three BRWs, three-state HMM comprising CRW and two BRWs (in each case filled black circles are initial positions and filled triangles are centres of attraction).*

and more directed movement, as expressed by smaller turning angles); see, for example, Morales *et al.* (2004). This kind of labelling of the states should, however, be made with caution; it is generally accepted that these states merely provide convenient proxies of an animal's actual behavioural state.

HMMs for animal movement data typically involve the assumption of contemporaneous conditional independence. That is, conditional on the current state, the step length and turning angle are assumed to be independent. This greatly facilitates the modelling of the data, yet does not impose overly restrictive assumptions. As we have seen in Section 9.4.2, the two component series will be (unconditionally) dependent. For example, in an encamped state an animal will, on average, make relatively short steps and many turns, while in an exploratory state an animal will, on average, make longer steps and fewer turns. Hence the states induce dependence between the component series, in this example

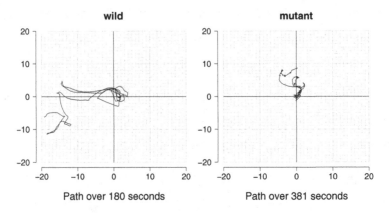

Figure 18.2 *Paths of two larvae of* Drosophila melanogaster.

leading to negative dependence between step length and turning angle; for an illustration, see Figure 18.3 below.

In practice it is then usually of interest to model the state transition probabilities, and hence state occupancy, as functions of environmental covariates, which can provide insights into how animals respond to their environment. For example, Patterson *et al.* (2009) found that southern bluefin tuna were more likely to occupy a 'resident' state, as opposed to a 'migratory' state, when water temperature was high. In other words, the movement distances tended to be greater in cooler waters. Morales *et al.* (2004) found that elk tended to be in an 'encamped' state when they were in open habitat (agricultural fields and opened forest). Corresponding analyses of animal movements can conveniently be carried out by practitioners using the **R** package `moveHMM` (Michelot *et al.*, 2015).

18.4 Case study I: A basic HMM for *Drosophila* movement

For the fruit fly *Drosophila melanogaster*, it is thought that locomotion can be largely summarized by the distribution of speed and direction change in each of two episodic states: 'forward peristalsis' (linear movement) and 'head swinging and turning' (Suster *et al.*, 2003). During linear movement, larvae maintain a high speed and a low direction change, in contrast to the low speed and high direction change characteristic of turning episodes. Given that the larvae apparently alternate thus between two states, an HMM in which both speed and turning rate are modelled according to two underlying states might be appropriate for describing the pattern of larval locomotion. By way of illustration we shall examine the movements of two of the experimental subjects of Suster (2000) (one wild larva, one mutant) whose positions were recorded once per second. The paths taken by the larvae are displayed in Figure 18.2.

We begin our analysis of the bivariate time series of speed and turning angles for *Drosophila* by examining a scatter plot of these quantities for each of the two subjects (top half of Figure 18.3). A smooth of these points is roughly horizontal, but the funnel shape of the plot in each case is conspicuous. We therefore plot also the speeds and absolute changes of direction, and we find, as one might expect from the funnel shape, that a smooth now has a clear downward slope.

We have also plotted in each of these figures a smooth of 10 000 points generated from the three-state model we describe later in this section. In only one of the four plots do the two lines differ appreciably, that of speed and absolute turning angle for the wild subject (lower left); there the line based on the model is the higher one. We defer further comment on the models to later in this section.

The structure of the HMMs fitted to this bivariate series is as follows. Conditional on the underlying state, the speed at time t and the turning angle at time t are assumed to be independent, with the former having a gamma distribution and the latter a von Mises distribution. A similar m-state model involves m CRWs as described in Section 18.3.2. In the three-state model there are in all 18 parameters to be estimated: six transition probabilities, two parameters for each of the three gamma distributions, and two parameters for each of the three von Mises distributions; more generally, there are $m^2 + 3m$ parameters for an m-state model of this kind.

These and other models for the *Drosophila* data were quite difficult to fit in that it was easy to become trapped at a local optimum of the log-likelihood which was not the global optimum.

Table 18.1 compares the two- and three-state models on the basis of likelihood, AIC and BIC, and indicates that AIC and BIC select three states, both for the wild subject and for the mutant. Figure 18.4 compares the sample and three-state model ACFs for absolute turning angle and for speed; for the absolute turning angle the model ACF was estimated by simulation, and for the speed it was computed by using the results of Exercise 3 of Chapter 2. Figure 18.5 depicts the marginal and state-dependent distributions of the three-state models. It is clear from the figure that the model for the wild subject is very different from that for the mutant, as regards both speed and change of direction (turning angle). In particular, the turning angle of the wild subject shows much higher concentration.

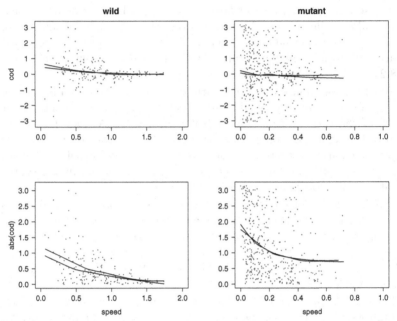

Figure 18.3 *Each panel shows the observations (turning angle, here abbreviated to c.o.d. for change for direction, against speed in the top panels, and absolute turning angles against speed in the bottom panels). In each case there are two nonparametric regression lines computed by the* **R** *function* loess*, one smoothing the observations and the other smoothing the 10 000 realizations generated from the fitted model.*

Table 18.1 *Comparison of two- and three-state bivariate HMMs for speed and direction change in two larvae of* Drosophila.

Subject	No. of states	No. of parameters	$-l$	AIC	BIC
wild	2	10	193.1711	406.3	438.3
	3	18	166.8110	**369.6**	**427.1**
mutant	2	10	332.3693	684.7	724.2
	3	18	303.7659	**643.5**	**714.5**

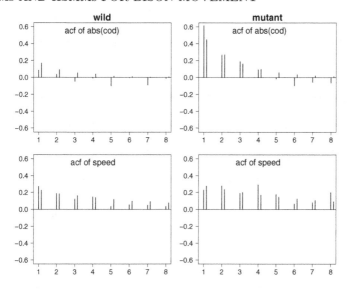

Figure 18.4 *Sample and three-state-HMM ACFs of absolute turning angle (top panels) and speed (bottom panels). The ACFs for the absolute turning angle under the model were computed by simulation of a series of length 50 000.*

18.5 Case study II: HMMs and HSMMs for bison movement

In Langrock, King *et al.* (2012), the movements of nine bison were analysed. Here we consider the locations of only one of those animals, recorded with GPS radio collars between October 2005 and April 2006 in Prince Albert National Park, Saskatchewan, Canada. Observations were made every 3 hours. From the locations we calculated turning angles (in radians) between successive movement directions, and the associated Euclidean step lengths (in kilometres).

Focusing on the case of two states, we initially fitted the same type of HMM as used in the fruit fly example. That is, with contemporaneous conditional independence assumed, a gamma distribution was considered for the step lengths, and a von Mises distribution for the turning angles. Figure 18.6 displays the marginal and state-dependent distributions of the fitted model. The shapes of the fitted gamma distributions differ substantially from those observed for the fruit fly, which is due to the large proportion of very small step lengths observed for the bison. In particular, neither of the fitted gamma distributions has a mode distinct from zero. However, the model identifies roughly the same general pattern that we saw for the fruit flies, with one state involving frequent reversals and mostly short steps, and the other state involving many fewer turns

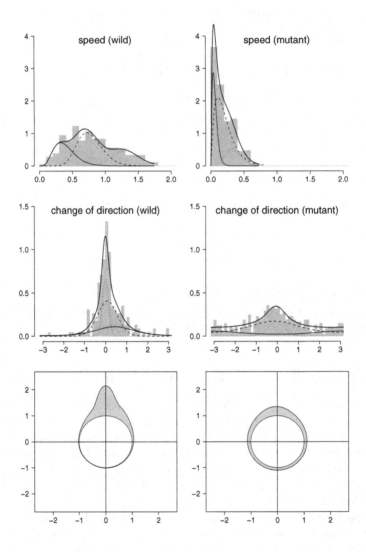

Figure 18.5 *Three-state (gamma–von Mises) HMMs for speed and turning angle in* Drosophila melanogaster: *wild subject (left) and mutant (right). The top panels show the marginal and the state-dependent (gamma) distributions for the speed, as well as a histogram of the speeds. The middle panels show the marginal and the state-dependent (von Mises) distributions for the turning angle, and a histogram of the turning angles. The bottom panels show the fitted marginal densities for turning angle.*

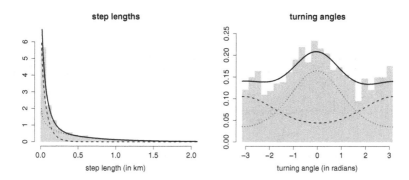

Figure 18.6 *Two-state (gamma–von Mises) HMM fitted to data on step lengths and turning angles observed for a bison. The left panel shows the marginal and the state-dependent (gamma) distributions for the step lengths, as well as a histogram of the step lengths. The right panel shows the marginal and the state-dependent (von Mises) distributions for the turning angles, as well as a histogram of the turning angles.*

and longer steps on average. In the fruit fly example, there was an additional intermediate state. The results are similar also to those obtained for elk in Morales *et al.* (2004), where the states were associated with 'encamped' and 'exploratory' behaviour. We will use this terminology below to refer to the two states. In the bison example, the encamped state is occupied about 45% of the time, according to the stationary distribution implied by the fitted model.

In order to check the appropriateness of geometric dwell-time distributions, we also fitted a stationary HSMM with shifted negative binomial distributions for the dwell times, as defined in (12.9). The sizes of the two state aggregates were $m_1 = 15$ and $m_2 = 8$. Since in the case $k_1 = k_2 = 1$ the shifted negative binomial distributions degenerate to geometric, the HMM is nested in this model. As values of the AIC we obtained 3718.23 and 3708.9 for the HMM and the HSMM, respectively. The transition probability matrix of the HMM is

$$\begin{pmatrix} 0.699 & 0.301 \\ 0.244 & 0.756 \end{pmatrix},$$

and Table 18.2 displays the parameters of the state-dependent distributions in the two models, which are very similar. Figure 18.7 displays the fitted dwell-time distributions for both HMM and HSMM. For the HSMM, the mode of the fitted dwell-time distribution in the exploratory state is 3, which shows deviation from the geometric distribution. In the

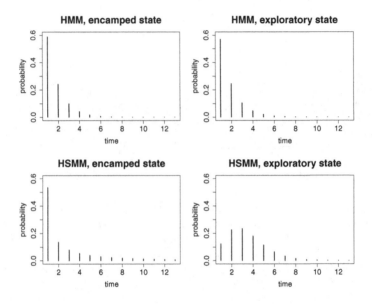

Figure 18.7 *Bison models: dwell-time distributions in the HMM and HSMM.*

Table 18.2 *Bison models: parameters of gamma and von Mises state-dependent distributions. The order of the parameters, from left to right, is: shape and scale parameters of the gamma distribution, directional mean and concentration of the von Mises.*

| | Encamped state | | | | Exploratory state | | | |
	gamma		von Mises		gamma		von Mises	
HMM	0.857	0.082	−3.101	0.437	0.721	0.711	−0.038	0.773
HSMM	0.826	0.091	−3.073	0.361	0.761	0.715	−0.029	0.804

encamped state of the HSMM, the fitted dwell-time distribution has a slightly heavier tail than does its HMM counterpart.

18.6 Case study III: Mixed HMMs for woodpecker movement

In this third case study, the movement of 97 family groups of woodpeckers is analysed by using mixed HMMs (see Chapter 13). Here we consider a subset of the data analysed by McKellar *et al.* (2015), involving one time series of locations in the horizontal plane for each of the 97 groups of birds observed in the study. (In McKellar *et al.* (2015), multiple time series for each group were considered.) Each group of birds was followed by a single observer equipped with a GPS set to record the geographic coordinates of its location automatically, once per minute. During observation sessions, woodpeckers moved from tree to tree, and although the group did not always occupy a single tree, birds remained in the same general areas and regularly made contact calls. The shortest series of those considered comprises only 15 locations, and the longest series 125 locations.

We consider the same type of multi-state random walk (i.e. HMM) as used in the fruit fly and bison case studies. A (state-dependent) von Mises distribution was used to model the turning angles. The groups sometimes remained stationary for some time, so there is a considerable proportion of zero step lengths in the observations. The step lengths were therefore assumed to have been generated by a (state-dependent) zero-inflated gamma distribution; an additional point mass, on zero, is included. For simplicity we again consider only the case of two states. Since this study involves longitudinal data, we considered, in addition to a baseline model in which parameters were assumed to be common to all component series, mixed HMMs with random effects in the state process (see Chapter 13), in order to allow for potentially heterogeneous movement patterns across woodpecker groups. We took the *component-specific* state transition probabilities to be

$$\mathbf{\Gamma}^{(k)} = \begin{pmatrix} \gamma_{11}^{(k)} & \gamma_{12}^{(k)} \\ \gamma_{21}^{(k)} & \gamma_{22}^{(k)} \end{pmatrix} = \begin{pmatrix} 1 - \varepsilon_1^{(k)} & \varepsilon_1^{(k)} \\ \varepsilon_2^{(k)} & 1 - \varepsilon_2^{(k)} \end{pmatrix},$$

with $k = 1, \ldots, 97$ indicating the component series, and the random variables $\varepsilon_1^{(k)}$ and $\varepsilon_2^{(k)}$ ($k = 1, \ldots, 97$) taking values in $[0, 1]$. We implemented three different specifications of the random effects $\varepsilon_i^{(k)}$, as described in Sections 13.3.1 and 13.3.2:

- independent continuous random effects, with $\varepsilon_i^{(k)}$ (for $i = 1, 2$) having a beta distribution (see p. 192);

- independent discrete random effects, with $\varepsilon_i^{(k)}$ taking one of q_i possible values (see p. 193);

- dependent discrete random effects (i.e. a single bivariate distribution, on q possible values) for $(\varepsilon_1^{(k)}, \varepsilon_2^{(k)})$ (see p. 194), allowing for dependence between the two random effects.

For any realization of $\mathbf{\Gamma}^{(k)}$, the associated stationary distribution was used as the initial distribution of the Markov chain. For comparison purposes, we also consider a baseline model in which the transition probabilities are assumed to be constant across component series. Table 18.3 displays the numbers of parameters, log-likelihood values and AIC values for the models considered. For the models with discrete random

Table 18.3 *Number of parameters (k), log-likelihood values and AIC values for the four models fitted to the woodpecker movement data.*

Model	Description	k	$-l$	AIC
1	no random effects	12	29 564.16	59 152.32
2	independent beta-distributed random effects	14	29 553.72	**59 135.44**
3	independent discrete random effects, $q_1 = q_2 = 2$	16	29 551.74	59 135.48
4	dependent discrete random effects, $q = 2$	15	29 554.93	59 139.86

effects, only the models with the optimal number of possible values for the random effects (optimal by the AIC) are listed. The AIC value is lowest for the model with beta-distributed random effects, although it is only marginally lower than that of the model with independent discrete-valued random effects and $q_1 = q_2 = 2$. Fitting the former model took several hours, whereas the latter took only a few minutes.

Figure 18.8 shows the marginal and state-dependent distributions of Model 1, the model without random effects. The model again identifies the same general pattern as we saw for the fruit flies and the bison, with one state involving short steps and many turns, and the other state involving longer steps and a higher directional persistence. Here we follow McKellar *et al.* (2015) and refer to these as the 'resting' and the 'foraging' state, respectively. According to the stationary distribution implied by the fitted baseline model, the resting state is occupied about 51% of the time. In the step-length distributions, the point masses on zero were estimated as 0.065 and 0.001 in the resting and foraging states, respectively.

It is striking that the 10 state-dependent parameters hardly differ between models. We quote only those from Model 1. In the resting state, the non-zero step lengths are gamma distributed with shape parameter 2.62 and scale parameter 2.31, zero step length has probability 0.065, and the turning angle has a von Mises distribution with directional mean -3.04 and concentration 0.317. In the foraging state, the distributions are: gamma $(1.97, 14.18)$, mass on zero 0.001, and von Mises $(-0.034, 1.39)$. Table 18.4 presents the transition parameters of the models. Figure 18.9 displays the fitted beta and (independent) discrete random-effects distributions. The point of this example is to illustrate the modelling of heterogeneity in longitudinal-data settings, and we do not enter here into a discussion of what the source of the heterogeneity might be. In

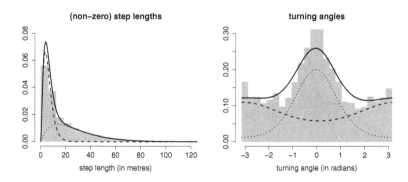

Figure 18.8 *Two-state basic HMM (gamma–von Mises, no random effects) fitted to data on step lengths and turning angles observed for 97 woodpecker groups. The left panel shows the marginal and the state-dependent (gamma) distributions for the non-zero step lengths, as well as a histogram of the non-zero step lengths. The right panel shows the marginal and the state-dependent (von Mises) distributions for the turning angles, as well as a histogram of the turning angles.*

Table 18.4 *Woodpecker movement: transition parameters for the four models.*

Model	
1	$\mathbf{\Gamma} = \begin{pmatrix} 0.815 & 0.185 \\ 0.196 & 0.804 \end{pmatrix}$
2	$\gamma_{12} \sim \text{Beta}(10.91, 43.86), \gamma_{21} \sim \text{Beta}(7.01, 26.41).$
3	$\Pr(\gamma_{12} = 0.123) = 0.314, \Pr(\gamma_{12} = 0.235) = 0.686,$ $\Pr(\gamma_{21} = 0.146) = 0.559, \Pr(\gamma_{21} = 0.295) = 0.441.$
4	$\Pr\big((\gamma_{12}, \gamma_{21}) = (0.269, 0.129)\big) = 0.277,$ $\Pr\big((\gamma_{12}, \gamma_{21}) = (0.173, 0.238)\big) = 0.723.$

practice, the heterogeneity is often not of interest, and is allowed for in the model only to ensure that the inference on other model components is valid (e.g. those that describe the effect of covariates on the transition probabilities). However, the quantification of variability across component series can sometimes also be of interest in itself.

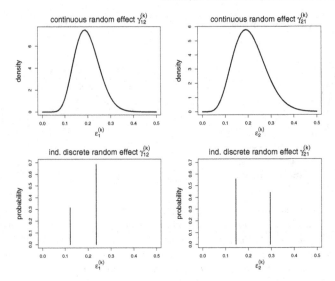

Figure 18.9 *Woodpecker example: Fitted beta (top panels) and independent discrete (bottom panels) random-effects distributions. The left panels give the distribution of the probability of switching from resting to foraging, and the right panels that of switching from foraging to resting.*

Exercises

1. Let μ be the directional mean of the circular random variable Θ, as defined on p. 228. Show that $\mathrm{E}\sin(\Theta - \mu) = 0$.

2. Let X have a von Mises distribution with parameters μ and κ. Show that

$$\mathrm{E}\cos(n(X - \mu)) = I_n(\kappa)/I_0(\kappa) \quad \text{and} \quad \mathrm{E}\sin(n(X - \mu)) = 0,$$

and hence that

$$\mathrm{E}\cos(nX) = \cos(n\mu)I_n(\kappa)/I_0(\kappa)$$

and

$$\mathrm{E}\sin(nX) = \sin(n\mu)I_n(\kappa)/I_0(\kappa).$$

More compactly,

$$\mathrm{E}(e^{inX}) = e^{in\mu}I_n(\kappa)/I_0(\kappa).$$

(Here I_n denotes the modified Bessel function of the first kind of integer order n, given in integral form by

$$I_n(\kappa) \;=\; (2\pi)^{-1}\int_{-\pi}^{\pi}\exp(\kappa\cos x)\cos(nx)\,\mathrm{d}x;$$

see, for example, equation (9.6.19) of Abramowitz *et al.* (1984).)

3. Again, let X have a von Mises distribution with parameters μ and κ. Deduce from the conclusions of Exercise 2 that

$$\tan \mu = E(\sin X)/E(\cos X).$$

4. Let $\{X_t\}$ be a stationary von Mises–HMM on m states, with the ith state-dependent distribution being von Mises (μ_i, κ_i), and with μ denoting the directional mean of X_t.

   Show that

$$E \sin^2(X_t - \mu) = \frac{1}{2}\left(1 - \sum_i \delta_i A_2(\kappa_i) \cos(2(\mu_i - \mu))\right),$$

   where, for positive integers n, $A_n(\kappa)$ is defined by $A_n(\kappa) = I_n(\kappa)/I_0(\kappa)$. Use the following steps:

$$E \sin^2(X_t - \mu) = \sum \delta_i E(\sin^2(X_t - \mu) \mid C_t = i);$$

$$\sin^2 A = (1 - \cos(2A))/2;$$

$$X_t - \mu = X_t - \mu_i + \mu_i - \mu;$$

$$E(\sin(2(X_t - \mu_i)) \mid C_t = i) = 0;$$

   and

$$E(\cos(2(X_t - \mu_i)) \mid C_t = i) = I_2(\kappa_i)/I_0(\kappa_i).$$

   Is it necessary to assume that μ is the directional mean of X_t?

5. Consider the HMM and the HSMM fitted to the bison movement data in Section 18.5. In both models an unrestricted von Mises distribution was used as the state-dependent distribution of the turning angle in each of the two states.

   (a) Suppose instead that in one of the two states of the HMM a uniform distribution on the circle is assumed: $f(x) = (2\pi)^{-1}$ for all $x \in (-\pi, \pi]$. This saves two parameters.

   Fit this modified HMM, and compare it with the HMM described in Section 18.5.

   (b) Fit a similarly modified HSMM, and compare it with the HSMM described in Section 18.5.

   (c) Which of the models considered has the lowest AIC?

CHAPTER 19

Wind direction at Koeberg

19.1 Introduction

South Africa's only nuclear power station is situated at Koeberg on the west coast, about 30 km north of Cape Town. Wind direction, wind speed, rainfall and other meteorological data are collected continuously by the Koeberg weather station with a view to their use in radioactive plume modelling, *inter alia*. Four years of data were made available by the staff of the Koeberg weather station, and this chapter describes an attempt to model the wind direction at Koeberg by means of HMMs.

The wind direction data consist of hourly values of average wind direction over the preceding hour at 35 m above ground level. The period covered is 1 May 1985 to 30 April 1989 inclusive. The average referred to is a vector average, which allows for the circular nature of the data, and is given in degrees. There are in all 35 064 observations; there are no missing values.

19.2 Wind direction classified into 16 categories

Although the hourly averages of wind direction were available in degrees, the first group of models fitted treated the observations as lying in one of the 16 conventional directions N, NNE, . . . , NNW, coded 1 to 16 in that order. This was done in order to illustrate the application of HMMs to time series of categorical observations, a special class of multinomial-like time series.

19.2.1 Three HMMs for hourly averages of wind direction

The first model fitted was a simple multinomial–HMM with two states and no seasonal components, the case $m = 2$ and $q = 16$ of the categorical model described in Section 9.3.2. There are 32 parameters to be estimated: two transition probabilities to specify the Markov chain, and 15 probabilities for each of the two states, subject to the sum of the 15 not exceeding one. The results are as follows. The underlying Markov chain

Table 19.1 *Koeberg wind data (hourly):* $1000\times$ *probabilities of each direction in each state for (from left to right) the simple two-state HMM, the simple three-state HMM, and the two-state HMM with cyclical components.*

		2-state HMM		3-state HMM			Cyclic HMM	
1	N	129	0	148	0	1	127	0
2	NNE	48	0	47	0	16	47	0
3	NE	59	1	16	0	97	57	2
4	ENE	44	26	3	0	148	27	40
5	E	6	50	1	0	132	4	52
6	ESE	1	75	0	0	182	1	76
7	SE	0	177	0	23	388	1	179
8	SSE	0	313	0	426	33	0	317
9	S	1	181	0	257	2	1	183
10	SSW	4	122	2	176	0	7	121
11	SW	34	48	20	89	0	59	26
12	WSW	110	8	111	28	0	114	3
13	W	147	0	169	2	0	145	0
14	WNW	130	0	151	0	0	128	0
15	NW	137	0	159	0	1	135	0
16	NNW	149	0	173	0	0	147	0

has transition probability matrix

$$\begin{pmatrix} 0.964 & 0.036 \\ 0.031 & 0.969 \end{pmatrix}$$

and stationary distribution $(0.462, 0.538)$, and the 16 probabilities associated with each of the two states are displayed in columns 3 and 4 of Table 19.1 and in Figure 19.1. The model identifies two distinct (and locally well-known) weather states: in state 1 the most likely direction is NNW, and in state 2 it is SSE. In both cases the probability falls away sharply on either side of the peak.

Two generalizations of this model were also fitted: firstly, a model based on a three-state Markov chain; and secondly, a model based on a two-state Markov chain but incorporating both a daily cycle and an annual cycle. The three-state HMM has 51 parameters: six to specify the Markov chain, and 15 probabilities associated with each of the states. Essentially this model splits state 2 of the two-state model into two new states, one of which peaks at SSE and the other at SE. The t.p.m. is

$$\begin{pmatrix} 0.957 & 0.030 & 0.013 \\ 0.015 & 0.923 & 0.062 \\ 0.051 & 0.077 & 0.872 \end{pmatrix}$$

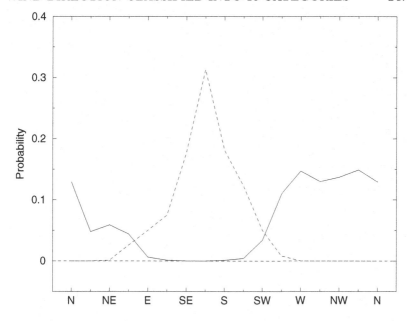

Figure 19.1 *Koeberg wind data (hourly): probabilities of each direction in the simple two-state HMM.*

and the stationary distribution is (0.400, 0.377, 0.223). The 16 probabilities associated with each of the three states are displayed in Figure 19.2 and columns 5–7 of Table 19.1.

As regards the model which adds daily and annual cyclical effects to the simple two-state HMM, it was decided to build these effects into the Markov chain rather than into the state-dependent probabilities. This was because, for a two-state chain, we can model cyclical effects parsimoniously by assuming that the two off-diagonal transition probabilities

$$\Pr(C_t \neq i \mid C_{t-1} = i) = {}_t\gamma_{i,3-i} \quad (i = 1, 2)$$

are given by an appropriate periodic function of t. We assume that logit ${}_t\gamma_{i,3-i}$ is equal to

$$\underbrace{a_i + b_i\cos(2\pi t/24) + c_i\sin(2\pi t/24)}_{\text{daily cycle}} + \underbrace{d_i\cos(2\pi t/8766) + e_i\sin(2\pi t/8766)}_{\text{annual cycle}}$$

(19.1)

for $i = 1, 2$ and $t = 2, 3, \ldots, T$. A similar model for each of the state-dependent probabilities in each of the two states would involve many more parameters. As discussed in Section 10.2.2, the estimation technique has to be modified when the underlying Markov chain is not

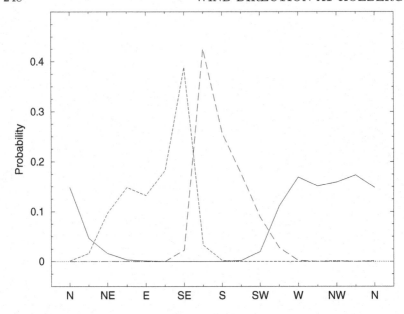

Figure 19.2 *Koeberg wind data (hourly): probabilities of each direction in the three-state HMM.*

assumed to be homogeneous. The estimates in this case were based on the initial state of the Markov chain being state 2. (Conditioning on state 1 yielded a slightly inferior value for the likelihood, and similar parameter estimates.)

The estimated parameters for the two off-diagonal transition probabilities are as follows:

i	a_i	b_i	c_i	d_i	e_i
1	−3.349	0.197	−0.695	−0.208	−0.401
2	−3.523	−0.272	0.801	0.082	−0.089

and the probabilities associated with each state are given in Table 19.1. The general pattern of the state-dependent probabilities is in this case very similar to that of the two-state HMM without cyclical components.

19.2.2 Model comparisons and other possible models

In Table 19.3 the three models described above are compared with each other and with a saturated 16-state Markov chain model, on the basis of AIC and BIC. The transition probabilities defining the Markov chain model were estimated by conditional maximum likelihood, as described in Section 1.3.4, and are displayed in Table 19.2.

Table 19.2 *Koeberg wind data (hourly):* 1000×*transition probability matrix of saturated Markov chain model.*

610	80	22	8	3	1	1	2	4	0	3	3	14	20	37	190
241	346	163	37	15	1	3	3	3	3	9	10	29	37	36	64
56	164	468	134	28	6	9	6	4	6	6	18	32	18	19	25
13	33	163	493	144	32	17	11	11	17	7	12	14	14	10	10
9	7	48	249	363	138	53	27	33	22	11	6	8	7	13	9
4	9	10	51	191	423	178	51	25	23	8	7	4	6	4	4
1	1	3	6	23	141	607	160	36	8	6	3	2	1	1	1
1	1	1	3	5	16	140	717	94	13	5	3	2	1	0	0
4	1	2	4	5	8	25	257	579	77	17	9	5	3	1	0
2	2	2	1	6	5	10	36	239	548	93	41	10	5	2	0
5	2	3	5	3	3	8	12	38	309	397	151	38	12	8	5
4	2	2	3	1	3	2	5	17	56	211	504	149	19	16	7
10	5	5	3	4	1	2	5	4	13	28	178	561	138	30	13
13	5	4	3	3	3	1	1	1	7	8	27	188	494	199	43
31	9	7	4	5	2	1	1	2	1	3	11	43	181	509	190
158	23	9	5	1	1	2	1	0	2	2	4	17	54	162	559

What is striking in Table 19.3 is that the likelihood of the saturated Markov chain model is so much higher than that of the HMMs that the large number of parameters of the Markov chain (240) is virtually irrelevant when comparisons are made by AIC or BIC. It is therefore interesting to compare certain properties of the Markov chain model with the corresponding properties of (say) the simple two-state HMM. The two models lead to almost identical distributions for the direction. However, the conditional distributions differ substantially.

For example, examination of $\Pr(X_{t+1} = 16 \mid X_t = 8)$, where X_t denotes the direction at time t, points to an important difference between the

Table 19.3 *Koeberg wind data (hourly): comparison of four models fitted.*

Model	k	$-l$	AIC	BIC
2-state HMM	32	75 832.1	151 728	151 999
3-state HMM	51	69 525.9	139 154	139 585
2-state HMM with cycles	40	75 658.5*	151 397	151 736
saturated Markov chain	240	48 301.7	**97 083**	**99 115**

*Conditional on state 2 being the initial state

models. For the Markov chain model this probability is zero, since no such transitions were observed. For the HMM it is, in the notation of equation (9.4),

$$\frac{\boldsymbol{\delta\Pi}(8)\boldsymbol{\Gamma\Pi}(16)\mathbf{1}'}{\boldsymbol{\delta\Pi}(8)\mathbf{1}'} = \frac{0.5375 \times 0.3131 \times 0.0310 \times 0.1494}{0.5375 \times 0.3131} = 0.0046.$$

Although small, this probability is not insignificant. The observed number of transitions from SSE (direction 8) was 5899. On the basis of the HMM one would therefore expect about 27 of these transitions to be to NNW. None were observed. Under the HMM, 180° switches in direction are quite possible. Every time the state changes (which happens at any given time point with probability in excess of 0.03), the most likely direction of wind changes by 180°. This is inconsistent with the observed gradual changes in direction; the matrix of observed transition counts is heavily dominated by diagonal and near-diagonal elements (and, because of the circular nature of the categories, elements in the corners furthest from the principal diagonal). Changes through 180° were rare.

We now consider whether a model which allows for higher-order dependence is superior to the (first-order) Markov chain model. A saturated second-order Markov chain model would have an excessive number of parameters, and the Pegram model for (say) a second-order Markov chain cannot reflect the property that, if a transition is made out of a given category, a nearby category is a more likely destination than a distant one.

The Raftery models (also known as MTD models: see p. 22) do not suffer from that disadvantage, and in fact it is very easy to find a Raftery model of lag 2 that is convincingly superior to the first-order Markov chain model. In the notation of Section 1.3.5, take $\mathbf{Q}$ to be the t.p.m. of the first-order Markov chain, as displayed in Table 19.2, and perform a simple line-search to find that value of λ_1 which maximizes the resulting conditional likelihood. This turns out to be 0.925. With these values for $\mathbf{Q}$ and λ_1 as starting values, the conditional likelihood was then maximized with respect to all 241 parameters, subject to the assumption that $0 \leq \lambda_1 \leq 1$. (This assumption makes it unnecessary to impose 16^3 pairs of nonlinear constraints on the maximization, and seems reasonable in the context of hourly wind directions showing a high degree of persistence.) The resulting value for λ_1 is 0.9125, and the resulting matrix $\mathbf{Q}$ does not differ much from its starting value. Table 19.4 shows that the Raftery model is superior to the Markov chain. Berchtold (2001) reports that this model cannot be improved on by another MTD model.

WIND DIRECTION AS A CIRCULAR VARIABLE

Table 19.4 *Koeberg wind data (hourly): comparison of first-order Markov chain with Raftery models of lag 2.*

Model	k	$-l$	AIC	BIC
saturated MC	240	48 301.7	97 083.4	99 115.0
Raftery model with starting values	241	48 087.8*	96 657.5	98 697.6
Raftery model fitted by max. likelihood	241	48 049.8*	**96 581.6**	**98 621.7**

*Conditioned on the first two states

19.3 Wind direction as a circular variable

We now revisit the Koeberg wind direction data, with a view to demonstrating some variations of HMMs that could be applied to data such as these. In particular, we do not confine ourselves here to use of the directions as classified into the 16 points of the compass, and we make use of the corresponding observations of wind speed in several different ways as a covariate in models for the direction, or for change in direction. Some of the models are not particularly successful, but it is not our intention here to confine our attention to models which are in some sense the best, rather to demonstrate the versatility of HMMs and to explore some of the questions which may arise in the fitting of more complex models.

19.3.1 Daily at hour 24: von Mises-HMMs

The analysis of wind direction that we have so far described is based entirely on the data as classified into the 16 directions N, NNE, ..., NNW. Although it illustrates the fitting of models to categorical series, it ignores the fact that the observations of (hourly average) direction are available in degrees, and it ignores the circular nature of the observations.

One of the strengths of the hidden Markov formulation is that almost any kind of data can be accommodated at observation level. Here we can exploit that flexibility by taking the state-dependent distribution from a family of distributions designed for circular data, the von Mises distributions. The probability density function of the von Mises distribution with parameters μ and κ is

$$f(x) = (2\pi I_0(\kappa))^{-1} \exp(\kappa \cos(x - \mu)), \quad \text{for } x \in (-\pi, \pi];$$

see also Sections 9.2 and 18.2.2.

In this section we shall fit to the directions data (in degrees) HMMs which have 1–4 states and von Mises state-dependent distributions. This

Table 19.5 *Koeberg wind data (daily at hour 24): comparison of models fitted to wind direction. m is the number of states in an HMM, k is in general the number of parameters in a model, and equals $m^2 + m$ for all but the Markov chain model. Note that the figures in the $-l$ column for the continuous models have to be adjusted before they can be compared with those for the discretized models.*

Model	m	k	$-l$	$-l$ (adjusted)	AIC	BIC
von Mises–HM	1	2	2581.312	3946.926	7897.9	7908.4
von Mises–HM	2	6	2143.914	3509.528	7031.1	7062.8
von Mises–HM	3	12	2087.133	3452.747	6929.5	6992.9
von Mises–HM	4	20	2034.133	3399.747	**6839.5**	**6945.2**
discretized von M–HM	1	2	3947.152		7898.3	7908.9
discretized von M–HM	2	6	3522.848		7057.7	7089.4
discretized von M–HM	3	12	3464.491		6953.0	7016.4
discretized von M–HM	4	20	3425.703		6891.4	6997.1
saturated Markov chain		240	3236.5		6953.0	8221.9

is perhaps a fairly obvious approach if one has available 'continuous' circular data (e.g. in integer degrees). We do this for the series of length 1461 formed by taking every 24th observation of direction. By taking every 24th observation we are focusing on the hour each day from 23:00 to 24:00, and modelling the average direction over that hour.

It is somewhat less obvious that it is still possible to fit von Mises–HMMs if one has available only the categorized directions; that is a second family of models which we present here. To fit such models, one uses in the likelihood computation not the von Mises density but the integral thereof over the interval corresponding to the category observed. For the purpose of comparison we present also a 16-state saturated Markov chain fitted to the same series.

A minor complication that arises if one wishes to compare the continuous log-likelihoods with the discrete is that one has to bear in mind that the direct use of a density in a continuous likelihood is just a convention. The density should in fact be integrated over the smallest interval that can contain the observation, or the integral approximated by the density multiplied by the appropriate interval length. The effect here is that, if one wishes to compare the continuous log-likelihoods with those relating to the 16 categories, one has to add $1461 \log(2\pi/16) = -1365.614$ to the continuous log-likelihood, or equivalently subtract it from minus the log-likelihood l. The resulting adjusted values of $-l$ appear in the column headed '$-l$ (adjusted)' in Table 19.5.

19.3.2 Modelling hourly change of direction

Given the strong persistence of wind direction which is apparent from the t.p.m. in Table 19.2, the most promising approach, however, seems to be to model the *change* in direction rather then the direction itself. We therefore describe here a variety of models for the change in (hourly average) direction from one hour to the next, both with and without the use of wind speed, lagged 1 hour, as a covariate. Observations of wind speed, in centimetres per second, were available for the same period as the observations of direction. The use of lagged wind speed as a covariate, rather than simultaneous, is motivated by the need for any covariate to be available at the time of forecast; here a 1-hour-ahead forecast is of interest.

19.3.3 Transition probabilities varying with lagged speed

Change of direction, like direction itself, is a circular variable, and first we model it by means of a von Mises–HMM with two, three, or four states. Then we introduce wind speed, lagged 1 hour, into the model as follows.

We use the transformation

$$\gamma_{ij} = \Pr(C_t = j \mid C_{t-1} = i, S_{t-1} = s) = \frac{e^{\tau_{ij}}}{\sum_{k=1}^{m} e^{\tau_{ik}}}, \quad j = 1, 2, \ldots m,$$

where

$$\tau_{ii} = \eta_i s$$

and S_{t-1} is the speed at time $t-1$. Equivalently, the transformation is, for row i,

$$\gamma_{ij} = \frac{e^{\tau_{ij}}}{e^{\eta_i s} + \sum_{k \neq i} e^{\tau_{ik}}}, \quad j \neq i,$$

with γ_{ii} determined by the row-sum constraint. This structure allows the speed at time $t-1$ to influence the probabilities of transition from state i to j between times $t-1$ and t.

In passing, we note that there would be no point in introducing an intercept η_{0i} as follows:

$$\tau_{ii} = \eta_{0i} + \eta_i s.$$

If we were to do so, it would follow that

$$\gamma_{ii} = \frac{e^{\eta_i s}}{e^{\eta_i s} + \sum_{k \neq i} e^{\tau_{ik} - \eta_{0i}}},$$

Table 19.6 *Koeberg wind data (hourly): comparison of von Mises–HM models fitted to* **change** *in wind direction. The covariate, if any, is here used to influence the transition probabilities. The number of states is* m, *and the number of parameters* k.

Covariate	m	k	$-l$	AIC	BIC
–	1	2	21 821.350	43 646.7	43 663.6
–	2	6	8 608.516	17 229.0	17 279.8
–	3	12	6 983.205	13 990.4	14 092.0
–	4	20	6 766.035	13 572.1	13 741.4
speed	2	8	6 868.472	13 752.9	13 820.7
speed	3	15	5 699.676	11 429.4	11 556.3
speed	4	24	5 476.199	11 000.4	11 203.6
$\sqrt{\text{speed}}$	2	8	6 771.533	13 559.1	13 626.8
$\sqrt{\text{speed}}$	3	15	5 595.228	11 220.5	11 347.4
$\sqrt{\text{speed}}$	4	24	5 361.759	**10 771.5**	**10 974.7**

and (for $j \neq i$) that

$$\gamma_{ij} = \frac{e^{\tau_{ij}-\eta_{0i}}}{e^{\eta_{is}} + \sum_{k\neq i} e^{\tau_{ik}-\eta_{0i}}}.$$

In all the corresponding transition probabilities, η_{0i} would be confounded with the parameters τ_{ij} and therefore non-identifiable.

The results are displayed in Table 19.6, along with results for the corresponding models which use the square root of speed, rather than speed itself, as covariate. In these models, the square root of speed is in general more successful as a covariate than is speed.

19.3.4 Concentration parameter varying with lagged speed

A quite different way in which the square root of speed, lagged 1 hour, could be used as a covariate is via the concentration parameter (κ) of the von Mises distributions used here as the state-dependent distributions. The intuition underlying this proposal is that the higher the speed, the more concentrated will be the change in wind direction in the following hour (unless the state changes). One possibility is this: we assume that, given $S_{t-1} = s$, the concentration parameter in state i is

$$\log \kappa_i = \zeta_{i0} + \zeta_{i1}\sqrt{s}. \tag{19.2}$$

However, a model which seems both more successful (as judged by likelihood) and more stable numerically is this. Given $S_{t-1} = s$, let the concentration parameter in state i (not its logarithm) be a linear function

Table 19.7 *Koeberg wind data (hourly): comparison of von Mises–HMMs fitted to change in wind direction. Here the covariate is used to influence the concentration parameter κ, as in equation (19.3). The number of states is m, and the number of parameters k.*

Covariate	m	k	$-l$	AIC	BIC
speed^2	1	3	10256.07	20518.1	20543.5
speed^2	2	8	4761.549	9539.1	9606.8
speed^2	3	15	4125.851	8281.7	8408.7
speed^2	4	24	3975.223	**7998.4**	**8201.6**

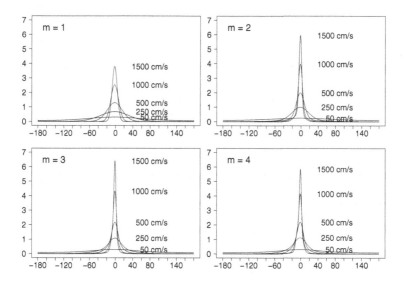

Figure 19.3 *Koeberg wind data (hourly), von Mises–HMMs of form (19.3): marginal distribution, for one to four states and several values of lagged speed, for change of direction (in degrees).*

of the square of the speed:

$$\kappa_i = \zeta_{i0} + \zeta_{i1}s^2. \tag{19.3}$$

To ensure that κ_i is positive we constrain ζ_{i0} and ζ_{i1} to be positive. Table 19.7 presents the log-likelihood values, AIC and BIC for four such models, and Figures 19.3 and 19.4 and Table 19.8 present some details of these models.

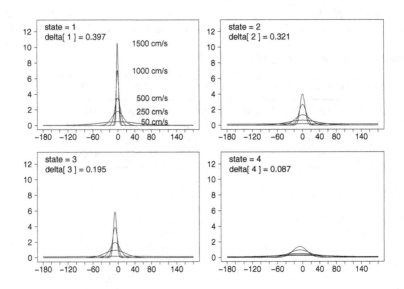

Figure 19.4 *Koeberg wind data (hourly), four-state von Mises–HMM of form (19.3): state-dependent distributions for change of direction (before multiplication by mixing probabilities δ_i).*

Table 19.8 *Koeberg wind data (hourly): parameters of four-state von Mises– HMM fitted to change in wind direction. Here the covariate is used to influence the concentration parameter κ, as in equation (19.3).*

$$\Gamma = \begin{pmatrix} 0.755 & 0.163 & 0.080 & 0.003 \\ 0.182 & 0.707 & 0.045 & 0.006 \\ 0.185 & 0.000 & 0.722 & 0.093 \\ 0.031 & 0.341 & 0.095 & 0.533 \end{pmatrix}$$

i	1	2	3	4
δ_i	0.397	0.321	0.195	0.087
μ_i	−0.0132	0.0037	−0.1273	−0.1179
ζ_{i0}	0.917	0.000	0.000	0.564
ζ_{i1}	31.01×10^{-5}	4.48×10^{-5}	9.61×10^{-5}	0.53×10^{-5}

We have in this section considered a multiplicity of models for wind direction, but more variations yet would be possible. One could, for instance, allow the state-dependent location parameters μ_i (not the concentrations) of the change in direction to depend on the speed lagged 1 hour, or allow the concentrations to depend on the speed in different ways in the different states; there is no reason *a priori* why a single relation such as (19.3) should apply to all m states. In any application one would have to be guided very much by the intended use of the model.

Exercises

1. (a) Write an **R** function to generate a series of observations from an m-state categorical HMM with q categories, as described in Section 9.3.2. (Hint: modify the code in Section A.1.5.)

   (b) Write functions to estimate the parameters of such a model. (Modify the code in Sections A.1.1–A.1.4.)

   (c) Generate a long series of observations using your code from (a), and estimate the parameters using your code from (b).

2. Use your code from Exercise 1(b) to fit two- and three-state models to the categorized wind directions data, and compare your models with those described in Section 19.2.1.

3. Generalize Exercise 1 to handle multinomial–HMMs, as described in Section 9.3.1, rather than merely categorical.

## CHAPTER 20

Models for financial series

Because that's where the money is.

attributed to Willie Sutton

20.1 Financial series I: A multivariate normal–HMM for returns on four shares

Time series of share returns generally display kurtosis in excess of 3 and little or no autocorrelation, even though the series of absolute or squared returns do display autocorrelation. These – and some other – phenomena are so widespread that they are termed 'stylized facts'. For discussion, see, for instance, Rydén *et al.* (1998), Cont (2001) or Bulla and Bulla (2007).

We describe a multivariate normal–HMM for the daily returns on the following shares on the Frankfurt stock exchange, for the 501 trading days from 4 March 2003 to 17 February 2005, both inclusive: Allianz (ALV), Deutsche Bank (DBK), DaimlerChrysler (DCX) and Siemens (SIE). The returns were computed as $100 \log(s_t/s_{t-1})$, where s_t is the price on day t. We assume that the four returns on a given day have a multivariate normal distribution selected from one of m such distributions by the underlying Markov chain. We do not, however, assume contemporaneous conditional independence; indeed, we impose no structure on the variance–covariance matrices of the multivariate normal state-dependent distributions. As usual, we assume longitudinal conditional independence. That is, given the states occupied, the four returns on day t are distributed independently of the returns on all other days. Such a model for p returns has in all $m\{m - 1 + p(p+3)/2\}$ parameters: $m^2 - m$ to determine $\boldsymbol{\Gamma}$, mp state-dependent means, and $mp(p + 1)/2$ state-dependent variances and covariances.

In order to use an unconstrained optimizer such as nlm to maximize the likelihood, it is necessary to parametrize the model in terms of unconstrained parameters. The transition probabilities can be transformed via the generalized logit transform (see p. 52). The means of the normal distributions are unconstrained. The m variance–covariance matrices are not, but they can conveniently be reparametrized as follows. A variance–covariance matrix $\boldsymbol{\Sigma}$ can be written as $\boldsymbol{\Sigma} = \mathbf{T}'\mathbf{T}$, where $\mathbf{T}$ is its unique

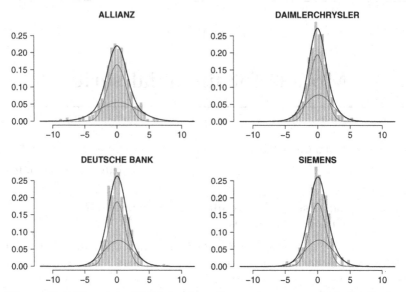

Figure 20.1 *Two-state HMM for four share returns: marginal distributions compared with histograms of the returns. The state-dependent distributions, multiplied by their mixing probabilities, are also shown.*

Cholesky upper-triangular 'square root' with positive diagonal elements. With a log-transform of the diagonal elements one has an unconstrained parametrization which ensures that $\boldsymbol{\Sigma}$ is symmetric and positive definite. This is the 'log-Cholesky parametrization' of Pinheiro and Bates (1996).

Two- and three-state multivariate normal–HMMs were fitted to the data. In contrast to the BIC, the AIC favours the 48-parameter three-state model ($-l = 3208.915$) over the 30-parameter two-state model ($-l = 3255.157$). The parameter estimates of the latter, which is depicted in Figure 20.1, include

$$\boldsymbol{\Gamma} = \left(\begin{array}{cc} 0.918 & 0.082 \\ 0.059 & 0.941 \end{array} \right), \quad \boldsymbol{\delta} = (0.418, 0.582).$$

The state-dependent means are:

	ALV	DBK	DCX	SIE
state 1	0.172	0.154	0.252	0.212
state 2	0.034	−0.021	0.021	0.011

$\boldsymbol{\Sigma}$, for states 1 and 2 respectively, is given by:

$$\begin{pmatrix} 9.162 & 3.981 & 4.496 & 4.542 \\ 3.981 & 4.602 & 2.947 & 3.153 \\ 4.496 & 2.947 & 4.819 & 3.317 \\ 4.542 & 3.153 & 3.317 & 4.625 \end{pmatrix},$$

$$\begin{pmatrix} 1.985 & 1.250 & 1.273 & 1.373 \\ 1.250 & 1.425 & 1.046 & 1.085 \\ 1.273 & 1.046 & 1.527 & 1.150 \\ 1.373 & 1.085 & 1.150 & 1.590 \end{pmatrix}.$$

The standard deviations are:

	ALV	DBK	DCX	SIE
state 1	3.027	2.145	2.195	2.151
state 2	1.409	1.194	1.236	1.261

The correlation matrices for states 1 and 2 respectively are:

$$\begin{pmatrix} 1.000 & 0.613 & 0.677 & 0.698 \\ 0.613 & 1.000 & 0.626 & 0.683 \\ 0.677 & 0.626 & 1.000 & 0.703 \\ 0.698 & 0.683 & 0.703 & 1.000 \end{pmatrix},$$

$$\begin{pmatrix} 1.000 & 0.743 & 0.731 & 0.773 \\ 0.743 & 1.000 & 0.709 & 0.721 \\ 0.731 & 0.709 & 1.000 & 0.738 \\ 0.773 & 0.721 & 0.738 & 1.000 \end{pmatrix}.$$

One notable feature of this model is the clear ordering by volatility that emerges for these shares. The two states are one of high and one of low volatility, for all four shares. There was also a clear ordering by volatility in the three-state model; the states were of high, intermediate and low volatility, for all four shares.

A property of the two-state model, but not the three-state model, is that the ranges of correlations within each state are narrow and non-overlapping: approximately 0.61–0.70 in state 1 (the more volatile state) and 0.71–0.77 in state 2. If these two correlation matrices had been approximately equal, we could usefully have examined the model in which they are assumed to be equal. It is also straightforward to fit a model in which all the correlations are equal within each state, or one in which they are equal both within and across states.

We now compare properties of the two-state model with the corresponding sample values. For all four shares, means and standard deviations agree extremely well. The same is not true of kurtosis. The model kurtoses, although above 3, are lower than the sample values (4.51, 3.97, 3.94, and 3.82, as opposed to 6.19, 4.69, 5.43, and 4.81). Much of this

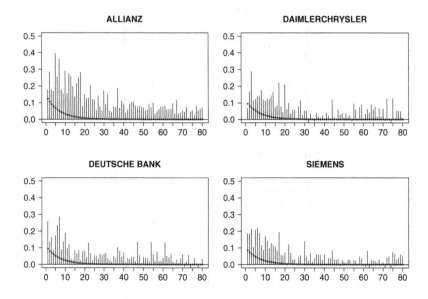

Figure 20.2 *Four share returns: sample ACF of squared returns, plus ACF of squared returns for two-state HMM (smooth curve).*

discrepancy is due to a few extreme returns. The kurtosis of Allianz, in particular, is much reduced if one caps absolute returns at 6%.

The model cross-correlations (at lag 0) between the individual returns match the sample values very well. The model ACFs of the individual return series are very low, as are the sample ACFs. But the ACFs of squared returns do not match well; see Figure 20.2. This is similar to the finding of Rydén *et al.* (1998) that HMMs could not capture the sample ACF behaviour of absolute returns. Bulla and Bulla (2007) discuss the use of hidden *semi*-Markov models rather than HMMs in order to represent the squared-return behaviour better.

20.2 Financial series II: Discrete state-space stochastic volatility models

Stochastic volatility (SV) models – which, like HMMs, are special cases of state-space models (SSMs) – provide an alternative way of modelling time series of share returns. These models have attracted much attention in the finance literature; see, for example, Shephard (1996) and Omori *et al.* (2007). SV models capture several of the stylized facts described above, including positive autocorrelation of squared returns, zero autocorrelation of the unsquared returns, and kurtosis in excess of

3. These models have precisely the same dependence structure as HMMs, but assume a continuous-valued state process, thus allowing for gradual changes in the volatility, which in the context of financial time series may be easier to interpret. The methods described in Chapter 11 can be used in order to fit SV models to data.

The first form of the model which we consider here, and the best-known, is the (Gaussian) SV model without leverage.

20.2.1 Stochastic volatility models without leverage

In such a model, the returns x_t on an asset $(t = 1, 2, \ldots, T)$ satisfy

$$x_t = \varepsilon_t \beta \exp(g_t/2), \quad g_{t+1} = \phi g_t + \eta_t, \quad (20.1)$$

where $|\phi| < 1$ and $\{\varepsilon_t\}$ and $\{\eta_t\}$ are independent sequences of independent normal random variables, with ε_t standard normal and $\eta_t \sim N(0, \sigma^2)$. (This is the 'alternative parametrization' in terms of β indicated by Shephard (1996, p. 22), although we use ϕ and σ where Shephard uses γ_1 and σ_η.) We shall later allow ε_t and η_t to be correlated, in which case the model will be said to accommodate leverage. This model has three parameters, β, ϕ and σ, and for identifiability reasons we constrain β to be positive. The model is simple, in that it has only three parameters, and some properties of the model are straightforward to establish. For instance, if $\{g_t\}$ is stationary, it follows that $g_t \sim N(0, \sigma^2/(1 - \phi^2))$. We shall assume that $\{g_t\}$ is indeed stationary.

But the principal difficulty in implementing the model in practice has been that direct evaluation of the likelihood of the observations – or forecast distributions – does not seem possible. Much ingenuity has been applied in the derivation and application of (*inter alia*) MCMC methods of estimating the parameters even in the case in which ε_t and η_t are assumed to be independent. By 1996 there was already a 'vast literature on fitting SV models' (Shephard, 1996, p. 35).

In Chapter 11 we discussed the idea of discretizing the state space of an SSM into a sufficiently large number of states to provide a good approximation to continuity, in order to allow for the application of well-established HMM techniques to estimate the parameters of the model. We now apply that idea to the estimation of SV models, as an alternative to the estimation methods described, for example, by Shephard (1996) and Kim, Shephard and Chib (1998). In particular, the ease of computation of the HMM likelihood enables one to fit models by numerical maximization of the likelihood, and to compute forecast distributions. The transition probability matrix of the Markov chain is structured in such a way that an increase in the number of states does not increase

the number of parameters; only the three parameters already listed are
used.

For the SV model described above, the likelihood of the observations
$\mathbf{x}^{(T)}$ is given by the T-fold multiple integral

$$p(\mathbf{x}^{(T)}) = \int \cdots \int p(\mathbf{x}^{(T)}, \mathbf{g}^{(T)}) \, d\mathbf{g}^{(T)},$$

the integrand of which can be decomposed as

$$p(g_1) \prod_{t=2}^{T} p(g_t \mid g_{t-1}) \prod_{t=1}^{T} p(x_t \mid g_t)$$

$$= \quad p(g_1) \, p(x_1 \mid g_1) \prod_{t=2}^{T} p(g_t \mid g_{t-1}) \, p(x_t \mid g_t),$$

where p is used as a general symbol for a density. As we have seen in
Chapter 11, by discretizing the range of g_t sufficiently finely, we can
approximate this integral by a multiple sum, and then evaluate that
sum recursively by the methods of HMMs.

In detail, we proceed as follows. Let the range of possible g_t-values be
split into m intervals $B_i = (b_{i-1}, b_i)$, $i = 1, 2, \ldots, m$. We denote by b_i^*
a representative point in B_i, such as the midpoint. The resulting T-fold
sum approximating the likelihood is

$$\sum_{i_1=1}^{m} \sum_{i_2=1}^{m} \cdots \sum_{i_T=1}^{m} p(g_1 \in B_{i_1}) \, n(x_1; 0, \beta^2 \exp(b_{i_1}^*))$$

$$\times \quad \prod_{t=2}^{T} \Pr(g_t \in B_{i_t} \mid g_{t-1} = b_{i_{t-1}}^*) \, n(x_t; 0, \beta^2 \exp(b_{i_t}^*)). \quad (20.2)$$

Here $n(\cdot; \mu, \sigma^2)$ is used to denote a normal density with mean μ and variance σ^2. With N denoting the corresponding (cumulative) distribution
function, and Φ the standard normal distribution function, we can write
the transition probability $\Pr(g_t \in B_{i_t} \mid g_{t-1} = b_{i_{t-1}}^*)$, corresponding to
a change of the log-volatility process from $b_{i_{t-1}}^*$ at time $t-1$ to some
value in B_{i_t} at time t, as

$$\gamma_{ij} = N(b_j; \phi b_i^*, \sigma^2) - N(b_{j-1}; \phi b_i^*, \sigma^2)$$

$$= \Phi((b_j - \phi b_i^*)/\sigma) - \Phi((b_{j-1} - \phi b_i^*)/\sigma).$$

We are in effect saying that, although g_t has (conditional) mean ϕg_{t-1},
we shall proceed as if that mean were $\phi \times$ the midpoint of the interval
in which g_{t-1} falls. Note that here we have applied a slightly different
numerical integration method from that presented in Chapter 11; the
difference between the two approaches is described in detail by Langrock,
MacDonald and Zucchini (2012). In practice, it does not matter which

of the two integration techniques is used, provided that m is sufficiently large.

Assuming stationarity of the approximating (m-state) HMM, expression (20.2) gives us the usual matrix expression for the likelihood:

$$\boldsymbol{\delta}\mathbf{P}(x_1)\boldsymbol{\Gamma}\mathbf{P}(x_2)\cdots\boldsymbol{\Gamma}\mathbf{P}(x_T)\mathbf{1}' = \boldsymbol{\delta}\boldsymbol{\Gamma}\mathbf{P}(x_1)\boldsymbol{\Gamma}\mathbf{P}(x_2)\cdots\boldsymbol{\Gamma}\mathbf{P}(x_T)\mathbf{1}',$$

$\boldsymbol{\delta} = \mathbf{1}(\mathbf{I} - \boldsymbol{\Gamma} + \mathbf{U})^{-1}$ being the stationary distribution implied by the t.p.m. $\boldsymbol{\Gamma} = (\gamma_{ij})$, and $\mathbf{P}(x_t)$ being the diagonal matrix with ith diagonal element equal to the normal density $n(x_t; 0, \beta^2 \exp(b_i^*))$.

It is then a routine matter to evaluate this approximate likelihood and maximize it with respect to the three parameters of the model, transformed to allow for the constraints $\beta > 0$, $|\phi| < 1$, and $\sigma > 0$. However, in practice one also has to decide what m is, what range of g_t-values to allow for (i.e. b_0 to b_m), whether the intervals (b_{i-1}, b_i) should (for example) be of equal length, and which value g_i^* to take as representative of (b_{i-1}, b_i). The accuracy of the approximation depends chiefly on the choice of m and the range (b_0, b_m). The range should of course cover the effective support of the p.d.f. of g_t, which one is approximating by the stationary distribution $\boldsymbol{\delta}$. Increasing the range beyond that erodes the accuracy. The estimate of $\boldsymbol{\delta}$ gives a simple indication of whether the selected range can be reduced to improve the accuracy for a given m. In our experience, m around 50 is usually adequate. The likelihood of an SV model can typically be calculated in a fraction of a second, even for T in the thousands and (say) $m = 150$, a value which renders the approximation excellent. In the applications we describe in Sections 20.2.2 and 20.2.4, we have used equally spaced intervals represented by their midpoints.

20.2.2 Application: FTSE 100 returns

The basic SV model was fitted to the daily returns on the FTSE 100 index for the period from 2 April 1986 to 6 May 1994. The return on day t was calculated as $100 \log(s_t/s_{t-1})$, where s_t is the index value at the close of day t. Using the technique described above, with a range of values of m and with g_t-values from -2.5 to 2.5, we fitted models to FTSE 100 returns for that period. Table 20.1 summarizes the findings and shows that the parameter estimates are reasonably stable by $m = 50$.

20.2.3 Stochastic volatility models with leverage

In the SV model without leverage, as described above, there is no feedback from past returns to the (log-)volatility process. But this is known to be too restrictive for many financial time series; see Yu (2005) and

Table 20.1 *SV model without leverage fitted to FTSE 100 returns, 2 April 1986 to 6 May 1994.*

m	β	ϕ	σ
20	0.866	0.964	0.144
50	0.866	0.963	0.160
100	0.866	0.963	0.162
500	0.866	0.963	0.163

the references therein. We therefore now discuss a second, more general, form of the model.

As before, the returns x_t on an asset $(t = 1, 2, \ldots, T)$ satisfy

$$x_t = \varepsilon_t \beta \exp(g_t/2), \quad g_{t+1} = \phi g_t + \eta_t, \qquad (20.3)$$

where $|\phi| < 1$, but now ε_t and η_t are permitted to be dependent. More specifically, for all t,

$$\begin{pmatrix} \varepsilon_t \\ \eta_t \end{pmatrix} \sim N(\mathbf{0}, \mathbf{\Sigma}) \quad \text{with } \mathbf{\Sigma} = \begin{pmatrix} 1 & \rho\sigma \\ \rho\sigma & \sigma^2 \end{pmatrix},$$

and the vectors $\begin{pmatrix} \varepsilon_t \\ \eta_t \end{pmatrix}$ are assumed independent.

This model has four parameters: β, ϕ, σ and ρ. We constrain β to be positive. If one does so, the model is equivalent to model (2) of Omori *et al.* (2007), via $\beta = \exp(\mu/2)$ and $g_t = h_t - \mu$. It is also – with notational differences – the discrete-time ASV1 model (2.2) of Yu (2005); note that our η_t corresponds to Yu's v_{t+1}. Yu contrasts the ASV1 specification with that of Jacquier, Polson and Rossi (2004), and concludes that the ASV1 version is preferable. The parameter ρ is said to measure leverage; it is expected to be negative, in order to accommodate an increase in volatility following a drop in returns. The structure of the model is conveniently represented by the directed graph in Figure 20.3.

The likelihood of the observations $\mathbf{x}^{(T)}$ is in this case also given by the multiple integral

$$\int \cdots \int p(\mathbf{x}^{(T)}, \mathbf{g}^{(T)}) \, d\mathbf{g}^{(T)}, \qquad (20.4)$$

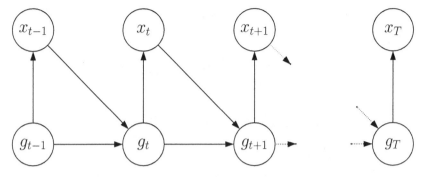

Figure 20.3 *Directed graph of stochastic volatility model with leverage.*

but here the integrand is decomposed as

$$p(g_1) \prod_{t=2}^{T} p(g_t \mid g_{t-1}, x_{t-1}) \prod_{t=1}^{T} p(x_t \mid g_t)$$

$$= p(g_1) p(x_1 \mid g_1) \prod_{t=2}^{T} p(g_t \mid g_{t-1}, x_{t-1}) p(x_t \mid g_t).$$

Notice the dependence of $p(g_t \mid g_{t-1}, x_{t-1})$ on x_{t-1}.

We can approximate this integral as well by discretizing the range of g_t and evaluating the sum recursively. But in order to approximate this likelihood, we need the conditional distribution of g_{t+1} given g_t and x_t, or equivalently – since $x_t = \varepsilon_t \beta \exp(g_t/2)$ – given g_t and ε_t.

This is the distribution of η_t given ε_t, except that ϕg_t is added to the mean. The distribution of η_t given ε_t is $\mathrm{N}(\rho\sigma\varepsilon_t, \sigma^2(1 - \rho^2))$; hence that of g_{t+1}, given g_t and ε_t, is

$$\mathrm{N}(\phi g_t + \rho\sigma\varepsilon_t, \sigma^2(1 - \rho^2)).$$

Writing this distribution in terms of the observations x_t shows that the required conditional distribution of g_{t+1} is

$$g_{t+1} \sim \mathrm{N}\left(\phi g_t + \frac{\rho\sigma x_t}{\beta \exp(g_t/2)}, \sigma^2(1 - \rho^2)\right).$$

Thus, in the approximate likelihood, $\Pr(g_{i_t} \in B_{i_t} \mid g_{t-1} = b^*_{i_{t-1}}, x_{t-1} =$

x) is given by

$$\gamma_{ij}(x) = N(b_j; \mu(b_i^*, x), \sigma^2(1 - \rho^2)) - N(b_{j-1}; \mu(b_i^*, x), \sigma^2(1 - \rho^2))$$

$$= \Phi\left(\frac{b_j - \mu(b_i^*, x)}{\sigma\sqrt{1 - \rho^2}}\right) - \Phi\left(\frac{b_{j-1} - \mu(b_i^*, x)}{\sigma\sqrt{1 - \rho^2}}\right),$$

where we define

$$\mu(b_i^*, x) = \phi b_i^* + \frac{\rho\sigma x}{\beta\exp(b_i^*/2)}.$$

In this case, therefore, the approximate likelihood is

$$\boldsymbol{\delta}\mathbf{P}(x_1)\boldsymbol{\Gamma}(x_1)\mathbf{P}(x_2)\boldsymbol{\Gamma}(x_2)\cdots\boldsymbol{\Gamma}(x_{T-1})\mathbf{P}(x_T)\mathbf{1}',$$

with $\boldsymbol{\delta}$ here determined by the distribution assumed for g_1, and $\boldsymbol{\Gamma}(x_t)$ the matrix with entries $\gamma_{ij}(x_t)$. This raises the question of what (if any) distribution for g_1 will produce stationarity in the process $\{g_t\}$, that is, in the case of the model with leverage. In the case of the model without leverage, that distribution is normal with mean zero and variance $\sigma^2/(1 - \phi^2)$, and it is not unreasonable to conjecture that that may also be the case here.

Here we know that, given g_t and ε_t,

$$g_{t+1} \sim N(\phi g_t + \rho\sigma\varepsilon_t, \sigma^2(1 - \rho^2)).$$

That is,

$$g_{t+1} = \phi g_t + \rho\sigma\varepsilon_t + \sigma\sqrt{1 - \rho^2}Z,$$

where Z is independently standard normal. If it is assumed that $g_t \sim N(0, \sigma^2/(1 - \phi^2))$, it follows that g_{t+1} is (unconditionally) normal with mean zero and variance given by $\phi^2(\sigma^2/(1 - \phi^2)) + \rho^2\sigma^2 + \sigma^2(1 - \rho^2) = \sigma^2/(1 - \phi^2)$. In the 'with leverage' case also, therefore, the stationary distribution for $\{g_t\}$ is $N(0, \sigma^2/(1 - \phi^2))$, and so that is the distribution we assume for g_1. For the vector $\boldsymbol{\delta}$ in the approximate likelihood we use the distribution of g_1, discretized into the intervals B_i.

20.2.4 Application: TOPIX returns

Using the daily opening prices of the Tokyo Stock Price Index (TOPIX) for the 1233 days from 30 December 1997 to 30 December 2002, both inclusive, we get a series of 1232 daily returns x_t with the summary statistics displayed in Table 20.2. This summary agrees completely with the statistics given by Omori *et al.* (2007) in their Table 4, although they state that they used closing prices. (We compute daily returns as $100\log(s_t/s_{t-1})$, where s_t is the price on day t, and we use the estimator with denominator T as the sample variance of T observations. The data were downloaded from http://index.onvista.de on 5 July 2006.)

STOCHASTIC VOLATILITY MODELS   269

Table 20.2 *Summary statistics of TOPIX returns, calculated from opening prices from 30 December 1997 to 30 December 2002, both inclusive.*

no. of returns	mean	std. dev.	max.	min.	+	−
1232	−0.02547	1.28394	5.37492	−5.68188	602	630

Table 20.3 *SV model with leverage fitted to TOPIX returns, opening prices 30 December 1997 to 30 December 2002, both inclusive, plus comparable figures from Table 5 of Omori* et al.

m	β	ϕ	σ	ρ
5	1.199	0.854	0.192	−0.609
10	1.206	0.935	0.129	−0.551
25	1.205	0.949	0.135	−0.399
50	1.205	0.949	0.140	−0.383
100	1.205	0.949	0.142	−0.379
200	1.205	0.949	0.142	−0.378

From Table 5 of Omori et al.*:*
posterior mean,

'unweighted'	1.2056	0.9511	0.1343	−0.3617
'weighted'	1.2052	0.9512	0.1341	−0.3578
95% interval	$(1.089, 1.318)$	$(0.908, 0.980)$	$(0.091, 0.193)$	$(−0.593, −0.107)$

Parametric bootstrap applied to the model with $m = 50$:

95% CI:	$(1.099, 1.293)$	$(0.826, 0.973)$	$(0.078, 0.262)$	$(−0.657, −0.050)$
correlations:				
β		0.105	−0.171	0.004
ϕ			−0.752	−0.192
σ				0.324

Again using a range of values of m, and with g_t ranging from $−2$ to 2, we have fitted an SV model with leverage to these data. The results are summarized in Table 20.3. All parameter estimates are reasonably stable by $m = 50$ and, as expected, the estimate of the leverage parameter ρ is consistently negative. The results for $m = 50$ agree well with the two sets of point estimates ('unweighted' and 'weighted') presented by Omori *et al.* in their Table 5, and for all four parameters our estimate is close to the middle of their 95% interval.

We also applied the parametric bootstrap, with bootstrap sample size

500, to our model with $m = 50$, in order to estimate the (percentile) bootstrap 95% confidence limits, the standard errors and the correlations of our estimators. The standard error of $\hat{\rho}$ (0.160) is the highest, and there is a high negative correlation (-0.752) between the estimators of ϕ and σ. Overall, our findings are consistent with the corresponding figures in Table 5 of Omori *et al.*

20.2.5 Non-standard stochastic volatility models

Alternative forms of feedback from the returns to the volatility process in an SV model are possible, and indeed have been proposed and implemented in a discrete state-space setting by Rossi and Gallo (2006). They attribute to Calvet and Fisher (2001, 2004) the first attempt to build accessible SV models based on high-dimensional regime switching. Fridman and Harris (1998) also present and implement a model involving such feedback; see their equation (7). Their route to the evaluation, and hence maximization, of the likelihood is via recursive numerical evaluation of the multiple integral which gives the likelihood; see equation (20.4) above.

In general, a convenient feature of the discretization strategy, which renders the HMM machinery applicable to SV models, is that it readily extends to a variety of non-standard formulations of SV models. Firstly, one can replace the normal distribution assumed for ε_t by some other distribution. In the approximate likelihood of the model without leverage, given in equation (20.2), the terms corresponding to the conditional density $p(x_t \mid g_t)$, $t = 1, \ldots, T$, are then simply

$$p(x_t \mid g_t = b_{i_t}) = \beta \exp(-b_{i_t}/2) p_\varepsilon(x_t \beta \exp(-b_{i_t}/2)),$$

where p_ε is the probability density function of the variables ε_t. SV models that assume the variables ε_t to follow a t distribution are particularly popular, since the basic model given in (20.1) tends to underestimate the probability of relatively extreme returns (Chib, Nardari and Shephard, 2002). The degrees of freedom of the t distribution can be estimated along with the other parameters, ϕ, β and σ. Secondly, it is straightforward to add additional parameters to the model, for example to capture a trend in the series of returns by estimating the mean of the distribution of ε_t, or to impose a lower bound on the variance of the observed process (see below). Thirdly, one can use as the volatility process some Markov process other than a Gaussian AR(1). A number of such extensions are discussed and applied by Langrock, MacDonald and Zucchini (2012). Here we give one concrete example.

20.2.6 A model with a mixture AR(1) volatility process

We consider a model formulation that generalizes the basic SV model in three ways. First, we replace the observation equation in the model (20.1) by

$$x_t = \varepsilon_t(\beta \exp(g_t/2) + \xi).$$

The additional parameter $\xi(\geq 0)$ is plausible on the grounds that some baseline volatility is always present. Of course, the model with $\xi = 0$ is nested in the model with $\xi \geq 0$. Second, we assume for ε_t a t distribution with ν degrees of freedom. The basic model, which assumes a standard normal distribution for ε_t, is a limiting case (as $\nu \to \infty$).

The third extension considered here concerns the log-volatility process. In the basic model, the persistence parameter ϕ is typically estimated to be only a little below 1. Consequently the sample paths of the standard Gaussian AR(1) model used to describe the log-volatility process are relatively smooth, in the sense that σ^2, the conditional variance of g_t given g_{t-1}, is much smaller than the marginal variance of g_t, namely $\sigma^2/(1-\phi^2)$. On the other hand, it can often be observed that the level of volatility is maintained for an extended period and then changes abruptly (cf. Figure 20.4). One possible way to model this type of behaviour is to generalize the model for the log-volatility process $\{g_t\}$ by using not one but a mixture of two Gaussian AR(1) processes. Assume that (given g_t) g_{t+1} is distributed either $N(\phi_1 g_t, \sigma_1^2)$ (with probability α) or $N(\phi_2 g_t, \sigma_2^2)$ (with probability $1 - \alpha$). Equivalently,

$$g_{t+1} = \begin{cases} \phi_1 g_t + \sigma_1 \eta_t & \text{with probability } \alpha, \\ \phi_2 g_t + \sigma_2 \eta_t & \text{with probability } 1 - \alpha, \end{cases}$$

with the innovations η_t being independent standard normal. This model allows for abrupt changes in the log-volatility process and thus offers additional flexibility. The basic model is again a nested special case (e.g. with $\alpha = 1$). The approximate likelihood of this non-standard SV model, after discretization of the log-volatility process, can easily be deduced in the same manner as for the basic SV model without leverage (see Exercise 2(a)). One could also consider using a mixture with more than two AR(1) components, but that generalization is not pursued here.

Wong and Li (2000) give the following necessary and sufficient condition for second-order stationarity of $\{g_t\}$:

$$\alpha \phi_1^2 + (1 - \alpha)\phi_2^2 < 1.$$

Note that it is possible for one of the AR(1) processes to be 'explosive' (e.g. $\phi_2 = 1.4$) without destroying the second-order stationarity of the

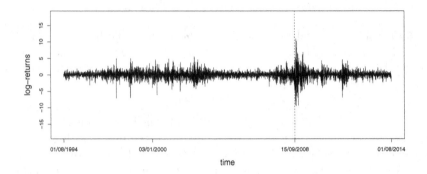

Figure 20.4 *Time series of log-returns on the S&P 500 index closing prices, 1 August 1994 to 1 August 2014. The dashed vertical line indicates the day of the Lehman Brothers bankruptcy filing.*

Table 20.4 *Parameter estimates for the non-standard SV model fitted to S&P 500 returns, from 1 August 1994 to 1 August 2014.*

β	ν	ξ	ϕ_1	ϕ_2	σ_1	σ_2	α
0.549	13.39	0.291	0.975	1.149	0.016	0.584	0.909

mixed process. The mean of $\{g_t\}$, if stationary, is 0 and the variance

$$\sigma_g^2 = \frac{\alpha\sigma_1^2 + (1-\alpha)\sigma_2^2}{1-(\alpha\phi_1^2 + (1-\alpha)\phi_2^2)}. \tag{20.5}$$

20.2.7 Application: S&P 500 returns

We fitted the non-standard SV model described in the previous subsection to the daily log-returns on the S&P 500 index for the period from 1 August 1994 to 1 August 2014; the time series is displayed in Figure 20.4. In the HMM-based estimation approach, we used $m = 200$ and restricted the range of g_t-values to the interval $[-7.5, 7.5]$.

Table 20.4 gives the parameter estimates. We find that one of the AR(1) components of the fitted model is non-stationary, although the mixture, and hence also the model for the observations, is stationary. Furthermore, the estimated baseline volatility, ξ, is clearly distinct from zero. Finally, the estimate of the degrees of freedom of the t conditional distribution indicates deviation from normality.

For comparison purposes, we also fitted the basic SV model given in equation (20.1). The basic model yielded an AIC value of 14 023.66 (BIC 14 043.23), whereas the non-standard model, with baseline volatility, t

conditional distribution and a mixture of two AR(1) components in the
log-volatility, yielded an AIC of 13 990.3 (BIC 14 042.5). By both AIC
and BIC, the non-standard model is preferable.

Exercises

1. For the basic SV model discussed in Section 20.2.1, assumed stationary:

   (a) show that the variance of $\{x_t\}$ is $\beta^2 \exp(0.5\sigma^2/(1-\phi^2))$;
   (b) show that the kurtosis of $\{x_t\}$ is $3\exp(\sigma^2/(1-\phi^2))$.

2. For the non-standard SV model discussed in Section 20.2.6:

   (a) give $\Pr(g_t \in B_{i_t} \mid g_{t-1} = b^*_{i_{t-1}})$, with the discretization exactly as
       described in Section 20.2.1;
   (b) use the formula

   $$\mathrm{Var}(g_{t+1}) = \mathrm{E}\big(\mathrm{Var}(g_{t+1} \mid g_t)\big) + \mathrm{Var}\big(\mathrm{E}(g_{t+1} \mid g_t)\big)$$

   to prove equation (20.5).

CHAPTER 21

Births at Edendale Hospital

21.1 Introduction

Haines, Munoz and van Gelderen (1989) have described the fitting of Gaussian ARIMA models to various discrete-valued time series related to births occurring during a 16-year period at Edendale Hospital in Natal,[*] South Africa. The data include monthly totals of mothers delivered and deliveries by various methods at the Obstetrics Unit of that hospital in the period from February 1970 to January 1986 inclusive. Except for the total deliveries, Haines *et al.* modelled only the final 8 years' observations, as we do here.

21.2 Models for the proportion Caesarean

One of the series considered by Haines *et al.*, to which they fitted two models, was the number of deliveries by Caesarean section. From their models they drew the conclusions (in respect of this particular series) that there is a clear dependence of present observations on past, and that there is a clear linear upward trend. In this section we describe the fitting of (discrete-valued) HM and Markov regression models to this series – Markov regression models, that is, in the sense in which that term is used by Zeger and Qaqish (1988). These models are of course rather different from those fitted by Haines *et al.* in that theirs, being based on the normal distribution, are continuous-valued. Furthermore, the discrete-valued models make it possible to model the proportion (as opposed to the number) of Caesareans performed in each month. Of the models proposed here, one type is 'observation-driven' and the other 'parameter-driven'; see Cox (1981) for these terms. The most important conclusion drawn from the discrete-valued models, and one which the Gaussian ARIMA models did not provide, is that there is a strong upward time trend in the proportion of the deliveries that are by Caesarean section.

The two models that Haines *et al.* fitted to the time series of Caesareans performed, and that they found to fit very well, may be described as follows. Let Z_t denote the number of Caesareans in month t (February 1978 being month 1 and t running up to 96), and let the process $\{a_t\}$ be

* Now KwaZulu-Natal.

275

Gaussian white noise, that is, uncorrelated random shocks distributed normally with mean zero and common variance σ_a^2. The first model fitted is the ARIMA$(0, 1, 2)$ model with constant term:

$$\nabla Z_t = \mu + a_t - \theta_1 a_{t-1} - \theta_2 a_{t-2}. \tag{21.1}$$

The maximum likelihood estimates of the parameters, with associated standard errors, are $\widehat{\mu} = 1.02 \pm 0.39$, $\widehat{\theta}_1 = 0.443 \pm 0.097$, $\widehat{\theta}_2 = 0.393 \pm 0.097$ and $\widehat{\sigma}_a^2 = 449.25$. The second model is an AR(1) with linear trend:

$$Z_t = \beta_0 + \beta_1 t + \phi Z_{t-1} + a_t, \tag{21.2}$$

with parameter estimates as follows: $\widehat{\beta}_0 = 120.2 \pm 8.2$, $\widehat{\beta}_1 = 1.14 \pm 0.15$, $\widehat{\phi} = 0.493 \pm 0.092$ and $\widehat{\sigma}_a^2 = 426.52$.

Both of these models, (21.1) and (21.2), provide support for the conclusion of Haines et al. that there is a dependence of present observations on past, and a linear upward trend. Furthermore, the models are nonseasonal; the Box–Jenkins methodology used found no seasonality in the Caesareans series. The X-11-ARIMA seasonal adjustment method employed in an earlier study (Munoz, Haines and van Gelderen, 1987) did, however, find some evidence, albeit weak, of a seasonal pattern in the Caesareans series similar to a pattern that was observed in the 'total deliveries' series. (This latter series shows marked seasonality, with a peak in September, and in Haines et al. (1989) it is modelled by the seasonal ARIMA model $(0, 1, 1) \times (0, 1, 1)_{12}$.)

It is of some interest to model the proportion, rather than the number, of Caesareans in each month. (See Figure 21.1 for a plot of this proportion for the years 1978–1986.)

It could be the case, for instance, that any trend, dependence or seasonality apparently present in the number of Caesareans is largely inherited from the total deliveries, and a constant proportion Caesarean is an adequate model. On the other hand, it could be the case that there is an upward trend in the proportion of the deliveries that are by Caesarean, and this accounts at least partially for the upward trend in the number of Caesareans. The two classes of model that we discuss in this section condition on the total number of deliveries in each month and seek to describe the principal features of the proportion Caesarean.

Now let n_t denote the total number of deliveries in month t. A model for $\{Z_t\}$ which allows for trend, dependence on previous observations and seasonality, in the proportion Caesarean, is as follows. Suppose that, conditional on the history $\mathbf{Z}^{(t-1)} = \{Z_s : s \le t-1\}$, Z_t is distributed

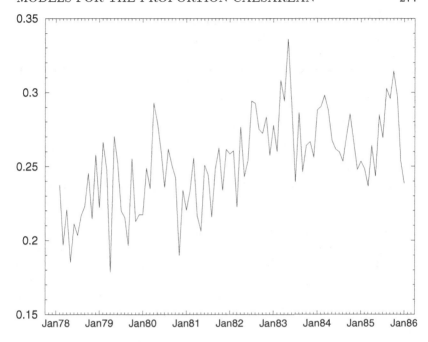

Figure 21.1 *Edendale births: monthly numbers of deliveries by Caesarean section, as a proportion of all deliveries, February 1978–January 1986.*

binomially with parameters n_t and $_tp$, where, for some positive integer q,

$$
\begin{aligned}
\operatorname{logit}_t p \;=\; & \alpha_1 + \alpha_2 t + \beta_1(Z_{t-1}/n_{t-1}) + \beta_2(Z_{t-2}/n_{t-2}) + \cdots \\
& + \beta_q(Z_{t-q}/n_{t-q}) + \gamma_1 \sin(2\pi t/12) + \gamma_2 \cos(2\pi t/12).
\end{aligned}
\tag{21.3}
$$

This is, in the terminology of Zeger and Qaqish (1988), a Markov regression model, and generalizes the model described by Cox (1981) as an 'observation-driven linear logistic autoregression', in that it incorporates trend and seasonality and is based on a binomial rather than a Bernoulli distribution. It is observation-driven in the sense that the distribution of the observation at a given time is specified in terms of the observations at earlier times. Clearly it is possible to add further terms to the above expression for $\operatorname{logit}_t p$ to allow for the effect of any further covariates, for example, the number or proportion of deliveries by various instrumental techniques.

It does not seem possible to formulate an unconditional maximum likelihood procedure to estimate the parameters α_1, α_2, β_1, ..., β_q, γ_1 and γ_2 of model (21.3). It is, however, straightforward to compute estimates of these parameters by maximizing a conditional likelihood. If, for instance, no observations earlier than Z_{t-1} appear in the model, the product

$$\prod_{t=1}^{96} \binom{n_t}{z_t} {}_t p^{z_t} (1 - {}_t p)^{n_t - z_t}$$

is the likelihood of $\{Z_t : t = 1, \ldots, 96\}$, conditional on Z_0. Maximizing this product with respect to $\alpha_1, \alpha_2, \beta_1, \gamma_1$ and γ_2 yields estimates of these parameters, and can be accomplished simply by performing a logistic regression of Z_t on t, Z_{t-1}/n_{t-1}, $\sin(2\pi t/12)$ and $\cos(2\pi t/12)$.

In the search for a suitable model the following explanatory variables were considered: t (i.e. the time in months, with February 1978 as month 1), t^2, the proportion Caesarean lagged 1, 2, 3 or 12 months, sinusoidal terms at the annual frequency (as in (21.3)), the calendar month, the proportion and number of deliveries by forceps or vacuum extraction, and the proportion and number of breech births. Model selection was performed by means of AIC and BIC. The **R** function used, glm, does not provide l, the log-likelihood, but it does provide the deviance, from which the log-likelihood can be computed (McCullagh and Nelder, 1989, p. 33). Here the maximum log-likelihood of a full model is -321.0402, from which it follows that $-l = 321.04 + \frac{1}{2} \times$ deviance.

The best models found with between one and four explanatory variables (other than the constant term) are listed in Table 21.1, as well as several other models that may be of interest. These other models are: the model with constant term only; the model with Z_{t-1}/n_{t-1}, the previous proportion Caesarean, as the only explanatory variable; and two models which replace the number of forceps deliveries as covariate by the proportion of instrumental deliveries. The last two models were included because the proportion of instrumental deliveries may seem a more sensible explanatory variable than the one it replaces; it will be observed, however, that the models involving the number of forceps deliveries are preferred by AIC and BIC. (By 'instrumental deliveries' we mean those which are either by forceps or by vacuum extraction.) Note, however, that both AIC and BIC indicate that one could still gain by inclusion of a fifth explanatory variable in the model.

The strongest conclusion we may draw from these models is that there is indeed a marked upward time trend in the proportion Caesarean. Secondly, there is positive dependence on the proportion Caesarean in the previous month. The negative association with the number (or proportion) of forceps deliveries is not surprising in view of the fact that de-

Table 21.1 *Edendale births: models fitted to the logit of the proportion Caesarean.*

Explanatory variables	Coefficients	Deviance	AIC	BIC
constant	−1.253	208.92	855.00	860.13
t (time in months)	0.003439			
constant	−1.594	191.70	839.78	847.47
t	0.002372			
previous proportion Caesarean	1.554			
constant	−1.445	183.63	833.71	843.97
t	0.001536			
previous proportion Caesarean	1.409			
no. forceps deliveries in month t	−0.002208			
constant	−1.446	175.60	**827.68**	**840.50**
t	0.001422			
previous proportion Caesarean	1.431			
no. forceps deliveries in month t	−0.002393			
October indicator	0.08962			
constant	−1.073	324.05	968.13	970.69
constant	−1.813	224.89	870.97	876.10
previous proportion Caesarean	2.899			
constant	−1.505	188.32	838.40	848.66
t	0.002060			
previous proportion Caesarean	1.561			
proportion instrumental				
deliveries in month t	−0.7507			
constant	−1.528	182.36	834.44	847.26
t	0.002056			
previous proportion Caesarean	1.590			
proportion instrumental				
deliveries in month t	−0.6654			
October indicator	0.07721			

Table 21.2 *Edendale births: the three two-state binomial–HMMs fitted to the proportion Caesarean. (The time in months is denoted by t, and February 1978 is month 1.)*

logit $_t p_i$	$-l$	AIC	BIC	γ_{12}	γ_{21}	α_1	α_2	β_1	β_2
α_i	420.322	848.6	858.9	0.059	0.086	−1.184	−0.960	–	–
$\alpha_i + \beta t$	402.350	**814.7**	**827.5**	0.162	0.262	−1.317	−1.140	0.003298	
$\alpha_i + \beta_i t$	402.314	816.6	832.0	0.161	0.257	−1.315	−1.150	0.003253	0.003473

livery by Caesarean and delivery by forceps are in some circumstances alternative techniques. As regards seasonality, the only possible seasonal pattern found in the proportion Caesarean is the positive 'October effect'. Among the calendar months only October stood out as having some explanatory power. As can be seen from Table 21.1, the indicator variable specifying whether the month was October was included in the 'best' set of four explanatory variables. A possible reason for an October effect is as follows. An overdue mother is more likely than others to give birth by Caesarean, and the proportion of overdue mothers may well be highest in October because the peak in total deliveries occurs in September.

Since the main conclusion emerging from the above logit-linear models is that there is a marked upward time trend in the proportion Caesarean, it is of interest also to fit HMMs with and without time trend. The HMMs we use in this application have two states and are defined as follows. Suppose $\{C_t\}$ is a stationary homogeneous Markov chain on state space $\{1, 2\}$, with transition probability matrix

$$\boldsymbol{\Gamma} = \begin{pmatrix} 1 - \gamma_{12} & \gamma_{12} \\ \gamma_{21} & 1 - \gamma_{21} \end{pmatrix}.$$

Suppose also that, conditional on the Markov chain, Z_t has a binomial distribution with parameters n_t and p_i, where $C_t = i$. A model without time trend assumes that p_1 and p_2 are constants and has four parameters. One possible model which allows p_i to depend on t has logit $_t p_i = \alpha_i + \beta t$ and has five parameters. A more general model yet, with six parameters, has logit $_t p_i = \alpha_i + \beta_i t$.

Maximization of the likelihood of the last 8 years' observations gives the three models appearing in Table 21.2 along with their associated log-likelihood, AIC and BIC values. It may be seen that, of the three models, that with a single time-trend parameter and a total of five parameters achieves the smallest AIC and BIC values.

In detail, that model is as follows, t being the time in months and February 1978 being month 1:

$$\Gamma = \begin{pmatrix} 0.838 & 0.162 \\ 0.262 & 0.738 \end{pmatrix},$$

$$\text{logit}_t p_1 = -1.317 + 0.003298t,$$

$$\text{logit}_t p_2 = -1.140 + 0.003298t.$$

The model can be described as consisting of a Markov chain with two moderately persistent states, along with their associated time-dependent probabilities of delivery being by Caesarean, the (upward) time trend being the same, on a logit scale, for the two states. State 1 is more likely than state 2, because the stationary distribution is $(0.618, 0.382)$, and has associated with it a lower probability of delivery being by Caesarean. For state 1 that probability increases from 0.212 in month 1 to 0.269 in month 96, and for state 2 from 0.243 to 0.305. The corresponding unconditional probability increases from 0.224 to 0.283. It may or may not be possible to interpret the states as (for instance) non-busy and busy periods in the Obstetrics Unit of the hospital, but without further information (e.g. on staffing levels) such an interpretation would be speculative.

It is true, however, that other models can reasonably be considered. One possibility, suggested by inspection of Figure 21.1, is that the proportion Caesarean was constant until January 1981, then increased linearly to a new level in about January 1983. Although we do not pursue such a model here, it is possible to fit an HMM incorporating this feature. (Models with change-points are discussed and used in Chapter 22.)

If one wishes to use the chosen model to forecast the proportion Caesarean at time 97 for a given number of deliveries, what is needed is the one-step-ahead forecast distribution of Z_{97}, that is, the distribution of Z_{97} conditional on $Z_1, \ldots, Z_{96}$. This is given by the likelihood of Z_1, $\ldots, Z_{97}$ divided by that of $Z_1, \ldots, Z_{96}$. More generally, the k-step-ahead forecast distribution, the conditional probability that $Z_{96+k} = z$, is given by a ratio of likelihoods, as described in Section 5.3.

The difference in likelihood between the HMMs with and without time trend is convincing evidence of an upward trend in the proportion Caesarean, and confirms the main conclusion drawn above from the logit-linear models. Although Haines *et al.* concluded that there is an upward trend in the number of Caesareans, it does not seem possible to draw any conclusion about the proportion Caesarean from their models, or from any other ARIMA models.

It is of interest also to compare the fit of the five-parameter HMM to the data with that of the logistic autoregressive models. Here it should be noted that the HMM produces a lower value of $-l$ (402.3) than does the logistic autoregressive model with four explanatory variables (408.8), without making use of the additional information used by the logistic autoregression. It does not use z_0, the number of Caesareans in January 1978, nor does it use information on forceps deliveries or the calendar month. This application demonstrates that HMMs are simple but useful tools for examining dependence on covariates (such as time) in the presence of serial dependence.

21.3 Models for the total number of deliveries

If one wishes to project the number of Caesareans, however, a model for the proportion Caesarean is not sufficient; one also needs a model for the total number of deliveries, which is a series of unbounded counts. The model of Haines *et al.* for the total deliveries was the seasonal ARIMA model $(0, 1, 1) \times (0, 1, 1)_{12}$ without constant term. For this series (unlike the others) they used all 16 years' data to fit the model, but we continue here to model only the final 8 years' observations.

First, five two-state Poisson–HMMs. were fitted to the monthly totals of deliveries (depicted in Figure 21.2). One was a model without covariates, that is,

$$\log{}_t\lambda_i = a_i. \tag{21.4}$$

The other four models for $\log{}_t\lambda_i$ were of the following forms:

$$\log{}_t\lambda_i = a_i + bt; \tag{21.5}$$
$$\log{}_t\lambda_i = a_i + bt + c\cos(2\pi t/12) + d\sin(2\pi t/12); \tag{21.6}$$
$$\log{}_t\lambda_i = a_i + bt + c\cos(2\pi t/12) + d\sin(2\pi t/12) + fn_{t-1}; \tag{21.7}$$
$$\log{}_t\lambda_i = a_i + bt + c\cos(2\pi t/12) + d\sin(2\pi t/12) + fn_{t-1} + gn_{t-2}. \tag{21.8}$$

Models (21.7) and (21.8) are examples of the incorporation of extra dependencies at observation level as described in Section 10.4; that is, they do not assume conditional independence of the observations $\{n_t\}$ given the Markov chain. But this does not significantly complicate the likelihood evaluation, as the state-dependent probabilities can just treat n_{t-1} and n_{t-2} in the same way as any other covariate. The models fitted are summarized in Table 21.3.

It may be seen that, of these five models, model (21.7), with a total of eight parameters, achieves the smallest AIC and BIC values. That model is as follows:

$$\log{}_t\lambda_i = 5.945/6.069 + 0.002538t - 0.05253\cos(2\pi t/12)$$
$$- 0.02287\sin(2\pi t/12) + 0.0005862n_{t-1}.$$

In passing, we may mention that models (21.5)–(21.8) were fitted by means of the **R** function constrOptim (for constrained optimization).

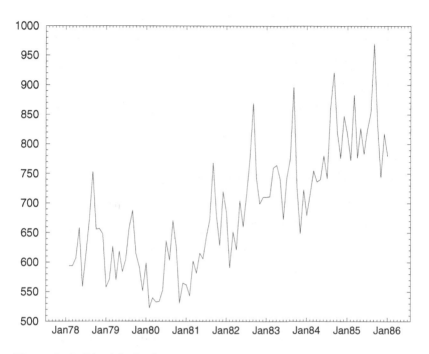

Figure 21.2 *Edendale births: monthly totals of deliveries, February 1978–January 1986.*

It is possible to refine these models for the monthly totals by allowing for the fact that the months are of unequal length, and by allowing for seasonality of frequency higher than annual, but these variations were not pursued.

A number of log-linear models were then fitted by glm. Table 21.4 compares the models fitted by glm to the total deliveries, and from that table it can be seen that of these models the BIC selects the model incorporating time trend, sinusoidal components at the annual frequency, and the number of deliveries in the previous month (n_{t-1}). The details of this model are as follows. Conditional on the history, the number of deliveries in month t (N_t) is distributed Poisson with mean $_t\lambda$, where

$$\log {}_t\lambda = 6.015 + 0.002436t - 0.03652\cos(2\pi t/12)$$
$$- 0.02164\sin(2\pi t/12) + 0.0005737n_{t-1}.$$

(Here, as before, February 1978 is month 1.)

Table 21.3 *Edendale births: summary of the five two-state Poisson–HMMs fitted to the number of deliveries. (The time in months is denoted by t, and February 1978 is month 1.)*

$\log{}_t\lambda_i$	$-l$	AIC BIC	γ_{12} γ_{21}	a_1 a_2	b	c d	f	g
(21.4)	642.023	1292.0 1302.3	0.109 0.125	6.414 6.659	– 	– –	– 	–
(21.5)	553.804	1117.6 1130.4	0.138 0.448	6.250 6.421	0.004873 	– –	– 	–
(21.6)	523.057	1060.1 1078.1	0.225 0.329	6.267 6.401	0.004217 	−0.0538 −0.0536	– 	–
(21.7)	510.840	**1037.7** **1058.2**	0.428 0.543	5.945 6.069	0.002538 	−0.0525 −0.0229	0.000586 	–
(21.8)	510.791	1039.6 1062.7	0.415 0.537	5.939 6.063	0.002511 	−0.0547 −0.0221	0.000561 	0.0000365

Table 21.4 *Edendale births: models fitted by* glm *to the log of the mean number of deliveries.*

Explanatory variables	Deviance	$-l$	AIC	BIC
t	545.9778	674.3990	1352.8	1357.9
t, sinusoidal terms*	428.6953	615.7577	1239.5	1249.8
t, sinusoidal terms, n_{t-1}	356.0960	579.4581	1168.9	**1181.7**
t, sinusoidal terms, n_{t-1}, n_{t-2}	353.9086	578.3644	**1168.7**	1184.1
sinusoidal terms, n_{t-1}	464.8346	633.8274	1275.7	1285.9
t, n_{t-1}	410.6908	606.7555	1219.5	1227.2
n_{t-1}	499.4742	651.1472	1306.3	1311.4

*This is a one-state model similar to the HMM (21.6), and its log-likelihood value can be used to some extent to check the code for the HMM. Similar comments apply to the other models listed above.

Both the HMMs and the log-linear autoregressions reveal time trend, seasonality, and dependence on the number of deliveries in the previous month. On the basis of both AIC and BIC the Poisson–HMM (21.7) is the best model.

21.4 Conclusion

The conclusion is therefore twofold. If a model for the number of Caesareans, given the total number of deliveries, is needed, the binomial–HMM with time trend is best of all of those considered (including various logit-linear autoregressive models). If a model for the total deliveries is needed (e.g. as a building-block in projecting the number of Caesareans), then the Poisson–HMM with time trend, 12-month seasonality and dependence on the number in the previous month is the best of those considered – contrary to our conclusion in MacDonald and Zucchini (1997), which was unduly pessimistic about the Poisson–HMMs.

The models chosen here suggest that there is a clear upward time trend in both the total deliveries and the proportion Caesarean, and seasonality in the total deliveries.

CHAPTER 22

Homicides and suicides in Cape Town, 1986–1991

22.1 Introduction

In South Africa, as in the USA, gun control is a subject of much public interest and debate. In a project intended to study the apparently increasing tendency for violent crime to involve firearms, Dr L.B. Lerer collected data relating to homicides and suicides from the South African Police mortuary in Salt River, Cape Town. Records relating to most of the homicide and suicide cases occurring in metropolitan Cape Town were kept at this mortuary. The remaining cases were dealt with at the Tygerberg Hospital mortuary. It is believed, however, that the exclusion of the Tygerberg data does not materially affect the conclusions.

The data consist of all the homicide and suicide cases appearing in the deaths registers relating to the 6-year period from 1 January 1986 to 31 December 1991. In each such case the information recorded included the date and cause of death. The five (mutually exclusive) categories used for the cause of death were: firearm homicide, non-firearm homicide, firearm suicide, non-firearm suicide, and 'legal intervention homicide'. (This last category refers to homicide by members of the police or army in the course of their work. In what follows, the word 'homicide', if unqualified, means homicide other than that resulting from such legal intervention.) Clearly some of the information recorded in the deaths registers could be inaccurate (e.g. a homicide recorded as a suicide, or a legal intervention homicide recorded as belonging to another category). This has to be borne in mind in drawing conclusions from the data.

22.2 Firearm homicides as a proportion of all homicides, suicides and legal intervention homicides

One question of interest that was examined by means of HMMs was whether there is an upward trend in the proportion of all the deaths recorded that are firearm homicides. This is of course quite distinct from the question of whether there is an upward trend in the *number* of firearm homicides. The latter kind of trend could be caused by an increase in the population exposed to risk of death, without there

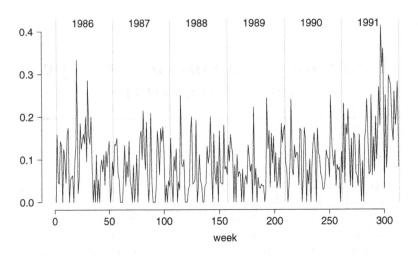

Figure 22.1 *Firearm homicides 1986–1991, as a proportion of all homicides, suicides and legal intervention homicides.*

being any other relevant change. This distinction is important because of the rapid urbanization which took place in the relevant years in South Africa and caused the population in and around Cape Town to increase dramatically.

Four models were fitted to the 313 weekly totals of firearm homicides (given the weekly totals of all the deaths recorded); for a plot of the firearm homicides in each week, as a proportion of all homicides, suicides and legal intervention homicides, see Figure 22.1.

The four models are: a two-state binomial–HMM with constant 'success probabilities' p_1 and p_2; a similar model with a linear time trend (the same for both states) in the logits of those probabilities; a model allowing differing time-trend parameters in the two states; and finally, a model which assumes that the success probabilities are piecewise constant with a single change-point at time 287 – that is, on 2 July 1991, 26 weeks before the end of the 6-year period studied. The time of the change-point was chosen because of the known upsurge of violence in some of the areas adjacent to Cape Town in the second half of 1991. Much of this violence was associated with the 'taxi wars', a dispute between rival groups of public transport operators.

The models were compared on the basis of AIC and BIC. The results are shown in Table 22.1. Broadly, the conclusion from the BIC is that a single (upward) time trend is better than either no trend or two trend

Table 22.1 *Comparison of several binomial–HMMs fitted to the weekly counts of firearm homicides, given the weekly counts of all homicides, suicides and legal intervention homicides.*

Model with	k	$-l$	AIC	BIC
p_1 and p_2 constant	4	590.258	1188.5	1203.5
one time-trend parameter	5	584.337	1178.7	1197.4
two time-trend parameters	6	579.757	1171.5	1194.0
change-point at time 287	6	573.275	**1158.5**	**1181.0**

parameters, but the model with a change-point is the best of the four. The details of this model are as follows. The underlying Markov chain has transition probability matrix

$$\begin{pmatrix} 0.658 & 0.342 \\ 0.254 & 0.746 \end{pmatrix}$$

and stationary distribution $(0.426, 0.574)$. The probabilities p_1 and p_2 are given by $(0.050, 0.116)$ for weeks 1–287, and by $(0.117, 0.253)$ for weeks 288–313. From this it appears that the proportion of the deaths that are firearm homicides was substantially greater by the second half of 1991 than it was earlier, and that this change is better accommodated by a discrete shift in the probabilities p_1 and p_2 than by gradual movement with time, at least gradual movement of the kind that we have incorporated into the models with time trend. One additional model was fitted: a model with change-point at the end of week 214. That week included 2 February 1990, on which day President de Klerk made a speech which is widely regarded as a watershed in South Africa's history. That model yielded a log-likelihood of -579.83, and AIC and BIC values of 1171.67 and 1194.14. Such a model is therefore inferior to the model with the change-point at the end of week 287.

22.3 The number of firearm homicides

In order to model the number (rather than the proportion) of firearm homicides, Poisson–HMMs were also fitted. The weekly counts of firearm homicides are shown in Figure 22.2. There is a marked increase in the level of the series at about week 287 (mid-1991), but another, less distinct, pattern is discernible in the values prior to that date. There seems to be persistence in those periods when the counts are high (e.g. around weeks 25 and 200); runs of relatively calm weeks seem to alternate with

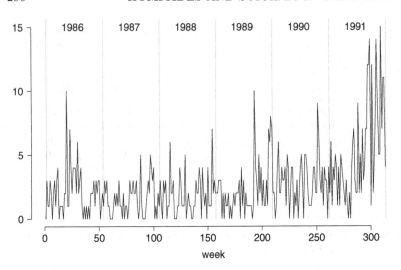

Figure 22.2 *Weekly counts of firearm homicides, 1986–1991.*

Table 22.2 *Comparison of several Poisson–HMMs fitted to weekly counts of firearm homicides.*

Model with	k	$-l$	AIC	BIC
λ_1 and λ_2 constant	4	626.637	1261.3	1276.3
log-linear trend	5	606.824	1223.6	1242.4
log-quadratic trend	6	602.273	**1216.5**	**1239.0**
change-point at time 287	6	605.559	1223.1	1245.6

runs of increased violence. This observation suggests that a two-state HMM might be an appropriate model.

The four models fitted in this case were: a two-state model with constant conditional means λ_1 and λ_2; a similar model with a single linear trend in the logs of those means; a model with a quadratic trend therein; and finally, a model allowing for a change-point at time 287. A comparison of these models is shown in Table 22.2.

The conclusion is that, of the four models, the model with a quadratic trend in the conditional means is best. In detail, that model is as follows. The underlying Markov chain has t.p.m. given by

$$\begin{pmatrix} 0.881 & 0.119 \\ 0.416 & 0.584 \end{pmatrix}$$

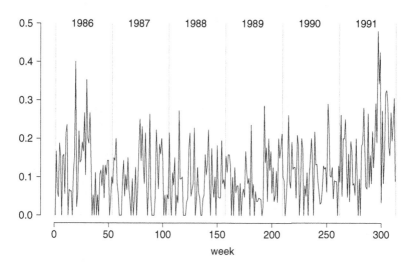

Figure 22.3 *Firearm homicides as a proportion of all homicides, 1986–1991. The week ending on 2 July 1991 is week 287.*

and stationary distribution $(0.777, 0.223)$. The conditional means are given by

$$\log {}_t\lambda_i = 0.4770/1.370 - 0.004858t + 0.00002665t^2,$$

where t is the week number and i the state. The fact that a smooth trend works better here than does a discrete shift may possibly be explained by population increase due to migration, especially towards the end of the 6-year period.

22.4 Firearm homicides as a proportion of all homicides, and firearm suicides as a proportion of all suicides

A question of interest that arises from the apparently increased proportion of firearm homicides is whether there is any similar tendency in respect of suicides. Here the most interesting comparison is between firearm homicides as a proportion of all homicides and firearm suicides as a proportion of all suicides. Plots of these two proportions appear as Figures 22.3 and 22.4. Binomial–HMMs of several types were used to model these proportions, and the results are given in Tables 22.3 and 22.4.

The chosen models for these two proportions are as follows. For the

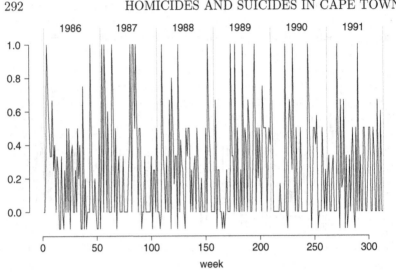

Figure 22.4 *Firearm suicides as a proportion of all suicides, 1986–1991. (Negative values indicate no suicides in that week.)*

firearm homicides the Markov chain has t.p.m.

$$\begin{pmatrix} 0.695 & 0.305 \\ 0.283 & 0.717 \end{pmatrix}$$

and stationary distribution $(0.481, 0.519)$. The probabilities p_1 and p_2 are given by $(0.060, 0.140)$ for weeks 1–287, and by $(0.143, 0.283)$ for weeks 288–313. The unconditional probability that a homicide involved the use of a firearm is therefore 0.102 before the change-point, and 0.216 thereafter. For the firearm suicides, the t.p.m. is

$$\begin{pmatrix} 0.854 & 0.146 \\ 0.117 & 0.883 \end{pmatrix},$$

and the stationary distribution is $(0.446, 0.554)$. The probabilities p_1 and p_2 are given by $(0.186, 0.333)$, and the unconditional probability that a suicide involves a firearm is 0.267.

A question worth considering, however, is whether time series models are needed at all for these proportions. Is it not perhaps sufficient to fit a one-state model, that is, a model which assumes independence of the consecutive observations but is otherwise identical to one of the time series models described above? Models of this type were therefore fitted both to the firearm homicides as a proportion of all homicides, and to the firearm suicides as a proportion of all suicides. The model compar-

Table 22.3 *Comparison of binomial–HMMs for firearm homicides, given all homicides.*

Model with	k	$-l$	AIC	BIC
p_1 and p_2 constant	4	590.747	1189.5	1204.5
one time-trend parameter	5	585.587	1181.2	1199.9
two time-trend parameters	6	580.779	1173.6	1196.0
change-point at time 287	6	575.037	**1162.1**	**1184.6**

Table 22.4 *Comparison of binomial–HMMs for firearm suicides, given all suicides.*

Model with	k	$-l$	AIC	BIC
p_1 and p_2 constant	4	289.929	**587.9**	**602.8**
one time-trend parameter	5	289.224	588.4	607.2
two time-trend parameters	6	288.516	589.0	611.5
change-point at time 287	6	289.212	590.4	612.9

isons are presented in Tables 22.5 and 22.6, in which the parameter p represents the probability that a death involves a firearm.

The conclusions that may be drawn from these models are as follows. For the homicides, the models based on independence are without exception clearly inferior to the corresponding HM time series models. There is sufficient serial dependence present in the proportion of the homicides involving a firearm to render inappropriate any analysis based on an assumption of independence. For the suicides the situation is reversed; the models based on independence are in general superior. There is in this case no convincing evidence of serial dependence, and time series models do not appear to be necessary. The 'best' model based on independence

Table 22.5 *Comparison of several 'independence' models for firearm homicides as a proportion of all homicides.*

Model with	k	$-l$	AIC	BIC
p constant	1	637.458	1276.9	1280.7
time trend in p	2	617.796	1239.6	1247.1
change-point at time 287	2	590.597	**1185.2**	**1192.7**

Table 22.6 *Comparison of several 'independence' models for firearm suicides as a proportion of all suicides.*

Model with	k	$-l$	AIC	BIC
p constant	1	291.166	**584.3**	**588.1**
time trend in p	2	290.275	584.6	592.0
change-point at time 287	2	291.044	586.1	593.6

assigns a value of 0.268 (= 223/833) to the probability that a suicide involves the use of a firearm, which is of course quite consistent with the value (0.267) implied by the chosen HMM.

To summarize the conclusions, therefore, we may say that the proportion of homicides that involve firearms does indeed seem to be at a higher level after June 1991, but that there is no evidence of a similar upward shift (or trend) in respect of the proportion of the suicides that involve firearms. There is evidence of serial dependence in the proportion of homicides that involve firearms, but not in the corresponding proportion of suicides.

In view of the finding that the proportion of homicides due to firearms seems to be higher after June 1991, it is interesting to see whether the monitoring technique introduced in Section 6.2.3 would have detected such a change if used over the final 2 years of the study period. The data for weeks 1–209 only (essentially the first 4 years, 1986–1989) were used to derive a model with constant probabilities p_1 and p_2 for the weekly numbers of firearm homicides as a proportion of all homicides. For each r from 210 to 313 the conditional distribution (under this model) of X_r given the full history $\mathbf{X}^{(r-1)}$ was then computed, and the extremeness or otherwise of the observation x_r was assessed with the help of a plot of forecast pseudo-residuals, which is shown as Figure 22.5.

The result is clear. Not one of the 78 weeks 210–287 produces a pseudo-residual segment lying outside the bands at 0.5% and 99.5%. Weeks 288–313 are very different, however. Weeks 297, 298, 299 and 305 all produce segments lying entirely above the 99.5% level, and within weeks 288–313 there are none even close to the 0.5% level. We therefore conclude that, although the data for 1990 and the first half of 1991 are reasonably consistent with a model based on the 4 years 1986–1989, the data for the second half of 1991 are not; after June 1991, firearm homicides are at a higher level, relative to all homicides, than is consistent with the model.

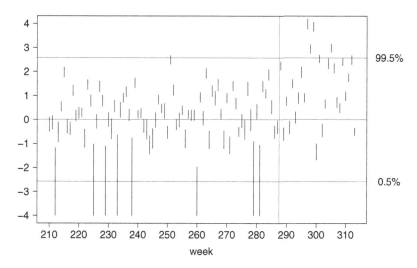

Figure 22.5 *Firearm homicides as a proportion of all homicides: forecast pseudo-residual segments computed from a model for weeks 1–209 only. The vertical line is immediately after week 287.*

22.5 Proportion in each of the five categories

As a final illustration of the application of HMMs to these data, we describe here two multinomial–HMMs for the weekly totals in each of the five categories of death. These models are of the kind introduced in Section 9.3.1. Each has two states. One model has constant 'success probabilities', and the other allows for a change in these probabilities at time 287. The model without change-point has 10 parameters: two to determine the Markov chain, and four independently determined probabilities for each of the two states. The model with change-point has 18 parameters, since there are eight independent success probabilities relevant to the period before the change-point, and eight after. For the model without change-point, $-l$ is 1810.059, and for the model with change-point it is 1775.610. The corresponding AIC and BIC values are 3640.1 and 3677.6 (without change-point), and 3587.2 and 3654.7 (with). The model with the change-point at time 287 is therefore preferred, and we give it in full here. The underlying Markov chain has t.p.m.

$$\begin{pmatrix} 0.541 & 0.459 \\ 0.097 & 0.903 \end{pmatrix}$$

and stationary distribution $(0.174, 0.826)$.

Table 22.7 displays, for the period up to the change-point and the

Table 22.7 *Multinomial–HMM with change-point at time 287. Probabilities associated with each category of death, before and after the change-point.*

Weeks 1–287

	Category 1	2	3	4	5
in state 1	0.124	0.665	0.053	0.098	0.059
in state 2	0.081	0.805	0.024	0.074	0.016
unconditional	0.089	0.780	0.029	0.079	0.023

Weeks 288–313

	Category 1	2	3	4	5
in state 1	0.352	0.528	0.010	0.075	0.036
in state 2	0.186	0.733	0.019	0.054	0.008
unconditional	0.215	0.697	0.018	0.058	0.013

Categories: 1, firearm homicide; 2, non-firearm homicide; 3, firearm suicide; 4, non-firearm suicide; 5, legal intervention homicide.

period thereafter, the probability of each category of death in state 1 and in state 2, and the corresponding unconditional probabilities. The most noticeable difference between the period before the change-point and the period thereafter is the sharp increase in the unconditional probability of category 1 (firearm homicide), with corresponding decreases in all the other categories.

Clearly the above discussion does not attempt to pursue all the questions of interest arising from these data that may be answered by the fitting of HM (or other) time series models. It is felt, however, that the models described, and the conclusions that may be drawn, are sufficiently illustrative of the technique to make clear its utility in such an application.

CHAPTER 23

A model for animal behaviour which incorporates feedback

23.1 Introduction

Animal behaviourists are interested in the causal factors that determine behavioural sequences: when animals perform particular activities, and under what circumstances they switch to alternative activities. It is accepted that observed behaviour results from the nervous system integrating information regarding the physiological state of the animal (e.g. the levels of nutrients in the blood) with sensory inputs (e.g. concerning the levels of nutrients in a food); see Barton Browne (1993). The combined physiological and perceptual state of the animal is termed the 'motivational state' (McFarland, 1999). MacDonald and Raubenheimer (1995) modelled behaviour sequences using an HMM whose unobserved underlying states were interpreted as motivational states. Their model captures an important aspect of the causal structure of behaviour, since an animal in a given motivational state (e.g. hungry) might perform not only the most likely behaviour for that state (feed) but also other behaviours (groom, drink, walk, etc.). There is not a one-to-one correspondence between motivational state and behaviour. And it is the runlength distributions of the motivational states that are of interest, rather than those of the observed behaviours. HMMs do not, however, take into account the fact that, in many cases, behaviour also influences motivational state; feeding, for example, leads to satiation.

We describe here a model, proposed by Zucchini, Raubenheimer and MacDonald (2008), that incorporates such feedbacks, and we apply it in order to model observed feeding patterns of caterpillars. We define a 'nutrient level', which is determined by the animal's recent feeding behaviour and which, in turn, influences the probability of transition to a different motivational state. Latent-state models, including HMMs, provide a means of grouping two or more behaviours, such as feeding and grooming, into 'activities'. Here the activities of interest are meal-taking, an activity characterized by feeding interspersed by brief pauses, and inter-meal intervals, in which the animal mainly rests but might also feed for brief periods.

The proposed model is not an HMM because the states do not form

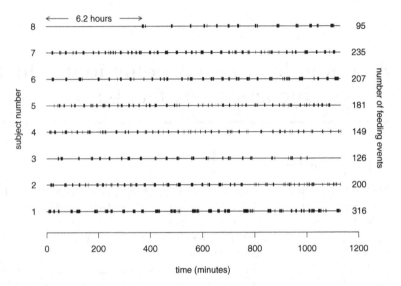

Figure 23.1 *Feeding behaviour of eight* Helicoverpa armigera *caterpillars ob-served at 1-minute intervals.*

a Markov chain. It can be regarded as a special-purpose extension of HMMs, and shares many of the features of HMMs. We therefore discuss the theoretical aspects of the model before moving on to an application.

In Section 23.9 we demonstrate the application of the model to data collected in an experiment in which eight caterpillars were observed at 1-minute intervals for almost 19 hours, and classified as feeding or not feeding. The data are displayed in Figure 23.1.

23.2 The model

Suppose an animal is observed at integer times $t = 1, 2, \ldots, T$, and classified as feeding at time t ($X_t = 1$) or not ($X_t = 0$). We propose the following model.

There are two possible (unobserved) motivational states, provisionally labelled 'hungry' (state 1) and 'sated' (state 2). The state process $\{C_t\}$ is a process in which the transition probabilities are driven by a process $\{N_t\}$ termed the 'nutrient level', which takes values in $[0, 1]$. We assume that N_t is some function of N_{t-1} and X_t. The development that follows is applicable more generally, but we shall restrict our attention to the exponential filter:

$$N_t = \lambda X_t + (1 - \lambda)N_{t-1}, \quad t = 1, 2, \ldots, T. \qquad (23.1)$$

In biological terms, N_t would correspond approximately with the levels of nutrients in the blood (Simpson and Raubenheimer, 1993). We take the transition probabilities

$$\gamma_{ij}(n_t) = \Pr(C_{t+1} = j \mid C_t = i, N_t = n_t), \quad t = 1, 2, \ldots, T - 1,$$

to be determined as follows by the nutrient level,

$$\text{logit } \gamma_{11}(n_t) = \alpha_0 + \alpha_1 n_t, \quad \text{logit } \gamma_{22}(n_t) = \beta_0 + \beta_1 n_t; \tag{23.2}$$

and by the row-sum constraints,

$$\gamma_{11}(n_t) + \gamma_{12}(n_t) = 1, \quad \gamma_{21}(n_t) + \gamma_{22}(n_t) = 1.$$

In equation (23.2) the logit could be replaced by some other monotonic function $g : (0, 1) \to \mathbb{R}$.

In state 1, the probability of feeding is always π_1, regardless of earlier motivational state, nutrient level or behaviour; similarly, π_2 in state 2. The behaviour X_t influences the nutrient level N_t, which in turn determines the transition probabilities of the state process and so influences the state occupied at the next time point, $t + 1$.

Our two fundamental assumptions are as follows, for $t = 2, 3, \ldots, T$ and $t = 1, 2, \ldots, T$ respectively:

$$\Pr(C_t \mid \mathbf{C}^{(t-1)}, \mathbf{N}^{(t-1)}, \mathbf{X}^{(t-1)}) = \Pr(C_t \mid C_{t-1}, N_{t-1}) \tag{23.3}$$

and

$$\Pr(X_t = 1 \mid \mathbf{C}^{(t)}, \mathbf{N}^{(t-1)}, \mathbf{X}^{(t-1)}) = \Pr(X_t = 1 \mid C_t)$$
$$= \begin{cases} \pi_1 & \text{if } C_t = 1 \\ \pi_2 & \text{if } C_t = 2. \end{cases} \tag{23.4}$$

We use here the usual notation $\mathbf{X}^{(t)} = (X_1, X_2, \ldots, X_t)$, and similarly $\mathbf{C}^{(t)}$, but define $\mathbf{N}^{(t)} = (N_0, N_1, N_2, \ldots, N_t)$. So $\mathbf{X}^{(0)}$ is an empty set of random variables, but $\mathbf{N}^{(0)}$ consists of N_0. Given the parameter λ, $\mathbf{N}^{(t-1)}$ is completely determined by N_0 and $\mathbf{X}^{(t-1)}$. Hence the above two assumptions can be written as

$$\Pr(C_t \mid \mathbf{C}^{(t-1)}, N_0, \mathbf{X}^{(t-1)}) = \Pr(C_t \mid C_{t-1}, N_{t-1}) \tag{23.5}$$

and

$$\Pr(X_t = 1 \mid \mathbf{C}^{(t)}, N_0, \mathbf{X}^{(t-1)}) = \begin{cases} \pi_1 & \text{if } C_t = 1 \\ \pi_2 & \text{if } C_t = 2. \end{cases} \tag{23.6}$$

This second form of assumptions (23.3) and (23.4) is useful in what follows in Section 23.3.

We expect α_1 to be negative, and β_1 positive. We expect π_1 to be close to 1 and π_2 to be close to 0. It is convenient to introduce here the notation

$$p_i(x) = \Pr(X_t = x \mid C_t = i) = \pi_i^x (1 - \pi_i)^{1-x}, \quad \text{for } x = 0, 1; \ i = 1, 2.$$

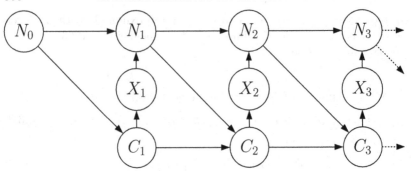

Figure 23.2 *Directed graph representing animal-behaviour model.*

With this notation, (23.6) becomes

$$\Pr(X_t = x \mid \mathbf{C}^{(t)}, N_0, \mathbf{X}^{(t-1)}) = p_{c_t}(x).$$

The model is conveniently represented by the directed graph in Figure 23.2. Notice that there is a path from C_1 to C_3 that does not pass through C_2; the state process is therefore not in general a Markov process. The following are treated as parameters of the model: α_0, α_1, β_0, β_1, π_1, π_2, and λ. In addition, we take $n_0 \in [0, 1]$ to be a parameter, albeit not one of any intrinsic interest.

One can also treat as a parameter the distribution of C_1 given N_0 (which we denote by the vector $\boldsymbol{\delta}$), or, more precisely, the probability $\delta_1 = \Pr(C_1 = 1 \mid N_0)$. It can, however, be shown that on maximizing the resulting likelihood one will simply have either $\widehat{\delta}_1 = 1$ or $\widehat{\delta}_1 = 0$. We follow this approach, but in the application described in Section 23.9 it turns out in any case that it is biologically reasonable to assume that $\delta_1 = 0$, that is, that the subject always starts in state 2.

23.3 Likelihood evaluation

Given a sequence of observations $\{\mathbf{x}^{(T)}\}$ assumed to arise from such a model, and given values of the parameters listed above, we need to be able to evaluate the likelihood, both to estimate parameters and to find the marginal and conditional distributions we shall use in this analysis. First we write the likelihood as a T-fold multiple sum, then we show that this sum can be efficiently computed via a recursion.

LIKELIHOOD EVALUATION

301

23.3.1 The likelihood as a multiple sum

We therefore seek

$$L_T = \Pr\left(\mathbf{X}^{(T)} \mid N_0\right) = \sum_{c_1,\dots,c_T} \Pr\left(\mathbf{C}^{(T)}, \mathbf{X}^{(T)} \mid N_0\right).$$

The summand $\Pr\left(\mathbf{C}^{(T)}, \mathbf{X}^{(T)} \mid N_0\right)$ may be decomposed as follows:

$$
\begin{aligned}
&\Pr\left(\mathbf{C}^{(T)}, \mathbf{X}^{(T)} \mid N_0\right) \\
&= \Pr(C_1 \mid N_0)\Pr(X_1 \mid C_1, N_0) \times \\
&\quad \prod_{t=2}^{T}\left(\Pr(C_t \mid \mathbf{C}^{(t-1)}, N_0, \mathbf{X}^{(t-1)})\Pr(X_t \mid \mathbf{C}^{(t)}, N_0, \mathbf{X}^{(t-1)})\right) \\
&= \Pr(C_1 \mid N_0)\Pr(X_1 \mid C_1)\prod_{t=2}^{T}\left(\Pr(C_t \mid C_{t-1}, N_{t-1})\Pr(X_t \mid C_t)\right).
\end{aligned}
$$

The first equality follows by repeated application of the definition of conditional probability, and the second from (23.5) and (23.6). We therefore conclude that

$$L_T = \sum_{c_1,\dots,c_T}\left(\delta_{c_1}\, p_{c_1}(x_1)\prod_{t=2}^{T}\left(\gamma_{c_{t-1},c_t}(n_{t-1})\,p_{c_t}(x_t)\right)\right).$$

23.3.2 Recursive evaluation

The likelihood is therefore a sum of the form

$$L = \sum_{c_1=1}^{2}\sum_{c_2=1}^{2}\cdots\sum_{c_T=1}^{2}\left(\alpha_1(c_1)\prod_{t=2}^{T}f_t(c_{t-1}, c_t)\right); \qquad (23.7)$$

that is, if here we define

$$\alpha_1(j) = \delta_j\, p_j(x_1)$$

and, for $t = 2, 3, \dots, T$,

$$f_t(i,j) = \gamma_{ij}(n_{t-1})\, p_j(x_t).$$

Multiple sums of the form (23.7) can in general be evaluated recursively; see Exercise 7 of Chapter 2. Indeed, this is the key property of hidden Markov likelihoods that makes them computationally feasible, and here makes it unnecessary to sum explicitly over the 2^T terms. (See Lange (2002, p. 120) for a discussion of such recursive summation as applied in the computation of likelihoods of pedigrees, a more complex problem than the one we need to consider.)

From Exercise 7(b) of Chapter 2 we conclude that

$$L = \alpha_1 \mathbf{F}_2 \mathbf{F}_3 \cdots \mathbf{F}_T \mathbf{1}',$$

where the 2×2 matrix $\mathbf{F}_t$ has (i, j) element equal to $f_t(i, j)$, and α_1 is the vector with jth element $\alpha_1(j)$. In the present context, $\alpha_1 = \delta \mathbf{P}(x_1)$, $\mathbf{F}_2 = \mathbf{\Gamma}(n_1)\mathbf{P}(x_2)$, and similarly $\mathbf{F}_3$, etc. Hence the likelihood of the model under discussion can be written as the matrix product

$$L_T = \delta \mathbf{P}(x_1)\mathbf{\Gamma}(n_1)\mathbf{P}(x_2)\mathbf{\Gamma}(n_2) \cdots \mathbf{\Gamma}(n_{T-1})\mathbf{P}(x_T)\mathbf{1}', \qquad (23.8)$$

where the vector δ is the distribution of C_1 given N_0, $\mathbf{P}(x_t)$ is the diagonal matrix with ith diagonal element $p_i(x_t)$, and $\mathbf{\Gamma}(n_t)$ is the matrix with (i, j) element $\gamma_{ij}(n_t)$.

As usual, precautions have to be taken against numerical underflow, but otherwise the matrix product (23.8) can be used as it stands to evaluate the likelihood. The computational effort is linear in T, the length of the series, in spite of the fact that there are 2^T terms in the sum.

23.4 Parameter estimation by maximum likelihood

Estimation may be carried out by direct numerical maximization of the log-likelihood. Since the parameters π_1, π_2, λ, n_0 and δ_1 are constrained to lie between 0 and 1, it is convenient to reparametrize the model in order to avoid these constraints. A variety of methods were used in this work to carry out the optimization and checking: the Nelder–Mead simplex algorithm, simulated annealing and methods of Newton type, as implemented by the R functions optim and nlm.

An alternative to direct numerical maximization would be to use the EM algorithm. In this particular model, however, there are no closed-form expressions for the parameter estimates given the complete data, that is, given the observations plus the states occupied at all times. The M step of the EM algorithm would therefore involve numerical optimization, and it seems circuitous to apply an algorithm which requires numerical optimization in each iteration, instead of only once. Whatever method is used, however, one has to bear in mind that there may well be multiple local optima in the likelihood.

23.5 Model checking

When we have fitted a model to the observed behaviour of an animal, we need to examine the model to assess its suitability. One way of doing so is as follows.

It is a routine calculation to find the forecast distributions under the fitted model, that is, the distribution of each X_t given the history $\mathbf{X}^{(t-1)}$.

The probabilities are the ratios of two likelihood values:

$$\Pr(X_t \mid \mathbf{X}^{(t-1)}) = \Pr(\mathbf{X}^{(t)})/\Pr(\mathbf{X}^{(t-1)}) = L_t/L_{t-1}.$$

We denote by $\widehat{p}_t$ the probability $\Pr(X_t = 1 \mid \mathbf{X}^{(t-1)})$ computed thus under the model. Since the joint probability function of $\mathbf{X}^{(T)}$ factorizes as

$$\Pr(\mathbf{X}^{(T)}) = \Pr(X_1)\Pr(X_2 \mid X_1)\Pr(X_3 \mid \mathbf{X}^{(2)}) \cdots \Pr(X_T \mid \mathbf{X}^{(T-1)}),$$

we have a problem of the following form. There are T binary observations x_t, assumed to be drawn independently, with $\mathrm{E}(x_t) = \widehat{p}_t$. We wish to test the null hypothesis $\mathrm{E}(x_t) = \widehat{p}_t$ (for all t), or equivalently

$$\mathrm{H}_0 : g(\mathrm{E}(x_t)) = g(\widehat{p}_t),$$

where g is the logit transform. We consider an alternative hypothesis of the form

$$\mathrm{H}_A : g(\mathrm{E}(x_t)) = f(g(\widehat{p}_t)),$$

where f is a smoothing spline (see, for example, Hastie and Tibshirani, 1990). Departure of f from the identity function constitutes evidence against the null hypothesis, and a plot of f against the identity function will reveal the nature of the departure.

23.6 Inferring the underlying state

A question that is of interest in many applications of latent-state models is this: what are the states of the latent process (here $\{C_t\}$) that are most likely (under the fitted model) to have given rise to the observation sequence? This is the decoding problem discussed in Section 5.4. More specifically, 'local decoding' of the state at time t refers to the determination of the state i_t which is most likely at that time,

$$i_t = \underset{i=1,2}{\operatorname{argmax}} \ \Pr(C_t{=}i \mid \mathbf{X}^{(T)}{=}\mathbf{x}^{(T)}).$$

In the context of feeding, local decoding might be of interest for determining the specific sets of sensory and metabolic events that distinguish meal-taking from inter-meal breaks (Simpson and Raubenheimer, 1993).

In contrast, global decoding refers to the determination of that sequence of states $c_1, c_2, \ldots, c_T$ which maximizes the conditional probability

$$\Pr(\mathbf{C}^{(T)}{=}\mathbf{c}^{(T)} \mid \mathbf{X}^{(T)}{=}\mathbf{x}^{(T)});$$

or equivalently, the joint probability

$$\Pr(\mathbf{C}^{(T)}, \mathbf{X}^{(T)}) = \delta_{c_1} \prod_{t=2}^{T} \gamma_{c_{t-1},c_t}(n_{t-1}) \prod_{t=1}^{T} p_{c_t}(x_t).$$

Global decoding can be carried out, both here and in other contexts, by means of the Viterbi algorithm; see Section 5.4.2. For the sake of completeness we present the details here, although little is needed beyond that which is in Section 5.4.2. Define

$$\xi_{1i} = \Pr(C_1=i, X_1=x_1) = \delta_i\, p_i(x_1),$$

and, for $t=2, 3, \ldots, T$,

$$\xi_{ti} = \max_{c_1,c_2,\ldots,c_{t-1}} \Pr(\mathbf{C}^{(t-1)}=\mathbf{c}^{(t-1)}, C_t=i, \mathbf{X}^{(t)}=\mathbf{x}^{(t)}).$$

It can then be shown that the probabilities ξ_{tj} satisfy the following recursion, for $t=2, 3, \ldots, T$:

$$\xi_{tj} = \left(\max_{i=1,2}\left(\xi_{t-1,i}\,\gamma_{ij}(n_{t-1})\right)\right) p_j(x_t). \qquad (23.9)$$

This provides an efficient means of computing the $T \times 2$ matrix of values ξ_{tj}, as the computational effort is linear in T. The required sequence of states $i_1, i_2, \ldots, i_T$ can then be determined recursively from

$$i_T = \underset{i=1,2}{\operatorname{argmax}}\ \xi_{Ti}$$

and, for $t=T-1, T-2, \ldots, 1$, from

$$i_t = \underset{i=1,2}{\operatorname{argmax}}\ \left(\xi_{ti}\,\gamma_{i,i_{t+1}}(n_t)\right).$$

23.7 Models for a heterogeneous group of subjects

The model described above is applicable to observations on a single subject. In this application $I = 8$ subjects were observed, that is, longitudinal data are available. As pointed out in Chapter 13, such data offer the opportunity to distinguish those features that are common to all subjects from those that differ across subjects. Common features can be modelled by assuming that some parameters are constant across subjects; parameters that clearly differ across subjects can be modelled either as fixed or as random effects. We shall consider each of these alternatives.

23.7.1 Models assuming some parameters to be constant across subjects

One extreme model for a group of I subjects is to apply complete pooling: to assume that, apart from nuisance parameters, the set of parameter *values* is the same for all subjects. We assume that the subjects are independent, and so the likelihood is just the product of the I individual likelihoods. It is a function of the seven parameters α_0, α_1, β_0, β_1, π_1, π_2 and λ, plus I values of n_0 and (and if they are treated as parameters) I values of δ_1; hence $7 + 2I$ parameters in all.

The other extreme model is that with no pooling: allowing each subject to have its own set of nine parameter values. In this case the comparable number of parameters is $9I$. Intermediate between these two cases are partially pooled models that assume that some but not all of the seven parameters listed are common to the I subjects. For instance, one might wish to assume that only the probabilities π_1 and π_2 are common to all subjects.

23.7.2 Mixed models

However, in drawing overall conclusions from a group, it may be useful to allow for between-subject variability by some means other than merely permitting parameters to differ between subjects. One way of doing so is to incorporate random effects; see Section 13.3.

The incorporation of a single subject-specific random effect into a model for a group of subjects is in principle straightforward; see Altman (2007) for a general discussion of the introduction of random effects into HMMs. For concreteness, suppose that the six parameters α_0, α_1, β_0, β_1, π_1 and π_2 can reasonably be supposed constant across subjects, but not λ. Instead we suppose that λ is generated by some density f with support $[0, 1]$.

Conditional on λ, the likelihood of the observations on subject i is

$$L_T(i, \lambda) = \boldsymbol{\delta}\mathbf{P}(x_{i1})\boldsymbol{\Gamma}(n_{i1})\mathbf{P}(x_{i2})\boldsymbol{\Gamma}(n_{i2})\cdots\boldsymbol{\Gamma}(n_{i,T-1})\mathbf{P}(x_{iT})\mathbf{1}',$$

where $\{x_{it} : t = 1,\ldots,T\}$ is the set of observations on subject i and similarly $\{n_{it}\}$ the set of values of the nutrient level. Unconditionally this likelihood is $\int_0^1 L_T(i, \lambda)\, f(\lambda)\, d\lambda$, and the likelihood of all I subjects is the product

$$L_T = \prod_{i=1}^{I} \int_0^1 L_T(i, \lambda)\, f(\lambda)\, d\lambda. \qquad (23.10)$$

Each evaluation of L_T therefore requires I numerical integrations, which can be performed in **R** by means of the function `integrate`, but slow the computation down considerably.

Incorporation of more than one random effect could proceed similarly, but would require the specification of the joint distribution of these effects, and the replacement of each of the one-dimensional integrals appearing in equation (23.10) by a multiple integral. The evaluation of such multiple integrals would of course make the computations even more time-consuming.

23.7.3 Inclusion of covariates

In some applications – although not the application which motivated this study – there may be covariate information available that could help to explain observed behaviour (e.g. dietary differences, or whether subjects are male or female). The important question then to be answered is whether such covariate information can efficiently be incorporated into the likelihood computation. The building-blocks of the likelihood are the transition probabilities and initial distribution of the latent process, and the probabilities π_1 and π_2 of the behaviour of interest in the two states. Any of these probabilities can be allowed to depend on covariates without greatly complicating the likelihood computation.

If we wished to introduce a (possibly time-dependent) covariate y_t into the probabilities π_i, here denoted $\pi_i(y_t)$, we could take logit $\pi_1(y_t)$ to be $a_1 + a_2 y_t$, and similarly logit $\pi_2(y_t)$. The likelihood evaluation would then present no new challenges, although the extra parameters would of course tend to slow down the optimization. If the covariate y_t were instead thought to affect the transition probabilities, we could define logit $\gamma_{11}(n_t, y_t)$ to be $\alpha_0 + \alpha_1 n_t + \alpha_2 y_t$, and similarly logit $\gamma_{22}(n_t, y_t)$.

23.8 Other modifications or extensions

Other potentially useful extensions are to increase the number of latent states above two, and to change the nature of the state-dependent distribution, for example, to allow for more than two behaviour categories or for a continuous behaviour variable.

23.8.1 Increasing the number of states

If more than two latent states are needed in the model, this can be accommodated, for example, by using in equation (23.2) a higher-dimensional analogue of the logit transform. Any such increase potentially brings with it a large increase in the number of parameters, however; if there are m states, there are $m^2 - m$ transition probabilities to be specified, and it might be necessary to impose some structure on the transition probabilities in order to reduce the number of parameters.

23.8.2 Changing the nature of the state-dependent distribution

In this work the observations are binary, and therefore so are the state-dependent distributions, that is, the conditional distributions of an observation given the underlying state. But there might well be more than two behaviour categories in some series of observations, or the observations might be continuous. That would require the use, in the likelihood

computation, of a different kind of distribution from the binary distribution used here. That is a simple matter, and indeed that flexibility at
observation level is one of the advantages of HM or similar models; almost any kind of data can be accommodated at observation level without
greatly complicating the likelihood computation. But since here the observations feed back into the nutrient level N_t, the feedback mechanism
would need to be correspondingly modified.

23.9 Application to caterpillar feeding behaviour

23.9.1 Data description and preliminary analysis

The model was applied to sequences of observations of eight final-instar
Helicoverpa armigera caterpillars, collected in an experiment designed to
quantify developmental changes in the pattern of feeding (Raubenheimer
and Barton Browne, 2000). The caterpillars were observed continuously
for 1132 minutes, during which time they were scanned at 1-minute
intervals and scored as either feeding or not feeding. In order to isolate
developmental changes from environmental effects, individually housed
caterpillars were fed semi-synthetic foods of homogeneous and constant
nutrient composition, and the recordings were made under conditions of
constant temperature and lighting. The caterpillars were derived from
a laboratory culture, and so had similar ancestry and developmental
histories. Figure 23.1 displays the data.

Some remarks can immediately be made. Firstly, despite the uniform
conditions of the experiment, there is considerable between-subject variation, both as regards the density of feeding events and the apparent
pattern thereof. The runlengths of the feeding events differ between subjects; see, for example, those of subjects 5 and 6. A striking feature is
that subject 8 began feeding only 6.2 hours after the start of the experiment. Closer examination of the original recordings made on this
subject revealed that it was initially eating its exuviae (moulted skin), a
behaviour which has been demonstrated to be of nutritional significance
(Mira, 2000). However, since the nutrient composition of the exuviae is
very different from that of the synthetic foods, in what follows only subjects 1–7 were included in the analysis. Also noticeable is the fact that
subject 3 stopped feeding more than 2 hours before the end of the experiment. This anomaly became apparent in the model-checking procedure
described below.

23.9.2 Parameter estimates and model checking

The first step in the model-fitting process was to fit a model separately
to each of the seven subjects, that is, to estimate for each one the nine

Table 23.1 *Parameter estimates and log-likelihood: individual models for subjects 1–7, and mixed model with six common parameters and random effect for* λ.

Subj.	$\widehat{\alpha}_0$	$\widehat{\alpha}_1$	$\widehat{\beta}_0$	$\widehat{\beta}_1$	$\widehat{\pi}_1$	$\widehat{\pi}_2$	$\widehat{\lambda}$	$\widehat{n}_0$	$-\log L$
1	5.807	−11.257	2.283	2.301	0.936	0.000	0.027	0.240	331.991
2	2.231	−5.284	−0.263	21.019	0.913	0.009	0.032	0.150	347.952
3	4.762	−10.124	2.900	15.912	0.794	0.004	0.080	0.740	225.166
4	2.274	−7.779	1.294	16.285	0.900	0.000	0.056	0.018	298.678
5	3.135	−7.271	1.682	10.908	0.911	0.006	0.097	1.000	332.510
6	3.080	−5.231	1.374	13.970	0.880	0.001	0.043	0.246	291.004
7	3.888	−9.057	0.617	13.341	0.976	0.003	0.054	0.375	315.188
									2142.488
mixed model	2.735	−5.328	2.127	7.083	0.919	0.003	$\widehat{\mu}=0.055$ $\widehat{\sigma}=0.051$		2230.090

parameters $\alpha_0, \alpha_1, \beta_0, \beta_1, \pi_1, \pi_2, \lambda, n_0, \delta_1$. In all cases δ_1 was estimated as zero, that is, the subject started in state 2. An explanation for this is that, until the post-moult skin has hardened, insects cannot use their mouthparts and so behave as if they were sated. In what follows we shall take it that all subjects start in state 2, and shall not further treat δ_1 as a parameter requiring estimation.

Table 23.1 displays the parameter estimates for each of subjects 1–7, and the corresponding values of minus the log-likelihood. Several features are apparent. For all subjects, $\widehat{\pi}_1$ is fairly close to 1, and $\widehat{\pi}_2$ very close to 0. All values of $\widehat{\alpha}_1$ are negative, and all values of $\widehat{\beta}_1$ positive. The estimates $\widehat{\alpha}_0, \widehat{\alpha}_1, \widehat{\beta}_0$ and $\widehat{\beta}_1$ differ substantially between subjects. (But it turns out that the resulting plots of transition probabilities are all similar to that for subject 1, shown as Figure 23.5.) The values of $\widehat{n}_0$ (essentially a nuisance parameter) differ substantially, but – more significantly – so also do the values of $\widehat{\lambda}$.

We now use subject 1 as illustrative. Figure 23.3 displays the observed feeding behaviour, the underlying motivational state sequence inferred by means of the Viterbi algorithm, and the nutrient level. Figure 23.4 presents an enlarged version of the observed feeding behaviour and inferred motivational state.

A point to note from the figures is the close correspondence between the series of feeding bouts and the inferred states. However, as is demonstrated by Figure 23.4, feeding bouts were interspersed with brief periods of non-feeding which did not break the continuity of the inferred state.

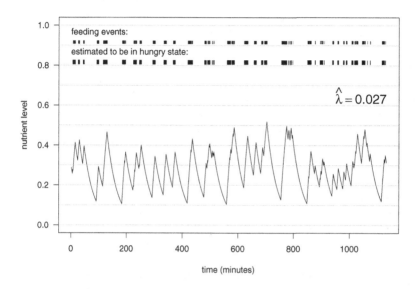

Figure 23.3 *Feeding behaviour, inferred motivational state and nutrient level for subject 1.*

The model thus succeeded in the aim of delimiting states according to the probability distributions of behaviours, rather than the occurrence of behaviours *per se*. The nutrient level for subject 1 ranges from about 0.1 to 0.5; for some other subjects the lower bound can reach zero. The parameter λ determines the (exponential) rate at which the nutrient level diminishes in the absence of feeding. The associated half-life is given by $\log(0.5)/\log(1-\lambda)$. Thus the estimated half-life for subject 1 is approximately 25 minutes.

Figure 23.5 displays the transition probabilities for subject 1 as a function of nutrient level. As expected, $\widehat{\gamma}_{11}$ decreases with increasing nutrient level, and $\widehat{\gamma}_{22}$ increases. Note also that $\widehat{\pi}_2$, the estimated probability of feeding when sated, is close to zero. In fact $\widehat{\pi}_2$ is less than 0.01 for all seven subjects, and less than 0.001 for two of them. The fact that these estimates are so close to the boundary of the parameter space has implications when one attempts to estimate standard errors. The standard

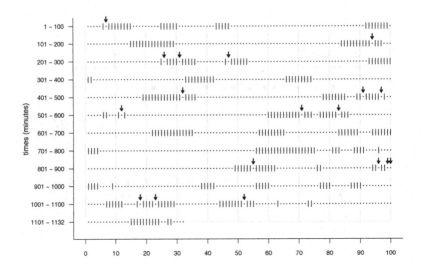

Figure 23.4 *Feeding behaviour of subject 1 (vertical lines), with arrows indicating times at which the subject was not feeding but was inferred to be in state 1, 'hungry'.*

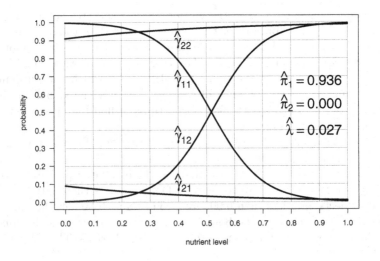

Figure 23.5 *Transition probabilities for subject 1.*

Table 23.2 *Runlength statistics. The seven columns are: subject no.; number of feeding runs, mean length of feeding runs, standard deviation of the length of feeding runs; number of (estimated) hungry runs, mean length of hungry runs, standard deviation of the length of hungry runs.*

Subject	Feeding runs			Estimated hungry runs		
	number	mean	s.d.	number	mean	s.d.
1	58	5.4	4.1	41	8.1	4.9
2	67	3.0	2.3	53	3.9	2.8
3	41	3.1	2.1	22	6.7	2.4
4	57	2.6	1.5	51	3.0	1.7
5	65	2.8	1.6	54	3.5	2.0
6	51	4.1	2.8	35	6.4	3.8
7	57	4.1	2.4	52	4.6	2.7

errors of the parameters for subject 1 were estimated by the parametric bootstrap, but are not included here.

Table 23.2 gives runlength statistics, for subjects 1–7, of the observed feeding sequences and the estimated sequences of the state 'hungry'. As expected, there are fewer runs for the latter (23% fewer on average) and the mean runlength is larger (45% on average), as is the standard deviation (20% on average).

In applying the model-checking technique of Section 23.5 to subjects 1–7, we were unable to reject H_0 in six cases. Using the chi-squared approximation to the distribution of deviance differences, we obtained p-values ranging from 0.30 to 0.82 in these six cases. In the case of subject 3 we concluded that the model fitted is unsatisfactory ($p = 0.051$). Except for subject 3, AIC would also select the hypothesized model.

Examining subject 3 more closely, we note that some of the parameter values are atypical; for example, the probability of feeding when hungry is atypically low, only about 0.8. The data for subject 3 revealed the unusual feature that there were no feeding events after about time 1000, that is, no feeding for more than 2 hours. This is clearly inconsistent with the earlier behaviour of this subject. This conclusion was reinforced by a plot of deviance residuals for subjects 1 and 3, along with spline smooths of these.

23.9.3 Runlength distributions

One of the key questions of biological interest that motivated this work was to assess the extent to which runs of feeding events differ from the runs in motivational state 1 (hungry), and similarly, runs of non-feeding

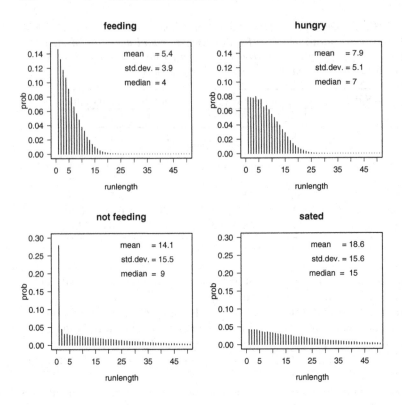

Figure 23.6 *Estimated runlength distributions for subject* 1.

events from runs in state 2 (sated). In an HMM the distributions of the runlengths in the two motivational states are geometric; that cannot be assumed to be the case here. Monte Carlo methods were used here to estimate the runlength distributions under the model.

For each of the seven subjects, a series of length 1 million was generated from the relevant fitted model, and the distribution of each of the four types of run estimated. Figure 23.6 displays plots and summary statistics of the four estimated distributions for subject 1.

The distribution of the feeding runlength clearly differs from that in the hungry state in the expected way. The probability of a runlength being 1 is almost twice as great for feeding runs as it is for hungry runs. The mean and median for the latter are greater. The duration of a hungry spell is often longer than one would conclude if one took feeding and hunger as synonymous.

Unlike the distribution of the sated runs, that of the non-feeding runs

Table 23.3 *Summary of models fitted jointly to the seven subjects. The number of parameters estimated is denoted by k. AIC selects the model with π_1 and π_2 common, and BIC the mixed model, which treats λ as a random effect.*

Model no.	Description	$-\log L$	k	AIC	BIC
1	no parameters common	2142.488	56	4396.976	4787.725
2	π_1 and π_2 common	2153.669	44	**4395.338**	4702.355
3	six parameters common	2219.229	20	4478.458	4618.011
4	seven parameters common	2265.510	14	4559.020	4656.707
5	mixed model (one random effect)	2230.090	15	4490.180	**4594.845**

has a marked peak at 1. This peak is attributable to hungry subjects often interrupting feeding for one time unit, rather than subjects being sated for a single time unit.

A further property worth noting is that the estimated distributions of hungry and sated runlengths do indeed have properties inconsistent with those of geometric distributions (on the positive integers). For hungry and sated runs the estimated means are 7.9 and 18.6 (respectively) and the s.d.s are 5.1 and 15.6. If they had geometric distributions with those means, the standard deviations would be higher: 7.4 and 18.1.

23.9.4 Joint models for seven subjects

Five different models were considered for the observations on all seven subjects:

1. a model with no parameters common to the subjects (no pooling);
2. a model with only π_1 and π_2 common to the subjects (partial pooling);
3. a model with the six parameters $\alpha_0, \alpha_1, \beta_0, \beta_1, \pi_1, \pi_2$ common to the subjects (partial pooling);
4. a model with the seven parameters $\alpha_0, \alpha_1, \beta_0, \beta_1, \pi_1, \pi_2, \lambda$ common to the subjects (complete pooling);
5. a model which incorporated a random effect for the parameter λ and common values for the six parameters $\alpha_0, \alpha_1, \beta_0, \beta_1, \pi_1, \pi_2$ (partial pooling with one random effect). The model used for λ, which is bounded by 0 and 1, was a normal distribution (mean μ and variance σ^2) restricted to the interval $[0, 1]$. This distribution was suggested by a kernel density estimate of λ-estimates from model 3.

Note that in each of the five cases we are still fitting the same type of model to each subject. The above five *joint* models differ because they are based on different assumptions about the values of the parameters for the seven subjects.

In all cases the subjects were taken to be independent of each other and to have started in state 2. As there were only seven subjects, models with more than one random effect were not fitted because one cannot expect to identify a joint distribution of random effects from only seven observations. In the case of model 5, it was necessary to replace a small part of the **R** code by code written in C in order to speed up the computations.

Table 23.3 displays in each case the log-likelihood attained, the number of parameters estimated, AIC and BIC. The model selected by the AIC is model 2, which has $\pi_1 = 0.913$ and $\pi_2 = 0.002$ for all subjects. The model selected by the BIC is model 5, the mixed model. The parameter estimates for that model appear in Table 23.1, along with those of the seven individual models. An interesting point to note is how much better the mixed model is than model 4. These two models differ only in their treatment of λ; model 4 uses a single fixed value, and the mixed model uses a random effect.

23.10 Discussion

The application of HMMs to animal behaviour (MacDonald and Raubenheimer, 1995) was initially limited to behaviours whose consequences do not readily alter motivational state.

The model presented here provides an extension which can allow for the important class of behaviours that are feedback-regulated (Toates, 1986). In the present example, the states 'hungry' and 'sated' as used above are of course an artefact of the model, and do not necessarily correspond to the accepted meanings of those terms. A different way of using the model described here would be to define state 1 as the state in which the probability of feeding is (say) 0.9, and similarly state 2 as that with probability (say) 0.1.

Irrespective of how the states are defined, the important point is that their delineation provides an objective means for exploring the physiological and environmental factors that determine the transitions between activities by animals. Such transitions are believed to play an important role in the evolution of behaviour (Sibly and McFarland, 1976), and the understanding of their causal factors is considered a central goal in the study of animal behaviour (Dewsbury, 1992).

The exponential filter (equation (23.1)) used here for the nutrient level seems plausible. It is, however, by no means the only possibility, and no fundamental difficulty arises if this component of the model is changed. Ideally, the filter should reflect the manner in which feeding affects the motivational process, which might be expected to vary with a range of factors such as the nutrient composition of the foods, the recent feeding history of the animals, their state of health, etc. (Simpson, 1990). In our example, the same filter applied reasonably well across the experimental animals, probably because they were standardized for age and developmental history, and were studied in uniform laboratory conditions. Interestingly, however, a study of foraging by wild grasshoppers revealed that the patterns of feeding were no less regular than those observed in tightly controlled laboratory conditions (Raubenheimer and Bernays, 1993), suggesting that there might be some uniformity of such characteristics as the decay function even in more complex ecological conditions.

The models introduced here, or variants thereof, are potentially of much wider applicability than to feeding behaviour. They may be applied essentially unchanged to any binary behaviour thought to have a feedback effect on some underlying state, and with some modification of the feedback mechanism the restriction to binary observations could be removed.

In overview, by allowing for feedback regulation we have extended the application of HM or similar models to a wider range of applications in the study of behaviour. We believe that these models hold potential for exploring the relationships among observed behaviours, the activities within which they occur, and the underlying causal factors.

CHAPTER 24

Estimating the survival rates of Soay sheep from mark–recapture–recovery data

Schafe können sicher weiden.

J.S. Bach, *Was mir behagt, ist nur die muntre Jagd* (Hunting Cantata BWV 208), 1713

24.1 Introduction

The population of Soay sheep (*Ovis aries*) on Hirta, a remote island off the west coast of Scotland, has been the subject of numerous studies of population dynamics, because of the unusually isolated environment and the relative ease with which individuals can be marked and recaptured. They have no natural predators, and the last human residents left Hirta in 1932, after which the sheep established a wild population.

During annual surveys, carried out in summer from 1985 to 2009, newborn sheep were tagged, and sightings of previously tagged live sheep were recorded. Only 62% of sightings (referred to as 'recaptures') involved physical capture, in which case the sheep was weighed; in the remaining 38% of recaptures the body mass was not recorded. Searches were made for dead sheep, but not all were found. The 'capture history' of each individual in the population indicates when the sheep was not seen (0), seen alive (1), or recovered dead (2). For example, the capture history

$$1 \ 0 \ 0 \ 1 \ 0 \ 2$$

indicates that the individual was seen alive (and marked) on the first occasion, not seen on occasions 2 and 3, recaptured on occasion 4, not seen on occasion 5, and recovered dead on occasion 6. The term 'recovered dead' on capture occasion t is reserved for individuals who were still alive at $t-1$; those that died before $t-1$ are termed 'long dead' and not recorded because they are assumed to be unidentified (due to erosion of the markings).

We will restrict our attention to the capture histories of the 1344 female sheep for which at least one recorded body mass is available.

The average number of observations per individual is 4.64, and 900 of them were recovered dead in the course of the observation period. Our objective here is to estimate the survival rate of Soay sheep and to investigate how this depends on their body mass, because the primary cause of mortality is starvation. The risk of starvation is highest for the young, and so the age of the sheep is also taken into account. It is not the objective here to investigate all the factors that are assumed to affect the survival rate. For details of the population dynamics of the Soay sheep, we refer to Clutton-Brock and Pemberton (2004).

Conditioning on the initial capture of each individual leads to open population models called Cormack–Jolly–Seber models (see Schwarz and Seber, 1999, for a review), and it is this class of models we consider here. These models are often used to investigate the effect of environmental and individual covariates on demographic indicators such as the survival probability (see, for instance, Catchpole $et\ al.$, 2000; Schofield and Barker, 2011). Statistical inference is fairly straightforward both in the case of environmental covariates and in that of individual-specific time-constant covariates. However, the case of individual-specific time-varying covariates, such as body mass, is much more difficult to deal with when some values of the covariates are missing (e.g. if there was no recapture). If the covariates take on a finite number of possible values then the Arnason–Schwarz model can be used (Schwarz $et\ al.$, 1993). For continuous-valued covariates, Langrock and King (2013) developed an HMM-based technique, in which the space of possible covariate values is discretized in a manner analogous to that described in Chapter 11.

We begin by presenting the HMM formulation for standard mark–recapture–recovery (MRR) data without covariate information (Section 24.2), and then extend this to include individual-specific time-varying continuous covariate information (Section 24.3). We illustrate the extension by investigating how the body mass of the Soay sheep affects their probability of survival (Section 24.4).

24.2 MRR data without use of covariates

MRR data are conveniently expressed as capture histories of the individual animals. We denote the capture history for an individual by $(X_1, \ldots, X_T)$, where

$$X_t = \begin{cases} 1 & \text{if the individual is observed alive,} \\ 2 & \text{if the individual is recovered dead at time } t, \\ 0 & \text{otherwise,} \end{cases}$$

and T is the number of capture occasions.

Following initial capture of the individual, the recorded values result from the combination of two distinct processes: one that gives the survival status (alive, recently dead, long dead) at time t and the second gives the observed value X_t, given the survival status. Thus, MRR data can be modelled using an HMM in which an underlying state process $\{C_t\}$ describes the survival state at time t, and a state-dependent process $\{X_t\}$ that models the observed recapture/recovery event, given the survival state. The survival states are coded as

$$C_t = \begin{cases} 1 & \text{if the individual is alive at time } t, \\ 2 & \text{if the individual is dead at time } t, \text{ but was alive at } t-1, \\ 3 & \text{if the individual died before time } t-1, \end{cases}$$

for $t = g, \ldots, T$, where g denotes the occasion on which the individual is initially observed and marked, here the year of birth. Note again that we distinguish between 'recently dead' individuals ($C_t = 2$) and 'long dead' individuals ($C_t = 3$), and we make the assumption, standard in MRR models, that only recently dead individuals can be classified as 'recovered dead'. A detailed discussion of state-space formulations for MRR data is given by Gimenez et al. (2007).

The likelihood of the observed capture histories is a function of survival, recapture and recovery probabilities. In particular, we set

$$\phi_t = \Pr(C_{t+1} = 1 \mid C_t = 1) \quad \text{(survival probability)},$$
$$p_t = \Pr(X_t = 1 \mid C_t = 1) \quad \text{(recapture probability)},$$
$$\lambda_t = \Pr(X_t = 2 \mid C_t = 2) \quad \text{(recovery probability)}.$$

For some capture histories $\{X_t\}$ it is possible to deduce the entire survival sequence $\{C_t\}$, in which case the likelihood is simply the product of survival and recapture/recovery probabilities. For example, from the history

$$1\ 0\ 0\ 1\ 0\ 2 \tag{24.1}$$

it can be deduced that the animal must have been alive on occasion 5 (despite the fact that it was not seen), as otherwise it could not have been found 'recently dead' on occasion 6. The corresponding sequence of survival states is therefore

$$1\ 1\ 1\ 1\ 1\ 2$$

so that the likelihood, conditional on the initial capture, is given by

$$\phi_1(1 - p_2)\phi_2(1 - p_3)\phi_3 p_4 \phi_4 (1 - p_5)(1 - \phi_5)\lambda_6.$$

However, in general it is not always possible to deduce the survival states. For example, from the history

$$1\ 0\ 1\ 0\ 0\ 0 \tag{24.2}$$

it cannot be determined whether or not the individual was alive on
occasions 4, 5 and 6. In the HMMs that we have considered so far, the
states are not observed at all. The model for this application differs in
that some of the states are observed (or can be deduced) and some are
not.

Let $S = \{t : C_t \text{ is known}\}$ denote the set of all occasions in $\{g, g +
1, \ldots, T\}$ at which the survival state of the individual is known, and S^c
the set of occasions in $\{g, g + 1, \ldots, T\}$ at which the survival state is
unknown, following initial capture. For the histories given in (24.1) and
(24.2), $S^c = \emptyset$ and $S^c = \{4, 5, 6\}$, respectively.

In general, conditional on the initial capture, the likelihood for a single
capture history can be written in the form

$$L_T = \sum_{\tau \in S^c} \sum_{C_\tau = 1}^{3} \prod_{t=g+1}^{T} \Pr(C_t \mid C_{t-1}) \Pr(X_t \mid C_t). \qquad (24.3)$$

The above expression sums over all possible survival histories that are
compatible with the observed capture history. Here

$$\Pr(C_t \mid C_{t-1}) = \begin{cases} \phi_{t-1} & C_t = 1; \ C_{t-1} = 1, \\ 1 - \phi_{t-1} & C_t = 2; \ C_{t-1} = 1, \\ 1 & C_t = 3; \ C_{t-1} \in \{2, 3\}, \\ 0 & \text{otherwise,} \end{cases}$$

and

$$\Pr(X_t \mid C_t) = \begin{cases} p_t & C_t = 1; \ X_t = 1, \\ 1 - p_t & C_t = 1; \ X_t = 0, \\ \lambda_t & C_t = 2; \ X_t = 2, \\ 1 - \lambda_t & C_t = 2; \ X_t = 0, \\ 1 & C_t = 3; \ X_t = 0, \\ 0 & \text{otherwise.} \end{cases}$$

As is usual for HMMs, the likelihood expression (24.3) can be written
as a matrix product

$$L_T = \delta \left(\prod_{t=g+1}^{T} \mathbf{\Gamma}_{t-1} \mathbf{P}(x_t) \right) \mathbf{1}', \qquad (24.4)$$

where

$$\mathbf{\Gamma}_t = \begin{pmatrix} \phi_t & 1 - \phi_t & 0 \\ 0 & 0 & 1 \\ 0 & 0 & 1 \end{pmatrix}$$

gives the transition probabilities for the survival process $\{C_t\}$ between times t and $t+1$, and

$$\mathbf{P}(x_t) = \begin{cases} \operatorname{diag}(1-p_t, 1-\lambda_t, 1) & \text{if } x_t = 0, \\ \operatorname{diag}(p_t, 0, 0) & \text{if } x_t = 1, \\ \operatorname{diag}(0, \lambda_t, 0) & \text{if } x_t = 2. \end{cases}$$

(The notation diag(...) denotes the diagonal matrix with the given diagonal.) Here the initial state distribution is $\boldsymbol{\delta} = (1,0,0)$ because we include in the study only individuals that were alive when first captured. The likelihood (24.4) is that of a *partially* hidden Markov model; one sums only over the unknown states, rather than over all possible state sequences. For basic MRR data, the likelihood can be calculated more efficiently by using sufficient statistics (see, for example, McCrea, 2012, and references therein). We have introduced the HMM form as a building-block of the extension to time-varying individual covariates discussed in the next section.

24.3 MRR data involving individual-specific time-varying continuous-valued covariates

We now consider the case in which the covariates are individual-specific, time-varying and continuous-valued, for example the condition of the individual as measured by proxies such as body mass or parasitic load. We restrict our attention to the case in which the survival probabilities depend on a single covariate. The extension to multiple covariates is straightforward in principle but it involves a large increase in computation time.

Let Y_t denote the value of the covariate of a given individual at time t, $t = g, \ldots, T$. It is only possible to record the value for the covariate when $X_t = 1$ but sometimes it is not recorded even then, for example if the animal is sighted alive but is not physically recaptured. For dead animals the value of the covariate is undefined.

As regards notation, we partition the set of times when the animal either is known to be, or could be, alive (i.e. when it was not known to be dead) into two subsets: $\mathcal{W}$ represents the times when Y_t was recorded and $\mathcal{V}$ the times when Y_t was not recorded. We define $\mathbf{Y}_\mathcal{V} = \{Y_t : t \in \mathcal{V}\}$. We use the general symbol f for all probability density functions.

We consider models in which the probability of survival from occasion t to $t+1$ is a function of Y_t, for example,

$$\operatorname{logit}(\phi_t) = \beta_0 + \beta_1 Y_t .$$

The covariate series is modelled by a first-order Markov process, $f(Y_t \mid Y_{t-1})$, for $t = g+1, \ldots, T$, with $f(Y_t \mid Y_{t-1}) = 1$ for t such that $C_t = 2$ or 3 (i.e. when an individual is dead). For example, Bonner and Schwarz

(2006) and King, Brooks and Coulson (2008) consider models of the type

$$Y_{t+1} \mid Y_t \sim N(Y_t + a_t, \sigma^2) \qquad (24.5)$$

with a_t varying over time, as well as extensions thereof that allow for additional complexities such as age-dependence. If the covariate is not recorded at the initial capture then we also need to specify a distribution, f_0, for the initial covariate value.

If the capture history of an individual is such that the survival state is known at all times, and the covariate values are also known at all times, then the likelihood is simply the product of survival probabilities, recapture probabilities and values of the conditional distribution $f(Y_t \mid Y_{t-1})$. However, not all the survival states are known, and also some covariate values are missing. The likelihood then involves summation over all possible survival state sequences (as in expression (24.3)) and also integration over all possible covariate sequences. If the covariate series is assumed to be a first-order Markov process, the likelihood (given the initial capture event) is of the following form:

$$L_T = \int \cdots \int \sum_{\tau \in S^c} \sum_{C_\tau = 1}^{3} f_0(Y_g)$$

$$\times \prod_{t=g+1}^{T} \Pr(C_t \mid C_{t-1}, Y_{t-1}) \Pr(X_t \mid C_t) f(Y_t \mid Y_{t-1}) \, d\mathbf{Y}_{\mathcal{V}}.$$

$$(24.6)$$

In general the integrals make this likelihood expression analytically intractable, which has led some authors to conclude that maximum likelihood estimation will usually be impossible (Bonner, Morgan and King, 2010). In the next section we demonstrate that it can be accurately approximated.

In Chapter 11 we described how HMMs can be used to approximate models having continuous-valued latent process, simply by discretizing the state space. The latent process considered here consists of survival process and covariate process, both of which are partially observed. The technique is nevertheless applicable. A difference between the models covered in Chapter 11 and the MRR models considered here is that we do not now assume that the states of the latent process are all unknown ('hidden'); here the survival status of the individual and the value of the covariate are sometimes observed. For such occasions we need not integrate (or sum) over those observed values.

As in Chapter 11, we define an essential range, that is, an interval (b_0, b_m) that covers all but an insignificant subset of the support of the marginal distribution of the covariate values. We partition the essential range into m intervals (e.g. $m = 100$) of equal length, $B_j = [b_{j-1}, b_j)$, $j = 1, \ldots, m$, and let b_j^* denote the midpoint of B_j. With such a partition it is possible to derive a multiple-sum expression approximating (24.6) which is analogous to (24.3) but rather more complicated. Such an expression appears as equation (2.6) of Langrock and King (2013).

Fortunately it is unnecessary to try to compute the likelihood in that way, because the likelihood can be given as a product of matrices. One begins by splitting the survival state 'alive' into m distinct states, corresponding to 'alive and with covariate value in B_j', $j = 1, \ldots, m$. The complete state space of the (partially) hidden process – now giving survival state and covariate value – comprises these m states plus 'recently dead' (state $m + 1$) and 'long dead' (state $m + 2$).

We define the $(m + 2) \times (m + 2)$ matrix

$$\Gamma_t^{(m)} = \begin{pmatrix} \phi_t(1)\Psi_t(1,1) & \cdots & \phi_t(1)\Psi_t(1,m) & 1 - \phi_t(1) & 0 \\ \vdots & \ddots & \vdots & \vdots & \vdots \\ \phi_t(m)\Psi_t(m,1) & \cdots & \phi_t(m)\Psi_t(m,m) & 1 - \phi_t(m) & 0 \\ 0 & & \cdots & 0 & 1 \\ 0 & & \cdots & 0 & 1 \end{pmatrix},$$

with

$$\Psi_t(i,j) = \begin{cases} f(Y_{t+1} \mid Y_t) & \text{if } t, t+1 \in \mathcal{W}, Y_t \in B_i, Y_{t+1} \in B_j, \\ f(Y_{t+1} \mid b_i^*) & \text{if } t \in \mathcal{V}, t+1 \in \mathcal{W}, Y_{t+1} \in B_j, \\ \Pr(Y_{t+1} \in B_j \mid Y_t) & \text{if } t \in \mathcal{W}, t+1 \in \mathcal{V}, Y_t \in B_i, \\ \Pr(Y_{t+1} \in B_j \mid b_i^*) & \text{if } t, t+1 \in \mathcal{V}, \\ 0 & \text{otherwise}, \end{cases}$$

and

$$\phi_t(i) = \begin{cases} \Pr(C_{t+1} = 1 \mid C_t = 1, Y_t) & \text{if } t \in \mathcal{W}, \ Y_t \in B_i, \\ \Pr(C_{t+1} = 1 \mid C_t = 1, b_i^*) & \text{if } t \in \mathcal{V}, \\ 0 & \text{otherwise}. \end{cases}$$

The product $\phi_t(i)\Psi_t(i,j)$ is the probability or density that the individual survives from time t to time $t + 1$, with the covariate value changing from a given value in the interval B_i at time t (the observed covariate value, if available, otherwise the representative value) to some value in the interval B_j at time $t + 1$ (again, the observed covariate value, if available).

The elements of $\Gamma_t^{(m)}$ are functions of the parameters of the model specified for the covariate process. For example, for the model given in

(24.5) one has

$$\Pr(Y_{t+1} \in B_j \mid b_i^*) = \Phi\left(\frac{b_j - (b_i^* + a_t)}{\sigma}\right) - \Phi\left(\frac{b_{j-1} - (b_i^* + a_t)}{\sigma}\right),$$

where Φ is the distribution function of the standard normal distribution.

We have increased the number of states from 3 to $m + 2$ and so the diagonal matrix of state-dependent probabilities is of dimension $(m + 2) \times (m + 2)$:

$$\mathbf{P}^{(m)}(x_t) = \begin{cases} \text{diag}(1 - p_t, \ldots, 1 - p_t, 1 - \lambda_t, 1) & \text{if } x_t = 0, \\ \text{diag}(p_t, \ldots, p_t, 0, 0) & \text{if } x_t = 1, \\ \text{diag}(0, \ldots, 0, \lambda_t, 0) & \text{if } x_t = 2. \end{cases}$$

(In each case above, the first m diagonal elements are equal.) Finally, in order to cover the case in which Y_g is unrecorded, we define the row vector $\boldsymbol{\delta}^{(m)}$ of length $m + 2$ with ith element

$$\delta_i^{(m)} = \begin{cases} \int_{b_{i-1}}^{b_i} f_0(z)\,\mathrm{d}z & \text{if } g \in \mathcal{V},\ i \in \{1, \ldots, m\}, \\ f_0(Y_g) & \text{if } g \in \mathcal{W},\ Y_g \in B_i, \\ 0 & \text{otherwise}, \end{cases}$$

where f_0 is the probability density function of Y_g.

Putting the above components together, the matrix formulation of the likelihood is

$$L_T = \boldsymbol{\delta}^{(m)} \left(\prod_{t=g+1}^{T} \boldsymbol{\Gamma}_{t-1}^{(m)} \mathbf{P}^{(m)}(x_t) \right) \mathbf{1}_{m+2}'. \qquad (24.7)$$

This matrix expression has the same structure as in the case without covariates; see (24.4).

The likelihood for multiple (independent) individuals is simply the product of the individual likelihoods computed using (24.7). It is then a routine matter to maximize this joint likelihood (numerically) with respect to the parameters. Approximate confidence intervals for the parameters can be obtained from the estimated Hessian, or by using a parametric bootstrap. Model selection, including selection of the underlying covariate process model, can be carried out by using a criterion such as the AIC.

24.4 Application to Soay sheep data

We now analyse the capture histories for the Soay sheep population described in Section 24.1. We assume that the survival probabilities are a function of body mass (in kilograms).

Table 24.1 *Number of parameters (k), minus log-likelihood and AIC values for models fitted to the Soay sheep data.*

	k	$-l$	AIC
model 1	70	10 221.85	**20 583.70**
model 2	61	10 350.66	20 823.32
model 3	24	10 309.28	20 666.56
model 4	47	10 260.68	20 615.36
model 5	66	10 404.45	20 940.90

Following Bonner *et al.* (2010), we distinguish four age groups: lambs (age < 1), yearlings (age $\in [1, 2)$), adults (age $\in [2, 7)$) and seniors (age ≥ 7). We assume a logistic-linear relationship between body mass and the survival probability, that is,

$$\text{logit}\,(\phi_t) = \beta_{a_t,0} + \beta_{a_t,1} Y_t.$$

For a given sheep, a_t indicates the age group (lamb, yearling, adult or senior) of the sheep at time t. The five models considered can be summarized as follows:

1. $Y_t = Y_{t-1} + \eta_{a_t} (\mu_{a_t} - Y_{t-1}) + \sigma_{a_t} \varepsilon_t$ (i.e. distinct covariate process parameters across age groups), recapture probability time-dependent, recovery probability time-dependent (70 parameters);

2. $Y_t = Y_{t-1} + \eta (\mu - Y_{t-1}) + \sigma \varepsilon_t$ (i.e. covariate process parameters fixed across age groups), recapture probability time-dependent, recovery probability time-dependent (61 parameters);

3. same covariate model as for Model 1, recapture probability constant, recovery probability constant (24 parameters);

4. same covariate model as for Model 1, recapture probability constant, recovery probability time-dependent (47 parameters);

5. $Y_t = Y_{t-1} + \mu_{a_t} + \sigma_{a_t} \varepsilon_t$, recapture probability time-dependent, recovery probability time-dependent (66 parameters).

In each case the ε_ts are independent and identically distributed standard normal random variables. Model 5 has a covariate process similar to those used by King *et al.* (2008) and Bonner *et al.* (2010).

Each of the models was fitted with $m = 50$ intervals for the discretization of the covariate process. The assumed essential range of covariate values was specified by $b_0 = 0.8 b_{\min}$ and $b_m = 1.2 b_{\max}$, where $b_{\min}$ and $b_{\max}$ denote the minimum and the maximum of the observed covariate values, respectively. For the given data, $b_{\min} = 2.9$ and $b_{\max} = 33.9$. A normal distribution was assumed for the initial covariate value. The log-likelihood and AIC values obtained for the five models listed are given in Table 24.1. In terms of the AIC, Model 1 is identified as the best of these models, and by a substantial margin.

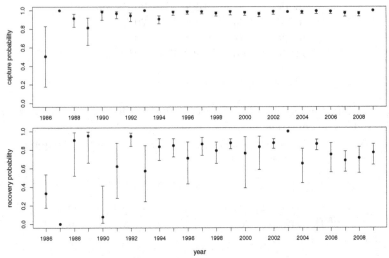

Figure 24.1 *Estimates of the yearly recapture and recovery probabilities ob-tained for Model 1. Points represent the ML estimates, and error bars indicate the 95% confidence intervals (only for those estimates that do not lie at the boundary of the parameter space).*

Figure 24.1 displays the estimated year-dependent recapture and re-covery probabilities for Model 1. The results generally match those of Bonner *et al.* (2010) well for the years common to the two analyses (i.e. 1986–2000), except for the first 2 years. This mismatch appears to be related to the use of slightly different data: for example, in our data set there are no recoveries in 1987, but Bonner *et al.* (2010) estimate a pos-itive recovery probability in that year. The recovery probabilities vary considerably over time, suggesting that models with constant recovery probabilities are inappropriate.

Figure 24.2 displays the estimated survival probabilities for Model 1 for the different age groups, in each case as a function of body mass. Pointwise confidence intervals based on the Hessian were obtained with the method described in Section 3.6.1. Again, the results are similar to those of Bonner *et al.* (2010). The survival probability increases with increasing body mass, with this effect strongest for lambs and seniors, and weakest for adults. The interval estimates are slightly narrower than

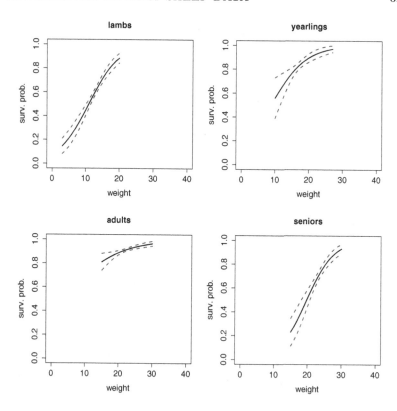

Figure 24.2 *Estimated survival probability as a function of the covariate body mass (kg), for the four different age groups (for Model 1). Solid lines give the maximum likelihood estimates, and dashed lines indicate the 95% pointwise confidence intervals.*

those obtained by Bonner *et al.*, which is not surprising, given that we consider a larger data set here.

In Figure 24.3, the observed body masses of sheep at ages 0–13 are compared to distributions of body masses (of live sheep) for these ages derived from Model 1. Similar plots were also examined for Models 2 and 5, and all three plots appear in Figure 3 of Langrock and King (2013). Model 1 appears to capture the development of the body mass over the years. However, as can be seen in Figure 3 of Langrock and King (2013), the 'diffusive' nature of the covariate process in Model 5 leads to increasingly wider interval estimates for body mass as age increases, with the intervals not capturing the observed quantiles well. Thus, as we have already seen from the AIC statistic, it appears that a model with

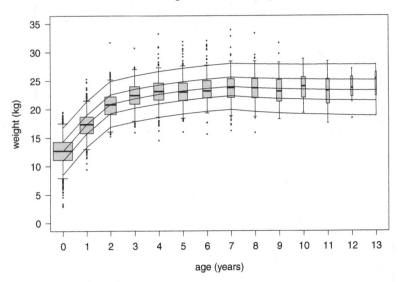

Figure 24.3 *Boxplot of observed body masses of sheep at ages 0–13, plus quantiles derived from Model 1. These quantiles (continuous lines) are for 5%, 25%, 50%, 75% and 95%. The box width is proportional to the square root of the corresponding number of observations.*

non-diffusive autoregressive covariate process is more appropriate in this application.

24.5 Conclusion

HMMs provide a framework for statistical inference in various MRR settings, and particularly when there are animal-specific time-varying covariates. Pradel (2005) gives a detailed account of the application of HMM techniques in scenarios with discrete individual-specific and time-varying covariates. For MRR studies that involve continuous individual-specific and time-varying covariates, the HMM-based approach discussed above (and originally proposed by Langrock and King, 2013) is the first method to implement maximum likelihood estimation in this context. Recently, Michelot *et al.* (2016) have extended this approach to allow for nonparametric estimation of the relationship between demographic parameters and covariates, using B-splines and a penalized maximum likelihood approach. Other recent developments related to the use of

HMMs for MRR data include models with multivariate states (Johnson *et al.*, 2016) and models with semi-Markov state process (King and Langrock, 2016).

APPENDIX A

Examples of R code

In this Appendix we present **R** functions which will perform basic analyses in respect of a Poisson–HMM, plus several examples of code which uses these functions to analyse the earthquakes series. It is intended that this code will be available on the web pages of the book, currently www.hmms-for-time-series.de, along with further examples of the use of these functions.

A.1 The functions

In Sections A.1.1–A.1.4 we give four functions which, for a Poisson–HMM:

- implement the transformations described in Section 3.3.1, and their inverse;

- compute minus the log-likelihood of a Poisson–HMM for given values of the working parameters; and

- estimate the (natural) parameters of the model by using numerical minimization of minus the log-likelihood.

The purpose of the transformation of the natural parameters (Poisson means, transition probabilities and, if appropriate, the initial distribution of the Markov chain) to working parameters is simply to convert a constrained optimization problem to an unconstrained one. The separation of the transformation functions from the estimation functions is not essential, but has the advantage of providing simple, easily checkable tools that can readily be modified to cope with constraints of a different form.

The code will cope with both the stationary and the non-stationary cases. The former case is when one assumes that the Markov chain is not merely homogeneous but stationary, and is the default. Note that, in the non-stationary case, the function pois.HMM.pn2pw cannot be expected to work if any component of delta is exactly zero. In practice this is not a problem, unless one explicitly provides such a delta as starting value.

These functions allow for the special case of a 'one-state HMM', and for the possibility that there are observations missing; missing data are treated as described in Section 2.3.4.

Notice, in particular, the form of the output of the code in Section A.1.4: a list consisting of

- the number of states (m);
- the estimates of λ, Γ and δ;
- the 'convergence code' of the optimizer `nlm`, an integer from 1 to 5; and
- the values of minus the log-likelihood, AIC and BIC.

This is the form in which all the functions given in Sections A.1.5–A.1.14 expect the input `mod` to be supplied.

A.1.1 Transforming natural parameters to working

```
 1  pois.HMM.pn2pw <- function(m,lambda,gamma,delta=NULL,
                                stationary=TRUE)
 2  {
 3    tlambda  <- log(lambda)
 4    if(m==1) return(tlambda)
 5    foo      <- log(gamma/diag(gamma))
 6    tgamma   <- as.vector(foo[!diag(m)])
 7    if(stationary) {tdelta   <- NULL}
 8      else {tdelta <- log(delta[-1]/delta[1])}
 9    parvect <- c(tlambda,tgamma,tdelta)
10    return(parvect)
11  }
```

A.1.2 Transforming working parameters to natural

```
 1  pois.HMM.pw2pn <- function(m,parvect,stationary=TRUE)
 2  {
 3    lambda        <- exp(parvect[1:m])
 4    gamma         <- diag(m)
 5    if (m==1) return(list(lambda=lambda,gamma=gamma,delta=1))
 6    gamma[!gamma] <- exp(parvect[(m+1):(m*m)])
 7    gamma         <- gamma/apply(gamma,1,sum)
 8    if(stationary){delta<-solve(t(diag(m)-gamma+1),rep(1,m))}
 9      else {foo<-c(1,exp(parvect[(m*m+1):(m*m+m-1)]))
10      delta<-foo/sum(foo)}
11    return(list(lambda=lambda,gamma=gamma,delta=delta))
12  }
```

A.1.3 Computing minus the log-likelihood from the working parameters

Notice that the vector `foo` is scaled at each iteration to have sum 1. When scaling is done in this way, the scaled likelihood that emerges at the end of the loop is automatically 1. Hence the log-likelihood we seek is just the final value of `lscale`, the sum of the logs of all the quantities by which the vector `foo` has been divided.

```
 1   pois.HMM.mllk  <- function(parvect,x,m,stationary=TRUE,...)
 2   {
 3    if(m==1) return(-sum(dpois(x,exp(parvect),log=TRUE)))
 4    n        <- length(x)
 5    pn       <- pois.HMM.pw2pn(m,parvect,stationary=stationary)
 6    foo      <- pn$delta*dpois(x[1],pn$lambda)
 7    sumfoo   <- sum(foo)
 8    lscale   <- log(sumfoo)
 9    foo      <- foo/sumfoo
10    for (i in 2:n)
11      {
12      if(!is.na(x[i])){P<-dpois(x[i],pn$lambda)}
13         else {P<-rep(1,m)}
14      foo     <- foo %*% pn$gamma*P
15      sumfoo  <- sum(foo)
16      lscale  <- lscale+log(sumfoo)
17      foo     <- foo/sumfoo
18      }
19    mllk      <- -lscale
20    return(mllk)
21   }
```

A.1.4 Computing the MLEs, given starting values for the natural parameters

```
 1   pois.HMM.mle <-
 2   function(x,m,lambda0,gamma0,delta0=NULL,stationary=TRUE,...)
 3   {
 4    parvect0  <- pois.HMM.pn2pw(m,lambda0,gamma0,delta0,
               stationary=stationary)
 5    mod       <- nlm(pois.HMM.mllk,parvect0,x=x,m=m,
               stationary=stationary)
 6    pn        <- pois.HMM.pw2pn(m=m,mod$estimate,stationary=stationary)
 7    mllk      <- mod$minimum
 8    np        <- length(parvect0)
 9    AIC       <- 2*(mllk+np)
10    n         <- sum(!is.na(x))
11    BIC       <- 2*mllk+np*log(n)
12    list(m=m,lambda=pn$lambda,gamma=pn$gamma,delta=pn$delta,
               code=mod$code,mllk=mllk,AIC=AIC,BIC=BIC)
13   }
```

A.1.5 Generating a sample

This function generates a realization, of length ns, of the HMM mod.

```
 1   pois.HMM.generate_sample  <- function(ns,mod)
 2   {
 3    mvect           <- 1:mod$m
 4    state           <- numeric(ns)
 5    state[1]        <- sample(mvect,1,prob=mod$delta)
 6    for (i in 2:ns) state[i] <- sample(mvect,1,
                             prob=mod$gamma[state[i-1],])
 7    x               <- rpois(ns,lambda=mod$lambda[state])
 8    return(x)
 9   }
```

A.1.6 Global decoding by the Viterbi algorithm

Given the model mod and observed series x, this function performs global decoding as described in Section 5.4.2.

```
1   pois.HMM.viterbi<-function(x,mod)
2   {
3    n                    <- length(x)
4    xi                   <- matrix(0,n,mod$m)
5    foo                  <- mod$delta*dpois(x[1],mod$lambda)
6    xi[1,]               <- foo/sum(foo)
7    for (i in 2:n)
8     {
9     foo<-apply(xi[i-1,]*mod$gamma,2,max)*dpois(x[i],mod$lambda)
10    xi[i,] <- foo/sum(foo)
11    }
12   iv<-numeric(n)
13   iv[n]      <-which.max(xi[n,])
14   for (i in (n-1):1)
15     iv[i] <- which.max(mod$gamma[,iv[i+1]]*xi[i,])
16   return(iv)
17  }
```

A.1.7 Computing log(forward probabilities)

Given data x and model mod, this function uses the recursion $\alpha_{t+1} = \alpha_t \mathbf{\Gamma P}(x_{t+1})$ (see Section 4.1.1) to find all the vectors of forward probabilities, in logarithmic form. A matrix (lalpha) is returned. Scaling of the same kind as used for likelihood computations is implemented.

```
1   pois.HMM.lforward<-function(x,mod)
2   {
3    n          <- length(x)
4    lalpha     <- matrix(NA,mod$m,n)
5    foo        <- mod$delta*dpois(x[1],mod$lambda)
6    sumfoo     <- sum(foo)
7    lscale     <- log(sumfoo)
8    foo        <- foo/sumfoo
9    lalpha[,1] <- lscale+log(foo)
10   for (i in 2:n)
11    {
12    foo        <- foo%*%mod$gamma*dpois(x[i],mod$lambda)
13    sumfoo     <- sum(foo)
14    lscale     <- lscale+log(sumfoo)
15    foo        <- foo/sumfoo
16    lalpha[,i] <- log(foo)+lscale
17    }
18   return(lalpha)
19  }
```

A.1.8 Computing log(backward probabilities)

Similarly, this function uses the recursion $\beta_t' = \mathbf{\Gamma P}(x_{t+1})\beta_{t+1}'$ (see Section 4.1.2) to find all the vectors of backward probabilities, in logarithmic form.

```
1   pois.HMM.lbackward<-function(x,mod)
2   {
3    n            <- length(x)
4    m            <- mod$m
5    lbeta        <- matrix(NA,m,n)
6    lbeta[,n]    <- rep(0,m)
7    foo          <- rep(1/m,m)
8    lscale       <- log(m)
9    for (i in (n-1):1)
10    {
11     foo          <- mod$gamma%*%(dpois(x[i+1],mod$lambda)*foo)
12     lbeta[,i]    <- log(foo)+lscale
13     sumfoo       <- sum(foo)
14     foo          <- foo/sumfoo
15     lscale       <- lscale+log(sumfoo)
16    }
17    return(lbeta)
18  }
```

A.1.9 Conditional probabilities

Equation (5.3) is used here to find, for all t, the conditional probabilities $\Pr\left(X_t = x \mid \mathbf{X}^{(-t)}\right)$, given the data x and model mod. The input xc specifies the range of x-values for which these probabilities are required. The use of the shifts lafact and lbfact is intended to prevent underflow in the exponentiations. The row vector foo is then normalized to have sum 1.

```
1   #==Conditional probability that observation at time t equals
2   #   xc, given all observations other than that at time t.
3   #   Note: xc is a vector and the result (dxc) is a matrix.
4   pois.HMM.conditional <- function(xc,x,mod)
5   {
6    n            <- length(x)
7    m            <- mod$m
8    nxc          <- length(xc)
9    dxc          <- matrix(NA,nrow=nxc,ncol=n)
10    Px           <- matrix(NA,nrow=m,ncol=nxc)
11    for (j in 1:nxc) Px[,j] <-dpois(xc[j],mod$lambda)
12    la           <- pois.HMM.lforward(x,mod)
13    lb           <- pois.HMM.lbackward(x,mod)
14    la           <- cbind(log(mod$delta),la)
15    lafact       <- apply(la,2,max)
16    lbfact       <- apply(lb,2,max)
17    for (i in 1:n)
18     {
19      foo          <-
20      (exp(la[,i]-lafact[i])%*%mod$gamma)*exp(lb[,i]-lbfact[i])
21      foo          <- foo/sum(foo)
22      dxc[,i]      <- foo%*%Px
23     }
24    return(dxc)
25  }
```

A.1.10 Pseudo-residuals

The function pois.HMM.conditional in Section A.1.9 is now used to
find ordinary normal pseudo-residuals as described in Sections 6.2.1 and
6.2.2. The form of the output is this (in the notation and terminology
of those sections). For each time t from 1 to n, the function provides
the lower and upper normal pseudo-residuals, z_t^- and z_t^+, and the mid-
pseudo-residual $z_t^m = \Phi^{-1}\left((u_t^- + u_t^+)/2\right)$. Plots such as those in the top
row of Figure 6.5 can then be produced.

```
1  pois.HMM.pseudo_residuals <- function(x,mod)
2  {
3   n          <- length(x)
4   cdists     <- pois.HMM.conditional(xc=0:max(x),x,mod)
5   cumdists <- rbind(rep(0,n),apply(cdists,2,cumsum))
6   ulo <- uhi <- rep(NA,n)
7   for (i in 1:n)
8     {
9     ulo[i]   <- cumdists[x[i]+1,i]
10    uhi[i]   <- cumdists[x[i]+2,i]
11    }
12   umi        <- 0.5*(ulo+uhi)
13   npsr       <- qnorm(rbind(ulo,umi,uhi))
14   return(npsr)
15 }
```

A.1.11 State probabilities

Here we compute probabilities $\Pr(C_t = i \mid \mathbf{X}^{(T)} = \mathbf{x}^{(T)})$, for $t \in \{1, 2, \ldots, T\}$. See Section 5.4.1, equation (5.6) in particular. The functions pois.HMM.lforward and pois.HMM.lbackward are used and, as
elsewhere, a shift (by c) is used to counteract underflow in the exponen-
tiation.

```
1  pois.HMM.state_probs <- function(x,mod)
2  {
3   n          <- length(x)
4   la         <- pois.HMM.lforward(x,mod)
5   lb         <- pois.HMM.lbackward(x,mod)
6   c          <- max(la[,n])
7   llk        <- c+log(sum(exp(la[,n]-c)))
8   stateprobs <- matrix(NA,ncol=n,nrow=mod$m)
9   for (i in 1:n) stateprobs[,i]<-exp(la[,i]+lb[,i]-llk)
10  return(stateprobs)
11 }
```

A.1.12 State prediction

Here we use equation (5.12) and the function pois.HMM.lforward to
compute probabilities $\Pr(C_{T+h} = i \mid \mathbf{X}^{(T)} = \mathbf{x}^{(T)})$, for a range of values
$h \in \mathbb{N}$.

```
1   # Note that state output 'statepreds' is a matrix even if h=1.
2   pois.HMM.state_prediction <- function(h=1,x,mod)
3   {
4    n            <- length(x)
5    la           <- pois.HMM.lforward(x,mod)
6    c            <- max(la[,n])
7    llk          <- c+log(sum(exp(la[,n]-c)))
8    statepreds <- matrix(NA,ncol=h,nrow=mod$m)
9    foo <- exp(la[,n]-llk)
10   for (i in 1:h){
11     foo<-foo%*%mod$gamma
12     statepreds[,i]<-foo
13     }
14   return(statepreds)
15  }
```

A.1.13 Local decoding

See Section 5.4.1. This is a straightforward application of the function
pois.HMM.state_probs in Section A.1.11.

```
1   pois.HMM.local_decoding <- function(x,mod)
2   {
3    n            <- length(x)
4    stateprobs <- pois.HMM.state_probs(x,mod)
5    ild          <- rep(NA,n)
6    for (i in 1:n) ild[i]<-which.max(stateprobs[,i])
7    ild
8   }
```

A.1.14 Forecast probabilities

Given x and mod, equation (5.5) is used here to find forecast probabilities
$\Pr(X_{T+h} = x \mid \mathbf{X}^{(T)} = \mathbf{x}^{(T)})$. The range of x-values for which these
probabilities are required is specified by the input xf (e.g. 0:50 or 0:45,
as in Section A.2.2) and the range of times for which they are required
by h.

```
1   # Note that the output 'dxf' is a matrix.
2   pois.HMM.forecast <- function(xf,h=1,x,mod)
3   {
4    n         <- length(x)
5    nxf       <- length(xf)
6    dxf       <- matrix(0,nrow=h,ncol=nxf)
7    foo       <- mod$delta*dpois(x[1],mod$lambda)
8    sumfoo    <- sum(foo)
9    lscale    <- log(sumfoo)
10   foo       <- foo/sumfoo
11   for (i in 2:n)
12     {
13     foo       <- foo%*%mod$gamma*dpois(x[i],mod$lambda)
14     sumfoo <- sum(foo)
15     lscale <- lscale+log(sumfoo)
16     foo       <- foo/sumfoo
17     }
```

```
18    for (i in 1:h)
19    {
20      foo    <- foo%*%mod$gamma
21      for (j in 1:mod$m) dxf[i,] <- dxf[i,] +
               foo[j]*dpois(xf,mod$lambda[j])
22    }
23    return(dxf)
24 }
```

A.2 Examples of code using the above functions

The purpose of this section is to provide a few examples of code which uses the functions in Section A.1, so it will be convenient to load those functions first. Further examples of the use of those functions will be provided on the web pages.

A.2.1 Fitting Poisson–HMMs to the earthquakes series

This code uses all the functions in Sections A.1.1–A.1.4 in order to fit to the earthquakes series both stationary and non-stationary models, with two, three and four states.

```
1  dat <- read.table(
2    "http://www.hmms-for-time-series.de/second/data/earthquakes.txt")
3  #(or set your own path)
4  x    <-dat[,2]
5  d    <-dat[,1]
6  n    <-length(x)
7  #==================================== fit 2-state HMM
8  m<-2
9  lambda0<-c(15,25)
10 gamma0<-matrix(
11 c(
12 0.9,0.1,
13 0.1,0.9
14 ),m,m,byrow=TRUE)
15 mod2s<-pois.HMM.mle(x,m,lambda0,gamma0,stationary=TRUE)
16 delta0<-c(1,1)/2
17 mod2h<-pois.HMM.mle(x,m,lambda0,gamma0,delta=delta0,stationary=FALSE)
18 mod2s; mod2h
19 #==================================== fit 3-state HMM
20 m<-3
21 lambda0<-c(10,20,30)
22 gamma0<-matrix(
23 c(
24   0.8,0.1,0.1,
25   0.1,0.8,0.1,
26   0.1,0.1,0.8
27 ),m,m,byrow=TRUE)
28 mod3s<-pois.HMM.mle(x,m,lambda0,gamma0,stationary=TRUE)
29 delta0 <- c(1,1,1)/3
30 mod3h<-pois.HMM.mle(x,m,lambda0,gamma0,delta=delta0,stationary=FALSE)
31 mod3s; mod3h
32 #==================================== fit 4-state HMM
33 m<-4
34 lambda0<-c(10,15,20,30)
35 gamma0<-matrix(
```

```
36    c(
37       0.85,0.05,0.05,0.05,
38       0.05,0.85,0.05,0.05,
39       0.05,0.05,0.85,0.05,
40       0.05,0.05,0.05,0.85
41    ),m,m,byrow=TRUE)
42  mod4s<-pois.HMM.mle(x,m,lambda0,gamma0,stationary=TRUE)
43  delta0<-c(1,1,1,1)/4
44  mod4h<-pois.HMM.mle(x,m,lambda0,gamma0,delta=delta0,stationary=FALSE)
45  mod4s; mod4h
```

A.2.2 Forecast probabilities

Here we demonstrate the use of pois.HMM.forecast, and plot the forecast distributions. Note that the output of that function is returned as a matrix.

```
1   #=== Use it for 1-step-ahead and plot the forecast distribution.
2   h<-1
3   xf<-0:50
4   forecasts<-pois.HMM.forecast(xf,h,x,mod3s)
5   fc<-forecasts[1,]
6   par(mfrow=c(1,1),las=1)
7   plot(xf,fc,type="h",
8   main=paste("Earthquake series: forecast distribution for", d[n]+1),
9   xlim=c(0,max(xf)),ylim=c(0,0.12),xlab="count",ylab="probability",lwd=3)
10
11  #=== Forecast 1-4 steps ahead and plot these.
12  h<-4
13  xf<-0:45
14  forecasts<-pois.HMM.forecast(xf,h,x,mod3s)
15
16  par(mfrow=c(2,2),las=1)
17  for (i in 1:4)
18     {
19  fc<-forecasts[i,]
20  plot(xf,fc,type="h",main=paste("Forecast distribution for", d[n]+i),
21  xlim=c(0,max(xf)),ylim=c(0,0.12),xlab="count",ylab="probability",lwd=3)
22     }
23
24  #=== Compute the marginal distribution (called "dstat" below)
25  #    for mod3h.
26  #=== This is also the long-term forecast.
27  m<-3.
28
29  lambda<-mod3h$lambda
30  delta<-solve(t(diag(m)-mod3h$gamma+1),rep(1,m))
31  dstat<-numeric(length(xf))
32  for (j in 1:m) dstat <- dstat + delta[j]*dpois(xf,lambda[j])
33
34  #=== Compare the 50-year-ahead forecast with the long-term forecast.
35  h<-50
36  xf<-0:45
37  forecasts<-pois.HMM.forecast(xf,h,x,mod3h)
38  fc<-forecasts[h,]
39  par(mfrow=c(1,1),las=1)
40  plot(xf,fc,type="h",
41  main=paste("Forecast distribution for", d[n]+h),
42  xlim=c(0,max(xf)),ylim=c(0,0.12),xlab="count",ylab="probability",lwd=3)
43  lines(xf,dstat,col="gray",lwd=3)
```

APPENDIX B

Some proofs

In this appendix we present proofs of five results, shown here in boxes, that are used principally in Section 4.1. None of these results is surprising, given the structure of an HMM. Indeed, a more intuitive, and less laborious, way to establish such properties is to invoke the separation properties of the directed graph of the model. More precisely, if one can establish that the sets of random variables $\mathbf{A}$ and $\mathbf{B}$ in a directed graphical model are 'd-separated' by the set $\mathbf{C}$, it will then follow that $\mathbf{A}$ and $\mathbf{B}$ are conditionally independent given $\mathbf{C}$; see Pearl (2000, pp. 16–18) or Bishop (2006, pp. 378 and 619). An account of the properties of HMMs that is similar to the approach we follow here is provided by Koski (2001, Chapter 13). We present the results for the case in which the random variables X_t are discrete. Analogous results hold in the continuous case.

B.1 A factorization needed for the forward probabilities

The first purpose of this appendix is to establish the following result, which we use in Section 4.1.1 in order to interpret $\alpha_t(i)$ as the forward probability $\Pr(\mathbf{X}^{(t)}, C_t = i)$. (Recall that $\alpha_t(i)$ was defined as the ith element of $\boldsymbol{\alpha}_t = \boldsymbol{\delta}\mathbf{P}(x_1) \prod_{s=2}^{t} \boldsymbol{\Gamma}\mathbf{P}(x_s)$; see equation (2.15).)

> For positive integers t,
> $$\Pr(\mathbf{X}^{(t+1)}, C_t, C_{t+1}) = \Pr(\mathbf{X}^{(t)}, C_t)\,\Pr(C_{t+1} \mid C_t)\,\Pr(X_{t+1} \mid C_{t+1}).$$
> $$\tag{B.1}$$

Throughout this appendix we assume that
$$\Pr(C_t \mid \mathbf{C}^{(t-1)}) = \Pr(C_t \mid C_{t-1})$$
and
$$\Pr(X_t \mid \mathbf{X}^{(t-1)}, \mathbf{C}^{(t)}) = \Pr(X_t \mid C_t).$$

In addition, we assume that these (and other) conditional probabilities are defined, in which case the probabilities that appear as denominators in what follows are strictly positive. The model may be represented, as usual, by the directed graph in Figure 2.2 on p. 30.

The tool we use throughout this appendix is the following factorization for the joint distribution of the set of random variables V_i in a directed graphical model, which appeared earlier as equation (2.5):

$$\Pr(V_1, V_2, \ldots, V_n) = \prod_{i=1}^{n} \Pr(V_i \mid \mathrm{pa}(V_i)), \qquad (\text{B.2})$$

where $\mathrm{pa}(V_i)$ denotes all the parents of V_i in the set V_1, V_2, $\ldots$, V_n. In our model, the only parent of X_k is C_k, and (for $k = 2, 3, \ldots$) the only parent of C_k is C_{k-1}; C_1 has no parent. The joint distribution of $\mathbf{X}^{(t)}$ and $\mathbf{C}^{(t)}$, for instance, is therefore given by

$$\Pr(\mathbf{X}^{(t)}, \mathbf{C}^{(t)}) = \Pr(C_1) \prod_{k=2}^{t} \Pr(C_k \mid C_{k-1}) \prod_{k=1}^{t} \Pr(X_k \mid C_k). \qquad (\text{B.3})$$

In order to prove equation (B.1), note that equation (B.3) and the analogous expression for $\Pr(\mathbf{X}^{(t+1)}, \mathbf{C}^{(t+1)})$ imply that

$$\Pr(\mathbf{X}^{(t+1)}, \mathbf{C}^{(t+1)}) = \Pr(C_{t+1} \mid C_t) \, \Pr(X_{t+1} \mid C_{t+1}) \, \Pr(\mathbf{X}^{(t)}, \mathbf{C}^{(t)}).$$

Now sum over $\mathbf{C}^{(t-1)}$; the result is equation (B.1). $\qquad\qquad\square$

Furthermore, (B.1) can be generalized as follows.

> For any (integer) $T \geq t + 1$,
> $$\Pr(\mathbf{X}_1^T, C_t, C_{t+1}) = \Pr(\mathbf{X}_1^t, C_t) \, \Pr(C_{t+1} \mid C_t) \, \Pr(\mathbf{X}_{t+1}^T \mid C_{t+1}).$$
> $$(\text{B.4})$$

(Recall the notation $\mathbf{X}_a^b = (X_a, X_{a+1}, \ldots, X_b)$.) Briefly, the proof of (B.4) proceeds as follows. First write $\Pr(\mathbf{X}_1^T, \mathbf{C}_1^T)$ as

$$\Pr(C_1) \prod_{k=2}^{T} \Pr(C_k \mid C_{k-1}) \prod_{k=1}^{T} \Pr(X_k \mid C_k),$$

then split each of the two products into $k \leq t$ and $k \geq t + 1$. Use the fact that

$$\Pr(\mathbf{X}_{t+1}^T, \mathbf{C}_{t+1}^T) = \Pr(C_{t+1}) \prod_{k=t+2}^{T} \Pr(C_k \mid C_{k-1}) \prod_{k=t+1}^{T} \Pr(X_k \mid C_k),$$

and sum $\Pr(\mathbf{X}_1^T, \mathbf{C}_1^T)$ over $\mathbf{C}_{t+2}^T$ and $\mathbf{C}_1^{t-1}$. $\qquad\qquad\square$

B.2 Two results needed for the backward probabilities

In this section we establish the two results used in Section 4.1.2 in order to interpret $\beta_t(i)$ as the backward probability $\Pr(\mathbf{X}_{t+1}^T \mid C_t = i)$.

The first of these is that,

> for $t = 0, 1, \ldots, T - 1$,
> $$\Pr(\mathbf{X}_{t+1}^T \mid C_{t+1}) = \Pr(X_{t+1} \mid C_{t+1}) \, \Pr(\mathbf{X}_{t+2}^T \mid C_{t+1}). \qquad (\text{B.5})$$

This is established by noting that

$$\Pr(\mathbf{X}_{t+1}^T, \mathbf{C}_{t+1}^T)$$

$$= \Pr(X_{t+1} \mid C_{t+1}) \left(\Pr(C_{t+1}) \prod_{k=t+2}^{T} \Pr(C_k \mid C_{k-1}) \prod_{k=t+2}^{T} \Pr(X_k \mid C_k) \right)$$

$$= \Pr(X_{t+1} \mid C_{t+1}) \Pr(\mathbf{X}_{t+2}^T, \mathbf{C}_{t+1}^T),$$

and then summing over $\mathbf{C}_{t+2}^T$ and dividing by $\Pr(C_{t+1})$. $\square$

The second result is that,

---

for $t = 1, 2, \ldots, T - 1$,

$$\Pr(\mathbf{X}_{t+1}^T \mid C_{t+1}) = \Pr(\mathbf{X}_{t+1}^T \mid C_t, C_{t+1}). \qquad \text{(B.6)}$$

This we prove as follows. The right-hand side of equation (B.6) is

$$\frac{1}{\Pr(C_t, C_{t+1})} \sum_{\mathbf{C}_{t+2}^T} \Pr(\mathbf{X}_{t+1}^T, \mathbf{C}_t^T),$$

which by (B.2) reduces to

$$\sum_{\mathbf{C}_{t+2}^T} \prod_{k=t+2}^{T} \Pr(C_k \mid C_{k-1}) \prod_{k=t+1}^{T} \Pr(X_k \mid C_k).$$

The left-hand side is

$$\frac{1}{\Pr(C_{t+1})} \sum_{\mathbf{C}_{t+2}^T} \Pr(\mathbf{X}_{t+1}^T, \mathbf{C}_{t+1}^T),$$

which reduces to the same expression. $\square$

B.3 Conditional independence of $\mathbf{X}_1^t$ and $\mathbf{X}_{t+1}^T$

Here we establish the conditional independence of $\mathbf{X}_1^t$ and $\mathbf{X}_{t+1}^T$ given C_t, used in Section 4.1.3 to link the forward and backward probabilities to the probabilities $\Pr(\mathbf{X}^{(T)} = \mathbf{x}^{(T)}, C_t = i)$. That is, we show that,

---

for $t = 1, 2, \ldots, T - 1$,

$$\Pr(\mathbf{X}_1^T \mid C_t) = \Pr(\mathbf{X}_1^t \mid C_t) \Pr(\mathbf{X}_{t+1}^T \mid C_t). \qquad \text{(B.7)}$$

To prove this, first note that

$$\Pr(\mathbf{X}_1^T, \mathbf{C}_1^T) = \Pr(\mathbf{X}_1^t, \mathbf{C}_1^t) \frac{1}{\Pr(C_t)} \Pr(\mathbf{X}_{t+1}^T, \mathbf{C}_t^T),$$

which follows by repeated application of equation (B.2). Then sum over $\mathbf{C}_1^{t-1}$ and $\mathbf{C}_{t+1}^T$. This yields

$$\Pr(\mathbf{X}_1^T, C_t) = \Pr(\mathbf{X}_1^t, C_t) \frac{1}{\Pr(C_t)} \Pr(\mathbf{X}_{t+1}^T, C_t),$$

from which the result is immediate.   □

References

Abramowitz, M., Stegun, I.A., Danos, M. and Rafelski, J. (eds) (1984). *Pocketbook of Mathematical Functions. Abridged Edition of Handbook of Mathematical Functions.* Verlag Harri Deutsch, Thun and Frankfurt am Main.

Aitchison, J. (1982). The statistical analysis of compositional data. *J. Roy. Statist. Soc. B* **44**, 139–177.

Aitkin, M. (1996). A general maximum likelihood analysis of overdispersion in generalized linear models. *Statist. and Computing* **6**, 251–262.

Albert, P.S. (1991). A two-state Markov mixture model for a time series of epileptic seizure counts. *Biometrics* **47**, 1371–1381.

Altman, R.M. (2007). Mixed hidden Markov models: an extension of the hidden Markov model to the longitudinal data setting. *J. Amer. Statist. Assoc.* **102**, 201–210.

Altman, R.M. and Petkau, J.A. (2005). Application of hidden Markov models to multiple sclerosis lesion count data. *Statist. Med.* **24**, 2335–2344.

Aston, J.A.D. and Martin, D.E.K. (2007). Distributions associated with general runs and patterns in hidden Markov models. *Ann. Appl. Statist.* **1**, 585–611.

Azzalini, A. and Bowman, A.W. (1990). A look at some data on the Old Faithful geyser. *Appl. Statist.* **39**, 357–365.

Barbu, V.S. and Limnios, N. (2008). *Semi-Markov Chains and Hidden Semi-Markov Models toward Applications: Their Use in Reliability and DNA Analysis.* Springer, New York.

Bartolucci, F. (2011). An alternative to the Baum-Welch recursions for hidden Markov models. arXiv:1201.0277v1 [math.ST].

Bartolucci, F. and De Luca, G. (2003). Likelihood-based inference for asymmetric stochastic volatility models. *Computat. Statist. & Data Analysis* **42**, 445–449.

Bartolucci, F., Farcomeni, A. and Pennoni, F. (2013). *Latent Markov Models for Longitudinal Data.* Chapman & Hall/CRC Press, Boca Raton, FL.

Barton Browne, L. (1993). Physiologically induced changes in resource-oriented behaviour. *Ann. Rev. Entomology* **38**, 1–25.

Baum, L.E. (1972). An inequality and associated maximization technique in statistical estimation for probabilistic functions of Markov processes. In *Proc. Third Symposium on Inequalities*, O. Shisha (ed.), 1–8. Academic Press, New York.

Baum, L.E., Petrie, T., Soules, G. and Weiss, N. (1970). A maximization technique occurring in the statistical analysis of probabilistic functions of Markov chains. *Ann. Math. Statist.* **41**, 164–171.

Bebbington, M.S. (2007). Identifying volcanic regimes using hidden Markov Models. *Geophys. J. Int.* **171**, 921–942.

Bellman, R. (1960). *Introduction to Matrix Analysis*. McGraw-Hill, New York.

Berchtold, A. (1999). The double chain Markov model. *Commun. Stat. Theory Meth.* **28**, 2569–2589.

Berchtold, A. (2001). Estimation in the mixture transition distribution model. *J. Time Series Anal.* **22**, 379–397.

Berchtold, A. and Raftery, A.E. (2002). The mixture transition distribution model for high-order Markov chains and non-Gaussian time series. *Statist. Sci.* **17**, 328–356.

Bisgaard, S. and Travis, L.E. (1991). Existence and uniqueness of the solution of the likelihood equations for binary Markov chains. *Statist. Prob. Letters* **12**, 29–35.

Bishop, C.M. (2006). *Pattern Recognition and Machine Learning*. Springer, New York.

Blackwell, P.G. (2003). Bayesian inference for Markov processes with diffusion and discrete components. *Biometrika* **90**, 613–627.

Bonner, S.J., Morgan, B.J.T. and King, R. (2010). Continuous covariates in mark-recapture-recovery analysis: a comparison of methods. *Biometrics* **66**, 1256–1265.

Bonner, S.J. and Schwarz, C.J. (2006). An extension of the Cormack–Jolly–Seber model for continuous covariates with application to *Microtus pennsylvanicus*. *Biometrics* **62**, 142–149.

Boys, R.J. and Henderson, D.A. (2004). A Bayesian approach to DNA sequence segmentation (with discussion). *Biometrics* **60**, 573–588.

Brockwell, A.E. (2007). Universal residuals: a multivariate transformation. *Statist. Prob. Letters* **77**, 1473–1478.

Brown, G.O. and Buckley, W.S. (2015). Experience rating with Poisson mixtures. *Ann. Actuarial Sci.* **9**, 304–321.

Bulla, J. and Berzel, A. (2008). Computational issues in parameter estimation for stationary hidden Markov models. *Computat. Statist.* **23**, 1–18.

Bulla, J. and Bulla, I. (2007). Stylized facts of financial time series and hidden semi-Markov models. *Computat. Statist. & Data Analysis* **51**, 2192–2209.

Bulla, J. and Bulla, I. (2013). hsmm: Hidden Semi Markov Models. R package version 0.4, URL http://CRAN.R-project.org/package=hsmm, accessed 22 July 2015.

Bulla, J., Bulla, I. and Nenadić, O. (2009). An R package for analyzing hidden semi-Markov models. *Computat. Statist. & Data Analysis* **54**, 611–619.

Bulla, J., Lagona, F., Maruotti, A. and Picone, M. (2012). A multivariate hidden Markov model for the identification of sea regimes from incomplete skewed and circular time series. *J. Agric. Biol. Envir. Statist.* **17**, 544–567.

Calvet, L. and Fisher, A.J. (2001). Forecasting multifractal volatility. *J. Econometrics* **105**, 27–58.

Calvet, L. and Fisher, A.J. (2004). How to forecast long-run volatility: regime switching and the estimation of multifractal processes. *J. Financial Econometrics* **2**, 49–83.

Camproux, A.C., Saunier, F., Chouvet, G., Thalabard, J.C. and Thomas, G.

(1996). A hidden Markov model approach to neuron firing patterns. *Biophys. J.* **75**, 2404–2412.

Cappé, O. (2001). Ten years of HMMs. Technical report, URL http://perso.telecom-paristech.fr/~cappe/docs/hmmbib.pdf, accessed 16 July 2014.

Cappé, O., Moulines, E. and Rydén, T. (2005). *Inference in Hidden Markov Models*. Springer, New York.

Carter, C.K. and Kohn, R. (1994). On Gibbs sampling for state space models. *Biometrika* **81**, 541–553.

Catchpole, E.A., Morgan, B.J.T., Coulson, T.N., Freeman, S.N. and Albon, S.D. (2000). Factors influencing Soay sheep survival. *J. Roy. Statist. Soc. C* **49**, 453–472.

Celeux, G. and Durand, J.-B. (2008). Selecting hidden Markov model state number with cross-validated likelihood. *Computat. Statist.* **23**, 541–564.

Celeux, G., Hurn, M. and Robert, C.P. (2000). Computational and inferential difficulties with mixture posterior distributions. *J. Amer. Statist. Assoc.* **95**, 957–970.

Chan, K.S. and Ledolter, J. (1995). Monte Carlo EM estimation for time series models involving counts. *J. Amer. Statist. Assoc.* **90**, 242–252.

Chib, S. (1996). Calculating posterior distributions and modal estimates in Markov mixture models. *J. Econometrics* **75**, 79–97.

Chib, S., Nardari, F. and Shephard, N. (2002). Markov chain Monte Carlo methods for stochastic volatility models. *J. Econometrics* **108**, 281–316.

Chopin, N. (2007). Inference and model choice for sequentially ordered hidden Markov models. *J. Roy. Statist. Soc. B* **69**, 269–284.

Churchill, G.A. (1989). Stochastic models for heterogeneous DNA sequences. *Bull. Math. Biol.* **51**, 79–94.

Clutton-Brock, T.H. and Pemberton, J.M. (2004). *Soay Sheep: Dynamics and Selection in an Island Population*. Cambridge University Press, Cambridge.

Codling, E.A., Plank, M.J. and Benhamou, S. (2008). Random walk models in biology. *J. Roy. Soc. Interface* **5**, 813–834.

Congdon, P. (2006). Bayesian model choice based on Monte Carlo estimates of posterior model probabilities. *Computat. Statist. & Data Analysis* **50**, 346–357.

Cont, R. (2001). Empirical properties of asset returns: stylized facts and statistical issues. *Quant. Finance* **1**, 223–236.

Cook, R.D. and Weisberg, S. (1982). *Residuals and Influence in Regression*. Chapman & Hall, London.

Cosslett, S.R. and Lee, L.-F. (1985). Serial correlation in latent discrete variable models. *J. Econometrics* **27**, 79–97.

Cox, D.R. (1981). Statistical analysis of time series: some recent developments. *Scand. J. Statist.* **8**, 93–115.

Cox, D.R. (1990). Role of models in statistical analysis. *Statist. Sci.* **5**, 169–174.

Cox, D.R. and Snell, E.J. (1968). A general definition of residuals (with discussion). *J. Roy. Statist. Soc. B* **30**, 248–275.

Davison, A.C. (2003). *Statistical Models*. Cambridge University Press, Cambridge.

Dempster, A.P., Laird, N.M. and Rubin, D.B. (1977). Maximum likelihood from incomplete data via the EM algorithm (with discussion). *J. Roy. Statist. Soc.* B **39**, 1–38.

DeRuiter, S.L., Langrock, R., Skirbutas, T., Goldbogen, J.A., Calambokidis, J., Friedlaender, A.S., Southall, B.L. (2016). A multivariate mixed hidden Markov model to analyze blue whale diving behaviour during controlled sound exposures. arXiv:1602.06570v1 [stat.AP].

Dewsbury, D.A. (1992). On the problems studied in ethology, comparative psychology, and animal behaviour. *Ethology* **92**, 89–107.

Diggle, P.J. (1993). Contribution to the discussion on the Meeting on the Gibbs Sampler and Other Markov Chain Monte Carlo Methods. *J. Roy. Statist. Soc.* B **55**, 67–68.

Draper, D. (2007). Contribution to the discussion of Raftery *et al.* (2007). In *Bayesian Statistics 8*, J.M. Bernardo, M.J. Bayarri, J.O. Berger, A.P. Dawid, D. Heckerman, A.F.M. Smith and M. West (eds), 36–37. Oxford University Press, Oxford.

Dunn, P.K. and Smyth, G.K. (1996). Randomized quantile residuals. *J. Comp. Graphical Statist.* **5**, 236–244.

Durbin, J. and Koopman, S.J. (1997). Monte Carlo maximum likelihood estimation for non-Gaussian state space models. *Biometrika* **84**, 669–684.

Durbin, R., Eddy, S.R., Krogh, A. and Mitchison, G. (1998). *Biological Sequence Analysis: Probabilistic Models of Proteins and Nucleic Acids.* Cambridge University Press, Cambridge.

Efron, B. and Tibshirani, R.J. (1993). *An Introduction to the Bootstrap.* Chapman & Hall, New York.

Ephraim, Y. and Merhav, N. (2002). Hidden Markov processes. *IEEE Trans. Inform. Th.* **48**, 1518–1569.

Feller, W. (1968). *An Introduction to Probability Theory and Its Applications, Volume 1*, third edition. Wiley, New York.

Ferguson, J.D. (1980). Variable duration models for speech. In *Proceedings of the Symposium on the Applications of Hidden Markov Models to Text and Speech*, J.D. Ferguson (ed.), 143–179. Institute for Defense Analyses, Communications Research Division, Princeton, NJ.

Fisher, N.I. (1993). *The Analysis of Circular Data.* Cambridge University Press, Cambridge.

Forney, G.D. (1973). The Viterbi algorithm. *Proc. IEEE* **61**, 268–278.

Fournier, D.A., Skaug, H.J., Ancheta, J., Ianelli, J., Magnusson, A., Maunder, M.N., Nielsen, A. and Sibert, J. (2012). AD Model Builder: using automatic differentiation for statistical inference of highly parameterized complex nonlinear models. *Optim. Methods Softw.* **27**, 233–249.

Franke, J. and Seligmann, T. (1993). Conditional maximum-likelihood estimates for INAR(1) processes and their application to modelling epileptic seizure counts. In *Developments in Time Series Analysis*, T. Subba Rao (ed.), 310–330. Chapman & Hall, London.

Fredkin, D.R. and Rice, J.A. (1992). Bayesian restoration of single-channel patch clamp recordings. *Biometrics* **48**, 427–448.

Fridman, M. and Harris, L. (1998). A maximum likelihood approach for non-Gaussian stochastic volatility models. *J. Bus. Econ. Statist.* **16**, 284–291.

Frühwirth-Schnatter, S. (2006). *Finite Mixture and Markov Switching Models*. Springer, New York.

Gill, P.E., Murray, W., Saunders, M.A. and Wright, M.H. (1986). User's Guide for NPSOL: a Fortran package for nonlinear programming. Report SOL 86-2, Department of Operations Research, Stanford University.

Gill, P.E., Murray, W. and Wright, M.H. (1981). *Practical Optimization*. Academic Press, London.

Gimenez, O., Rossi, V., Choquet, R., Dehais, C., Doris, B., Varella, H., Vila, J.-P. and Pradel, R. (2007). State-space modelling of data on marked individuals. *Ecol. Model.* **206**, 431–438.

Gould, S.J. (1997). *The Mismeasure of Man*, revised and expanded edition. Penguin Books, London.

Granger, C.W.J. (1982). Acronyms in time series analysis (ATSA). *J. Time Series Anal.* **3**, 103–107.

Green, P.J. (1995). Reversible jump Markov chain Monte Carlo computation and Bayesian model determination. *Biometrika* **82**, 711–732.

Grimmett, G.R. and Stirzaker, D.R. (2001). *Probability and Random Processes*, third edition. Oxford University Press, Oxford.

Grünewälder, S., Broekhuis, F., Macdonald, D.W., Wilson, A.M., McNutt, J.W., Shawe-Taylor, J. and Hailes, S. (2012). Movement activity based classification of animal behaviour with an application to data from cheetah (*Acinonyx jubatus*). *PLOS ONE* **7**, e49120.

Guédon, Y. (2003). Estimating hidden semi-Markov chains from discrete sequences. *J. Comp. Graphical Statist.* **12**, 604–639.

Guédon, Y. (2005). Hidden hybrid Markov/semi-Markov chains. *Computat. Statist. & Data Analysis* **49**, 663–688.

Guttorp, P. (1995). *Stochastic Modeling of Scientific Data*. Chapman & Hall, London.

Haines, L.M., Munoz, W.P. and van Gelderen, C.J. (1989). ARIMA modelling of birth data. *J. Applied Statist.* **16**, 55–67.

Hamilton, J.D. (1994). *Time Series Analysis*. Princeton University Press, Princeton, NJ.

Haney, D.J. (1993). Methods for analyzing discrete-time, finite state Markov chains. Ph.D. dissertation, Department of Statistics, Stanford University.

Harte, D. (2014). R package 'HiddenMarkov', version 1.8-1. URL `http://homepages.maxnet.co.nz/davidharte/SSLib/`, accessed 10 January 2015.

Harvey, A.C. (1989). *Forecasting, Structural Time Series Models and the Kalman Filter*. Cambridge University Press, Cambridge.

Hasselblad, V. (1969). Estimation of finite mixtures of distributions from the exponential family. *J. Amer. Statist. Assoc.* **64**, 1459–1471.

Hastie, T.J. and Tibshirani, R.J. (1990). *Generalized Additive Models*. Chapman & Hall, London.

Hastie, T., Tibshirani, R.J. and Friedman, J. (2009). *The Elements of Statistical Learning: Data Mining, Inference and Prediction*, second edition.

Springer, New York.

Holzmann, H., Munk, A., Suster, M.L. and Zucchini, W. (2006). Hidden Markov models for circular and linear-circular time series. *Environ. Ecol. Stat.* **13**, 325–347.

Hopkins, A., Davies, P. and Dobson, C. (1985). Mathematical models of patterns of seizures: their use in the evaluation of drugs. *Arch. Neurol.* **42**, 463–467.

Hughes, J.P. (1993). A class of stochastic models for relating synoptic atmospheric patterns to local hydrologic phenomena. Ph.D. dissertation, University of Washington.

Hughes, J.P., Guttorp, P. and Charles, S.P. (1999). A non-homogeneous hidden Markov model for precipitation occurrence. *J. Roy. Statist. Soc.* C **48**, 15–30.

Ihaka, R. and Gentleman, R. (1996). R: a language for data analysis and graphics. *J. Comp. Graphical Statist.* **5**, 299–314.

Israel, R.B., Rosenthal, J.S. and Wei, J.Z. (2001). Finding generators for Markov chains via empirical transition matrices, with applications to credit ratings. *Math. Finance* **11**, 245–265.

Jackson, C.H., Sharples, L.D., Thompson, S.G., Duffy, S.W. and Couto, E. (2003). Multistate Markov models for disease progression with classification error. *The Statistician* **52**, 193–209.

Jacquier, E., Polson, N.G. and Rossi, P.E. (2004). Bayesian analysis of stochastic volatility models with fat-tails and correlated errors. *J. Econometrics* **122**, 185–212.

Johnson, D.S., Laake, J.L., Melin, S.R. and DeLong, R.L. (2016). Multivariate state hidden Markov models for mark-recapture data. *Statist. Sci.*, in press.

Johnson, M.T. (2005). Capacity and complexity of HMM duration modeling techniques. *IEEE Signal Processing Letters* **12**, 407–410.

Jonsen, I.D., Flemming, J.M. and Myers, R.A. (2005). Robust state-space modeling of animal movement data. *Ecology* **86**, 2874–2880.

Jordan, M.I. (2004). Graphical models. *Statist. Sci.* **19**, 140–155.

Juang, B.H. and Rabiner, L.R. (1991). Hidden Markov models for speech recognition. *Technometrics* **33**, 251–272.

Kim, S., Shephard, N. and Chib, S. (1998). Stochastic volatility: likelihood inference and comparison with ARCH models. *Rev. Econ. Studies* **65**, 361–393.

King, R., Brooks, S.P. and Coulson, T. (2008). Analyzing complex capture-recapture data in the presence of individual and temporal covariates and model uncertainty. *Biometrics* **64**, 1187–1195.

King, R. and Langrock, R. (2016). Semi-Markov Arnason–Schwarz models. *Biometrics*, in press. doi:10.1111/biom.12446.

Kitagawa, G. (1987). Non-Gaussian state-space modeling of nonstationary time series (with discussion). *J. Amer. Statist. Assoc.* **82**, 1032–1063.

Kitagawa, G. (1996). Monte Carlo filter and smoother for non-Gaussian nonlinear state space models. *J. Comp. Graph. Stat.* **5**, 1–25.

Koski, T. (2001). *Hidden Markov Models for Bioinformatics.* Kluwer Academic Publishers, Dordrecht.

Kulkarni, V.G. (2010). *Modeling and Analysis of Stochastic Systems*, second edition. Chapman & Hall/CRC Press, Boca Raton, FL.

Kundu, A., He, Y. and Bahl, P. (1989). Recognition of handwritten word: First and second order hidden Markov model based approach. *Pattern Recognit.* **22**, 283–297.

Laake, J.L. (2013). Capture-recapture analysis with hidden Markov models. Alaska Fish. Sci. Cent. processed report, 2013-04.

Lange, K. (1995). A quasi-Newton acceleration of the EM algorithm. *Statistica Sinica* **5**, 1–18.

Lange, K. (2002). *Mathematical and Statistical Methods for Genetic Analysis*, second edition. Springer, New York.

Lange, K. (2004). *Optimization.* Springer, New York.

Lange, K. and Boehnke, M. (1983). Extensions to pedigree analysis V. Optimal calculation of Mendelian likelihoods. *Hum. Hered.* **33**, 291–301.

Langrock, R. (2011). Some applications of nonlinear and non-Gaussian state-space modelling by means of hidden Markov models. *J. Appl. Statist.* **38**, 2955–2970.

Langrock, R. and King, R. (2013). Maximum likelihood estimation of mark-recapture-recovery models in the presence of continuous covariates. *Ann. Appl. Stat.* **7**, 1709–1732.

Langrock, R., King, R., Matthiopoulos, J., Thomas, L., Fortin, D. and Morales, J.M. (2012). Flexible and practical modeling of animal telemetry data: hidden Markov models and extensions. *Ecology* **93**, 2336–2342.

Langrock, R., Kneib, T., Sohn, A. and DeRuiter, S.L. (2015). Nonparametric inference in hidden Markov models using P-splines. *Biometrics* **71**, 520–528.

Langrock, R., MacDonald, I.L. and Zucchini, W. (2012). Some nonstandard stochastic volatility models and their estimation using structured hidden Markov models. *J. Empirical Finance* **19**, 147–161.

Langrock, R., Swihart, B.J., Caffo, B.S., Punjabi, N.M. and Crainiceanu, C.M. (2013). Combining hidden Markov models for comparing the dynamics of multiple sleep electroencephalograms. *Statist. Med.* **32**, 3342–3356.

Langrock, R. and Zucchini, W. (2011). Hidden Markov models with arbitrary state dwell-time distributions. *Computat. Statist. & Data Analysis* **55**, 715–724.

Laverty, W.H., Miket, M.J. and Kelly, I.W. (2002). Application of hidden Markov models on residuals: an example using Canadian traffic accident data. *Percept. Mot. Skills* **94**, 1151–1156.

Le, N.D., Leroux, B.G. and Puterman, M.L. (1992). Reader reaction: Exact likelihood evaluation in a Markov mixture model for time series of seizure counts. *Biometrics* **48**, 317–323.

Leisch, F. (2004). FlexMix: A general framework for finite mixture models and latent class regression in R. *J. Statistical Software* **11**. URL http://www.jstatsoft.org/v11/i08/.

Leroux, B.G. and Puterman, M.L. (1992). Maximum-penalized-likelihood estimation for independent and Markov-dependent mixture models. *Biometrics* **48**, 545–558.

Levinson, S.E., Rabiner, L.R. and Sondhi, M.M. (1983). An introduction to the application of the theory of probabilistic functions of a Markov process to automatic speech recognition. *Bell System Tech. J.* **62**, 1035–1074.

Lindsey, J.K. (2004). *Statistical Analysis of Stochastic Processes in Time.* Cambridge University Press, Cambridge.

Linhart, H. and Zucchini, W. (1986). *Model Selection.* Wiley, New York.

Little, R.J.A. (2009). Selection and pattern-mixture models. In *Longitudinal Data Analysis*, G. Fitzmaurice, M. Davidian, G. Verbeke and G. Molenberghs (eds), 409–431. Chapman & Hall/CRC, Boca Raton, FL.

Little, R.J.A. and Rubin, D.B. (2002). *Statistical Analysis with Missing Data*, second edition. Wiley, New York.

Liu, S., Wu, H. and Meeker, W.Q. (2015). Understanding and addressing the unbounded 'likelihood' problem. *Amer. Statistician* **69**, 191–200.

Lloyd, E.H. (1980). *Handbook of Applicable Mathematics, Vol. 2: Probability.* Wiley, New York.

Lystig, T.C. and Hughes, J.P. (2002). Exact computation of the observed information matrix for hidden Markov models. *J. Comp. Graphical Statist.* **11**, 678–689.

McClintock, B.T., King, R., Thomas, L., Matthiopoulos, J., McConnell, B.J. and Morales, J.M. (2012). A general modeling framework for animal movement and migration using multi-state random walks. *Ecol. Monogr.* **82**, 335–349.

McCrea, R.S. (2012). Sufficient statistic likelihood construction for age- and time-dependent multi-state joint recapture and recovery data. *Statist. Prob. Letters* **82**, 357–359.

McCullagh, P. and Nelder, J.A. (1989). *Generalized Linear Models*, second edition. Chapman & Hall, London.

MacDonald, I.L. and Raubenheimer, D. (1995). Hidden Markov models and animal behaviour. *Biometrical J.* **37**, 701–712.

MacDonald, I.L. and Zucchini, W. (1997). *Hidden Markov and Other Models for Discrete-Valued Time Series.* Chapman & Hall, London.

McFarland, D. (1999). *Animal Behaviour: Psychobiology, Ethology and Evolution*, third edition. Longman Scientific and Technical, Harlow.

McKellar, A.E., Langrock, R., Walters, J.R. and Kesler, D.C. (2015). Using mixed hidden Markov models to examine behavioral states in a cooperatively breeding bird. *Behav. Ecol.* **26**, 148–157.

McLachlan, G.J. and Krishnan, T. (2008). *The EM Algorithm and Extensions*, second edition. Wiley, Hoboken, NJ.

McLachlan, G.J. and Peel, D. (2000). *Finite Mixture Models.* Wiley, New York.

Mark, T., Lemon, K.N., Vandenbosch, M., Bulla, J. and Maruotti, A. (2013). Capturing the evolution of customer-firm relationships: how customers become more (or less) valuable over time. *J. Retailing* **89**, 231–245.

Marsh, L. and Jones, R. (1988). The form and consequences of random walk movement models. *J. Theor. Biol.* **133**, 113–131.

Maruotti, A. (2011). Mixed hidden Markov models for longitudinal data: an overview. *Int. Stat. Rev.* **79**, 427–454.

Maruotti, A. and Rydén, T. (2009). A semiparametric approach to hidden Markov models under longitudinal observations. *Statist. and Computing* **19**,

381–393.

Michelot, T., Langrock, R., Kneib, T. and King, R. (2016). Maximum penalized likelihood estimation in semiparametric capture-recapture models. *Biometrical J.* **58**, 222–239.

Michelot, T., Langrock, R., Patterson, T.A. and Rexstad, E. (2015). moveHMM: animal movement modelling using hidden Markov models. R package version 1.1, URL http://CRAN.R-project.org/package=moveHMM, accessed 1 January 2016.

Mira, A. (2000). Exuviae eating: a nitrogen meal? *J. Insect Physiol.* **46**, 605–610.

Monahan, J.F. (2011). *Numerical Methods of Statistics*, second edition. Cambridge University Press, New York.

Morales, J.M., Haydon, D.T., Frair, J.L., Holsinger, K.E. and Fryxell, J.M. (2004). Extracting more out of relocation data: building movement models as mixtures of random walks. *Ecology* **85**, 2436–2445.

Morales, J.M., Moorcroft, P.R., Matthiopoulos, J., Frair, J.L., Kie, J.G., Powell, R.A., Merrill, E.H. and Haydon, D.T. (2010). Building the bridge between animal movement and population dynamics. *Phil. Trans. R. Soc. B* **365**, 2289–2301.

Munoz, W.P., Haines, L.M. and van Gelderen, C.J. (1987). An analysis of the maternity data of Edendale Hospital in Natal for the period 1970–1985. Part 1: Trends and seasonality. Internal report, Edendale Hospital.

Newton, M.A. and Raftery, A.E. (1994). Approximate Bayesian inference with the weighted likelihood bootstrap (with discussion). *J. Roy. Statist. Soc.* B **56**, 3–48.

Neykov, N., Neytchev, P., Zucchini, W. and Hristov, H. (2012). Linking atmospheric circulation to daily precipitation patterns over the territory of Bulgaria. *Environ. Ecol. Stat.* **19**, 249–267.

Nicolas, P., Bize, L., Muri, F., Hoebeke, M., Rodolphe, F., Ehrlich, S.D., Prum, B. and Bessières, P. (2002). Mining *Bacillus subtilis* chromosome heterogeneities using hidden Markov models. *Nucleic Acids Res.* **30**, 1418–1426.

Omori, Y., Chib, S., Shephard, N. and Nakajima, J. (2007). Stochastic volatility with leverage: fast and efficient likelihood inference. *J. Econometrics* **140**, 425–449.

Patterson, T.A., Thomas, L., Wilcox, C., Ovaskainen, O. and Matthiopoulos, J. (2008). State-space models of individual animal movement. *Trends Ecol. Evol.* **23**, 87–94.

Patterson, T.A., Basson, M., Bravington, M.V. and Gunn, J.S. (2009). Classifying movement behaviour in relation to environmental conditions using hidden Markov models. *J. Anim. Ecol.* **78**, 1113–1123.

Pearl, J. (2000). *Causality: Models, Reasoning and Inference.* Cambridge University Press, Cambridge.

Pedersen, M.W., Patterson, T.A., Thygesen, U.H. and Madsen, H. (2011). Estimating animal behaviour and residency from movement data. *Oikos* **120**, 1281–1290.

Pegram, G.G.S. (1980). An autoregressive model for multilag Markov chains. *J. Appl. Prob.* **17**, 350–362.

Pinheiro, J.C. and Bates, D.M. (1996). Unconstrained parametrizations for variance-covariance matrices. *Statist. and Computing* **6**, 289–296.

Pradel, R. (2005). Multievent: an extension of multistate capture-recapture models to uncertain states. *Biometrics* **61**, 442–447.

R Core Team (2015). *R: A language and environment for statistical computing*. R Foundation for Statistical Computing, Vienna. URL http://www.R-project.org/.

Rabiner, L.R. (1989). A tutorial on hidden Markov models and selected applications in speech recognition. *Proc. IEEE* **77**, 257–286.

Raftery, A.E. (1985a). A model for high-order Markov chains. *J. Roy. Statist. Soc.* B **47**, 528–539.

Raftery, A.E. (1985b). A new model for discrete-valued time series: autocorrelations and extensions. *Rassegna di Metodi Statistici ed Applicazioni* **3–4**, 149–162.

Raftery, A.E., Newton, M.A., Satagopan, J.M. and Krivitsky, P.N. (2007). Estimating the integrated likelihood via posterior simulation using the harmonic mean identity (with discussion). In *Bayesian Statistics 8*, J.M. Bernardo, M.J. Bayarri, J.O. Berger, A.P. Dawid, D. Heckerman, A.F.M. Smith and M. West (eds), 1–45. Oxford University Press, Oxford.

Raftery, A.E. and Tavaré, S. (1994). Estimation and modelling repeated patterns in high order Markov chains with the mixture transition distribution model. *Appl. Statist.* **43**, 179–199.

Raubenheimer, D. and Barton Browne, L. (2000). Developmental changes in the patterns of feeding in fourth- and fifth-instar *Helicoverpa armigera* caterpillars. *Physiol. Entomology* **25**, 390–399.

Raubenheimer, D. and Bernays, E.A. (1993). Patterns of feeding in the polyphagous grasshopper *Taeniopoda eques*: a field study. *Anim. Behav.* **45**, 153–167.

Richardson, S. and Green, P.J. (1997). On Bayesian analysis of mixtures with an unknown number of components (with discussion). *J. Roy. Statist. Soc.* B **59**, 731–792.

Robert, C.P. and Casella, G. (1999). *Monte Carlo Statistical Methods*. Springer, New York.

Robert, C.P., Rydén, T. and Titterington, D.M. (2000). Bayesian inference in hidden Markov models through the reversible jump Markov chain Monte Carlo method. *J. Roy. Statist. Soc.* B **62**, 57–75.

Robert, C.P. and Titterington, D.M. (1998). Reparameterization strategies for hidden Markov models and Bayesian approaches to maximum likelihood estimation. *Statist. and Computing* **8**, 145–158.

Rosenblatt, M. (1952). Remarks on a multivariate transformation. *Ann. Math. Statist.* **23**, 470–472.

Rossi, A. and Gallo, G.M. (2006). Volatility estimation via hidden Markov models. *J. Empirical Finance* **13**, 203–230.

Ruiz-Cárdenas, R., Krainski, E.T. and Rue, H. (2012). Direct fitting of dynamic models using integrated nested Laplace approximations – INLA. *Computat. Statist. & Data Analysis* **56**, 1808–1828.

Russell, M.J. and Cook, A.E. (1987). Experimental evaluation of duration modelling techniques for automatic speech recognition. In *IEEE Proceedings of the International Conference on Acoustics, Speech and Signal Processing*, Dallas, Texas, 2376–2379.

Rydén, T. (2008). EM versus Markov chain Monte Carlo for estimation of hidden Markov models: a computational perspective. *Bayesian Analysis* **3**, 659–688.

Rydén, T., Teräsvirta, T. and Åsbrink, S. (1998). Stylized facts of daily return series and the hidden Markov model. *J. Appl. Econometr.* **13**, 217–244.

Sansom, J. and Thomson, P. (2001). Fitting hidden semi-Markov models to breakpoint rainfall data. *J. Appl. Prob.* **38A**, 142–157.

Schilling, W. (1947). A frequency distribution represented as the sum of two Poisson distributions. *J. Amer. Statist. Assoc.* **42**, 407–424.

Schimert, J. (1992). A high order hidden Markov model. Ph.D. dissertation, University of Washington.

Schliehe-Diecks, S., Kappeler, P.M. and Langrock, R. (2012). On the application of mixed hidden Markov models to multiple behavioural time series. *Interface Focus* **2**, 180–189.

Schofield, M.R. and Barker, R.J. (2011). Full open capture-recapture models with individual covariates. *J. Agric. Biol. Envir. Statist.* **16**, 253–268.

Scholz, F.W. (2006). Maximum likelihood estimation. In *Encyclopedia of Statistical Sciences*, second edition, S. Kotz, N. Balakrishnan, C.B. Read, B. Vidakovic and N.L. Johnson (eds), 4629–4639. Wiley, Hoboken, NJ.

Schwarz, C.J. and Seber, G.A.F. (1999). Estimating animal abundance: review III. *Statist. Sci.* **14**, 427–456.

Schwarz, C.J., Schweigert, J.F. and Arnason, A.N. (1993). Estimating migration rates using tag-recovery data. *Biometrics* **49**, 177–193.

Scott, D.W. (1992). *Multivariate Density Estimation: Theory, Practice and Visualization*. Wiley, New York.

Scott, S.L. (2002). Bayesian methods for hidden Markov models: Recursive computing in the 21st century. *J. Amer. Statist. Assoc.* **97**, 337–351.

Scott, S.L., James, G.M. and Sugar, C.A. (2005). Hidden Markov models for longitudinal comparisons. *J. Amer. Statist. Assoc.* **100**, 359–369.

Shephard, N.G. (1996). Statistical aspects of ARCH and stochastic volatility. In *Time Series Models: In Econometrics, Finance and Other Fields*, D.R. Cox, D.V. Hinkley and O.E. Barndorff-Nielsen (eds), 1–67. Chapman & Hall, London.

Sibly, R.M. and McFarland, D. (1976). On the fitness of behavior sequences. *American Naturalist* **110**, 601–617.

Silverman, B.W. (1985). Some aspects of the spline smoothing approach to non-parametric regression curve fitting (with discussion). *J. Roy. Statist. Soc.* B **47**, 1–52.

Silverman, B.W. (1986). *Density Estimation for Statistics and Data Analysis*. Chapman & Hall, London.

Simpson, S.J. (1990). The pattern of feeding. In *A Biology of Grasshoppers*, R.F. Chapman and T. Joern (eds), 73–103. Wiley, New York.

Simpson, S.J. and Raubenheimer, D. (1993). The central role of the haemolymph in the regulation of nutrient intake in insects. *Physiol. En-*

tomology **18**, 395–403.

Singh, G.B. (2003). Statistical modeling of DNA sequences and patterns. In *Introduction to Bioinformatics: A Theoretical and Practical Approach*, S.A. Krawetz and D.D. Womble (eds), 357–373. Humana Press, Totowa, NJ.

Smyth, P., Heckerman, D. and Jordan, M.I. (1997). Probabilistic independence networks for hidden Markov probability models. *Neural Computation* **9**, 227–269.

Speed, T.P. (2008). Terence's stuff: my favourite algorithm. *IMS Bulletin* **37**(9), 14.

Spreij, P. (2001). On the Markov property of a finite hidden Markov chain. *Statist. Prob. Letters* **52**, 279–288.

Srikanthan, R. and McMahon, T.A. (2001). Stochastic generation of annual, monthly and daily climate data: A review. *Hydrol. Earth Syst. Sc.* **5**, 653–670.

Stadie, A. (2002). Überprüfung stochastischer Modelle mit Pseudo–Residuen. Ph.D. dissertation, Universität Göttingen.

Suster, M.L. (2000). Neural control of larval locomotion in *Drosophila melanogaster*. Ph.D. thesis, University of Cambridge.

Suster, M.L., Martin, J.R., Sung, C. and Robinow, S. (2003). Targeted expression of tetanus toxin reveals sets of neurons involved in larval locomotion in *Drosophila*. *J. Neurobiology* **55**, 233–246.

Thorndike, F. (1926). Applications of Poisson's probability summation. *Bell System Tech. J.* **5**, 604–624.

Timmermann, A. (2000). Moments of Markov switching models. *J. Econometrics* **96**, 75–111.

Titterington, D.M., Smith, A.F.M. and Makov, U.E. (1985). *Statistical Analysis of Finite Mixture Distributions*. Wiley, New York.

Toates, F. (1986). *Motivational Systems*. Cambridge University Press, Cambridge.

Turner, R. (2008). Direct maximization of the likelihood of a hidden Markov model. *Computat. Statist. & Data Analysis* **52**, 4147–4160.

van Belle, G. (2002). *Statistical Rules of Thumb*. Wiley, New York.

Visser, I. (2010). Book review. *J. Math. Psychology* **54**, 509–511.

Visser, I., Raijmakers, M.E.J. and Molenaar, P.C.M. (2002). Fitting hidden Markov models to psychological data. *Scientific Programming* **10**, 185–199.

Visser, I., and Speekenbrink, M. (2010). depmixS4: An R package for hidden Markov models. *J. Stat. Software* **36**, 1–21.

Viterbi, A.J. (1967). Error bounds for convolutional codes and an asymptotically optimal decoding algorithm. *IEEE Trans. Inform. Th.* **13**, 260–269.

Wang, P. and Puterman, M.L. (1999a). Markov Poisson regression models for discrete time series. Part 1: Methodology. *J. Applied Statist.* **26**, 855–869.

Wang, P. and Puterman, M.L. (1999b). Markov Poisson regression models for discrete time series. Part 2: Applications. *J. Applied Statist.* **26**, 871–882.

Wasserman, L. (2000). Bayesian model selection and model averaging. *J. Math. Psychology* **44**, 92–107.

Weisberg, S. (1985). *Applied Linear Regression*, second edition. Wiley, New York.

Welch, L.R. (2003). Hidden Markov models and the Baum–Welch algorithm. *IEEE Inform. Th. Soc. Newsl.* **53**, 1, 10–13.

Whitaker, L. (1914). On the Poisson law of small numbers. *Biometrika* **10**, 36–71.

Wittmann, B.K., Rurak, D.W. and Taylor, S. (1984). Real-time ultrasound observation of breathing and body movements in foetal lambs from 55 days gestation to term. Abstract presented at the XI Annual Conference, Society for the Study of Foetal Physiology, Oxford.

Wong, C.S. and Li, W.K. (2000). On a mixture autoregressive model. *J. Roy. Statist. Soc.* B **62**, 95–115.

Woolhiser, D.A. (1992). Modeling daily precipitation – progress and problems. In *Statistics in the Environmental and Earth Sciences*, A.T. Walden and P. Guttorp (eds), 71–89. Edward Arnold, London.

Yu, J. (2005). On leverage in a stochastic volatility model. *J. Econometrics* **127**, 165–178.

Yu, S.-Z. (2010). Hidden semi-Markov models. *Artificial Intelligence* **174**, 215–243.

Zeger, S.L. (1988). A regression model for time series of counts. *Biometrika* **75**, 621–629.

Zeger, S.L. and Qaqish, B. (1988). Markov regression models for time series: a quasi-likelihood approach. *Biometrics* **44**, 1019–1031.

Zucchini, W. (2000). An introduction to model selection. *J. Math. Psychology* **44**, 41–61.

Zucchini, W. and Guttorp, P. (1991). A hidden Markov model for space-time precipitation. *Water Resour. Res.* **27**, 1917–1923.

Zucchini, W. and MacDonald, I.L. (1998). Hidden Markov time series models: some computational issues. In *Computing Science and Statistics, Volume 30*, S. Weisberg (ed.), 157–163. Interface Foundation of North America, Fairfax Station, VA.

Zucchini, W. and MacDonald, I.L. (1999). Illustrations of the use of pseudo-residuals in assessing the fit of a model. In *Statistical Modelling. Proceedings of the 14th International Workshop on Statistical Modelling, Graz, July 19–23, 1999*, H. Friedl, A. Berghold, G. Kauermann (eds), 409–416.

Zucchini, W., Raubenheimer, D. and MacDonald, I.L. (2008). Modeling time series of animal behavior by means of a latent-state model with feedback. *Biometrics* **64**, 807–815.

Author index

Abramowitz, M., 243, 345
Aitchison, J., 52, 345
Aitkin, M., 193, 345
Albert, P.S., 201, 345
Albon, S.D., 318, 347
Altman, R.M., 78, 191, 193, 195, 200, 305, 345
Ancheta, J., 162, 348
Arnason, A.N., 318, 355
Åsbrink, S., 141, 200, 259, 262, 355
Aston, J.A.D., 149, 213, 345
Azzalini, A., 22, 213, 219, 345

Bach, J.S., 317
Bahl, P., 200, 351
Barbu, V.S., 165, 345
Barker, R.J., 318, 355
Bartolucci, F., 30, 149, 162, 191, 345
Barton Browne, L., 297, 307, 345, 354
Basson, M., 227, 232, 353
Bates, D.M., 260, 354
Baum, L.E., 35, 69, 345
Bebbington, M.S., 200, 346
Bellman, R., xxi, 346
Benhamou, S., 230, 347
Berchtold, A., 22, 151, 250, 346
Bernays, E.A., 315, 354
Berzel, A., 77, 346
Bessières, P., 151, 353
Bisgaard, S., 20, 27, 215, 346
Bishop, C.M., 341, 346
Bize, L., 151, 353
Blackwell, P.G., 227, 346
Boehnke, M., 35, 38, 351
Bonner, S.J., 321, 322, 325–327, 346
Bowman, A.W., 22, 213, 219, 345
Boys, R.J., 151, 346

Bravington, M.V., 227, 232, 353
Brockwell, A.E., 106, 346
Broekhuis, F., 94, 349
Brooks, S.P., 322, 325, 350
Brown, G.O., 23, 346
Buckley, W.S., 23, 346
Bulla, I., 166, 181, 184, 259, 262, 346
Bulla, J., 77, 166, 181, 184, 200, 259, 262, 346, 352

Caffo, B.S., 138, 200, 351
Calambokidis, J., 188, 196, 348
Calvet, L., 270, 346
Camproux, A.C., 200, 346
Cappé, O., xviii, 30, 56, 78, 97, 111, 156, 199, 347
Carter, C.K., 162, 347
Casella, G., 76, 354
Catchpole, E.A., 318, 347
Celeux, G., 97, 120, 347
Chan, K.S., 156, 347
Charles, S.P., 187, 207, 350
Chib, S., 76, 111, 113, 262, 263, 266, 268–270, 347, 350, 353
Chopin, N., 120, 347
Choquet, R., 319, 349
Chouvet, G., 200, 346
Churchill, G.A., 200, 347
Clutton-Brock, T.H., 318, 347
Codling, E.A., 230, 347
Congdon, P., 111, 116, 347
Cont, R., 259, 347
Cook, A.E., 167, 354
Cook, R.D., 213, 347
Cosslett, S.R., 35, 347
Coulson, T.N., 318, 322, 325, 347, 350
Couto, E., 53, 126, 350

Cox, D.R., 104, 109, 275, 277, 347
Crainiceanu, C.M., 138, 200, 351

Danos, M., 243, 345
Davies, P., 201, 350
Davison, A.C., 33, 347
De Luca, G., 162, 345
Dehais, C., 319, 349
DeLong, R.L., 329, 350
Dempster, A.P., 69, 70, 348
DeRuiter, S.L., 135, 188, 196, 348,
     351
Dewsbury, D.A., 314, 348
Diggle, P.J., 213, 348
Dobson, C., 201, 350
Doris, B., 319, 349
Draper, D., 115, 348
Duffy, S.W., 53, 126, 350
Dunn, P.K., 104, 106, 348
Durand, J.-B., 97, 347
Durbin, J., 162, 348
Durbin, R., 48, 199, 348

Eddy, S.R., 48, 199, 348
Efron, B., 58, 59, 348
Ehrlich, S.D., 151, 353
Ephraim, Y., xviii, 30, 348

Farcomeni, A., 30, 191, 345
Feller, W., 14, 17, 85, 348
Ferguson, J.D., 166, 348
Fisher, A.J., 270, 346
Fisher, N.I., 136, 228, 348
Flemming, J.M., 227, 350
Forney, G.D., 89, 348
Fortin, D., 200, 227, 235, 351
Fournier, D.A., 162, 348
Frair, J.L., 227, 231, 232, 237, 353
Franke, J., 201, 348
Fredkin, D.R., 85, 348
Freeman, S.N., 318, 347
Fridman, M., 270, 349
Friedlaender, A.S., 188, 196, 348
Friedman, J., 13, 14, 349
Frühwirth-Schnatter, S., 10, 42, 56,
     111, 112, 349
Fryxell, J.M., 227, 231, 232, 237, 353

Gallo, G.M., 270, 354
Gentleman, R., 5, 350
Gill, P.E., 50, 349
Gimenez, O., 319, 349
Goldbogen, J.A., 188, 196, 348
Gould, S.J., 109, 349
Granger, C.W.J., xxi, 349
Green, P.J., 120, 349, 354
Grimmett, G.R., xviii, 14, 17, 26, 349
Grünewälder, S., 94, 349
Guédon, Y., 166, 167, 173, 181, 183,
     184, 349
Gunn, J.S., 227, 232, 353
Guttorp, P., 85, 140, 187, 200, 207,
     349, 350, 357

Hailes, S., 94, 349
Haines, L.M., 275, 276, 281, 282,
     349, 353
Hamilton, J.D., 150, 349
Haney, D.J., 22, 349
Harris, L., 270, 349
Harte, D., 124–126, 349
Harvey, A.C., 156, 349
Hasselblad, V., 24, 349
Hastie, T.J., 13, 14, 303, 349
Haydon, D.T., 227, 231, 232, 237,
     353
He, Y., 200, 351
Heckerman, D., 35, 356
Henderson, D.A., 151, 346
Hoebeke, M., 151, 353
Holsinger, K.E., 227, 231, 232, 237,
     353
Holzmann, H., 227, 350
Hopkins, A., 201, 350
Hristov, H., 207, 353
Hughes, J.P., 58, 148, 187, 207, 350,
     352
Hurn, M., 120, 347

Ianelli, J., 162, 348
Ihaka, R., 5, 350
Israel, R.B., 62, 63, 127, 350

Jackson, C.H., 53, 126, 350
Jacquier, E., 266, 350

James, G.M., 114, 355
Johnson, D.S., 329, 350
Johnson, M.T., 166, 167, 350
Jones, R., 229, 352
Jonsen, I.D., 227, 350
Jordan, M.I., 33, 35, 350, 356
Juang, B.H., 85, 200, 350

Kappeler, P.M., 191, 355
Kelly, I.W., 200, 351
Kesler, D.C., 238, 240, 352
Kie, J.G., 227, 353
Kim, S., 263, 350
King, R., 162, 200, 227, 230, 235,
        318, 322, 323, 325–329,
        346, 350–353
Kitagawa, G., 156, 162, 350
Kneib, T., 135, 328, 351, 353
Kohn, R., 162, 347
Koopman, S.J., 162, 348
Koski, T., 341, 350
Krainski, E.T., 162, 354
Krishnan, T., 77, 352
Krivitsky, P.N., 115, 121, 354
Krogh, A., 48, 199, 348
Kulkarni, V.G., 165, 182, 351
Kundu, A., 200, 351

Laake, J.L., 200, 329, 350, 351
Lagona, F., 200, 346
Laird, N.M., 69, 70, 348
Lange, K., 24, 35, 38, 301, 351
Langrock, R., 135, 138, 156, 158,
        160, 162, 166, 167, 181,
        188, 191, 196, 200, 207,
        227, 232, 235, 238, 240,
        264, 270, 318, 323,
        327–329, 348, 350–353, 355
Laverty, W.H., 200, 351
Le, N.D., 201–202, 351
Ledolter, J., 156, 347
Lee, L.-F., 35, 347
Leisch, F., 10, 351
Lemon, K.N., 200, 352
Leroux, B.G., 30, 48, 73, 76,
        201–202, 351
Levinson, S.E., 73, 351

Li, W.K., 271, 357
Limnios, N., 165, 345
Lindsey, J.K., xviii, 352
Linhart, H., 98, 352
Little, R.J.A., 40, 70, 352
Liu, S., 12, 352
Lloyd, E.H., 21, 352
Lystig, T.C., 58, 352

McClintock, B.T., 230, 352
McConnell, B.J., 230, 352
McCrea, R.S., 321, 352
McCullagh, P., 278, 352
McFarland, D., 297, 314, 352, 355
McKellar, A.E., 238, 240, 352
McLachlan, G.J., 10, 17, 56, 77, 352
McMahon, T.A., 207, 356
McNutt, J.W., 94, 349
Macdonald, D.W., 94, 349
MacDonald, I.L., 52, 101, 141, 148,
        156, 158, 160, 162, 201,
        264, 270, 285, 297, 314,
        351, 352, 357
Madsen, H., 162, 353
Magnusson, A., 162, 348
Makov, U.E., 24, 356
Mark, T., 200, 352
Marsh, L., 229, 352
Martin, D.E.K., 149, 213, 345
Martin, J.R., 232, 356
Maruotti, A., 191, 193, 200, 346, 352
Matthiopoulos, J., 200, 227, 230,
        235, 351–353
Maunder, M.N., 162, 348
Meeker, W.Q., 12, 352
Melin, S.R., 329, 350
Merhav, N., xviii, 30, 348
Merrill, E.H., 227, 353
Michelot, T., 232, 328, 353
Miket, M.J., 200, 351
Mira, A., 307, 353
Mitchison, G., 48, 199, 348
Molenaar, P.C.M., 78, 200, 356
Monahan, J.F., 57, 353
Moorcroft, P.R., 227, 353
Morales, J.M., 200, 227, 230–232,
        235, 237, 351–353

Morgan, B.J.T., 318, 322, 325–327, 346, 347
Moulines, E., xviii, 30, 56, 78, 97, 111, 156, 347
Mumford, D.B., 155
Munk, A., 227, 350
Munoz, W.P., 275, 276, 281, 282, 349, 353
Muri, F., 151, 353
Murray, W., 50, 349
Myers, R.A., 227, 350

Nakajima, J., 262, 266, 268–270, 353
Nardari, F., 270, 347
Nelder, J.A., 278, 352
Nenadić, O., 166, 181, 346
Newton, M.A., 115, 121, 353, 354
Neykov, N., 207, 353
Neytchev, P., 207, 353
Nicolas, P., 151, 353
Nielsen, A., 162, 348

Omori, Y., 262, 266, 268–270, 353
Ovaskainen, O., 227, 353

Patterson, T.A., 162, 227, 232, 353
Pearl, J., 121, 341, 353
Pedersen, M.W., 162, 353
Peel, D., 10, 17, 56, 352
Pegram, G.G.S., 22, 149, 153, 218, 353
Pemberton, J.M., 318, 347
Pennoni, F., 30, 191, 345
Petkau, J.A., 78, 200, 345
Petrie, T., 69, 345
Picone, M., 200, 346
Pinheiro, J.C., 260, 354
Plank, M.J., 230, 347
Polson, N.G., 266, 350
Powell, R.A., 227, 353
Pradel, R., 319, 328, 349, 354
Prum, B., 151, 353
Punjabi, N.M., 138, 200, 351
Puterman, M.L., 30, 48, 73, 76, 147, 201–202, 351, 356

Qaqish, B., 275, 277, 357

R Core Team, 5, 354
Rabiner, L.R., 73, 85, 94, 200, 350, 351, 354
Rafelski, J., 243, 345
Raftery, A.E., 22, 115, 121, 149, 153, 218, 346, 353, 354
Raijmakers, M.E.J., 78, 200, 356
Raubenheimer, D., 297, 299, 303, 307, 314, 315, 352, 354, 355, 357
Rexstad, E., 232, 353
Rice, J.A., 85, 348
Richardson, S., 120, 354
Robert, C.P., 76, 111, 119, 120, 221, 347, 354
Robinow, S., 232, 356
Rodolphe, F., 151, 353
Rosenblatt, M., 106, 354
Rosenthal, J.S., 62, 63, 127, 350
Rossi, A., 270, 354
Rossi, P.E., 266, 350
Rossi, V., 319, 349
Rubin, D.B., 69, 70, 348, 352
Rue, H., 162, 354
Ruiz-Cárdenas, R., 162, 354
Rurak, D.W., 76, 357
Russell, M.J., 167, 354
Rydén, T., xviii, 30, 56, 78, 97, 111, 120, 141, 156, 193, 200, 259, 262, 347, 352, 354, 355

Sansom, J., 166, 181, 182, 355
Satagopan, J.M., 115, 121, 354
Saunders, M.A., 50, 349
Saunier, F., 200, 346
Schilling, W., 24, 355
Schimert, J., 149, 355
Schliehe-Diecks, S., 191, 355
Schofield, M.R., 318, 355
Scholz, F.W., 11, 355
Schwarz, C.J., 318, 321, 346, 355
Schweigert, J.F., 318, 355
Scott, D.W., 213, 355
Scott, S.L., 76, 111, 114, 116, 355
Seber, G.A.F., 318, 355
Seligmann, T., 201, 348
Sharples, L.D., 53, 126, 350

Shawe-Taylor, J., 94, 349
Shephard, N.G., 262, 263, 266,
    268–270, 347, 350, 353, 355
Sibert, J., 162, 348
Sibly, R.M., 314, 355
Silverman, B.W., 213, 355
Simpson, S.J., 299, 303, 315, 355
Singh, G.B., 27, 28, 356
Skaug, H.J., 162, 348
Skirbutas, T., 188, 196, 348
Smith, A.F.M., 24, 356
Smyth, G.K., 104, 106, 348
Smyth, P., 35, 356
Snell, E.J., 104, 347
Sohn, A., 135, 351
Sondhi, M.M., 73, 351
Soules, G., 69, 345
Southall, B.L., 188, 196, 348
Speed, T.P., 65, 356
Speekenbrink, M., 123, 356
Spreij, P., 40, 356
Srikanthan, R., 207, 356
Stadie, A., 101, 104, 356
Stegun, I.A., 243, 345
Stirzaker, D.R., xviii, 14, 17, 26, 349
Sugar, C.A., 114, 355
Sung, C., 232, 356
Suster, M.L., 227, 232, 233, 350, 356
Sutton, W., 259
Swihart, B.J., 138, 200, 351

Tavaré, S., 22, 149, 354
Taylor, S., 76, 357
Teräsvirta, T., 141, 200, 259, 262,
    355
Thalabard, J.C., 200, 346
Thomas, G., 200, 346
Thomas, L., 200, 227, 230, 235,
    351–353
Thompson, S.G., 53, 126, 350
Thomson, P., 166, 181, 182, 355
Thorndike, F., 24, 356
Thygesen, U.H., 162, 353
Tibshirani, R.J., 13, 14, 58, 59, 303,
    348, 349
Timmermann, A., 42, 356

Titterington, D.M., 24, 76, 111, 119,
    120, 221, 354, 356
Toates, F., 314, 356
Travis, L.E., 20, 27, 215, 346
Turner, R., 58, 78, 356

van Belle, G., 220, 356
van Gelderen, C.J., 275, 276, 281,
    282, 349, 353
Vandenbosch, M., 200, 352
Varella, H., 319, 349
Vila, J.-P., 319, 349
Visser, I., 78, 123, 199, 200, 356
Viterbi, A.J., 89, 356

Walters, J.R., 238, 240, 352
Wang, P., 147, 356
Wasserman, L., 98, 356
Wei, J.Z., 62, 63, 127, 350
Weisberg, S., 213, 347, 356
Weiss, N., 69, 345
Welch, L.R., 70, 357
Whitaker, L., 24, 357
Wilcox, C., 227, 353
Wilson, A.M., 94, 349
Wittmann, B.K., 76, 357
Wong, C.S., 271, 357
Woolhiser, D.A., 207, 357
Wright, M.H., 50, 349
Wu, H., 12, 352

Yu, J., 265, 266, 357
Yu, S.-Z., 166, 357

Zeger, S.L., 156, 275, 277, 357
Zucchini, W., 52, 97, 98, 101, 140,
    141, 148, 156, 158, 160,
    162, 166, 167, 181, 187,
    200, 201, 207, 227, 264,
    270, 285, 297, 350–353, 357

Subject index

ACF
earthquakes, 29, 55, 62, 100
epileptic seizures, 202, 203
general HMM, 42
Old Faithful geyser, 214–217
Poisson–HMM, 34, 43
squared share returns, 262
AIC and BIC, 97–100
bison movement, 237
caterpillar feeding, 314
Drosophila locomotion, 233, 234
earthquakes, 99, 152
Edendale births, 278, 280, 283
epileptic seizures, 202
homicides and suicides, 288–293
Koeberg wind direction,
248–255
Old Faithful geyser, 219, 221,
222
share returns, 260
woodpecker movement, 240
Animal movement, 227–243
ARMA models, 275, 276, 282

Backward probabilities, 65–69, 86,
342, 343
Baum–Welch algorithm, 65
Bayes factor, 114
Bernoulli–HMM, 35, 134
daily rainfall occurrence, 208
Old Faithful geyser, 216–220
runlength distribution, 142–144
Bessel function, 229, 242
Beta distribution, 135
BIC, *see* AIC and BIC
Binary series
caterpillar feeding, 307
Old Faithful geyser, 214

Binomial–HMM, 134
Edendale births, 280–282
homicides and suicides, 288, 291
with change-point, 288
Bison, 235–238
Bivariate normal–HMM
Old Faithful geyser, 223–225

Categorical HMMs, 137
Categorical series
homicides, 295
Koeberg wind direction,
245–250
mark–recapture–recovery data,
317–329
Caterpillar feeding, 307
global decoding, 308
mixed model, 313
runlengths, 312
Chapman–Kolmogorov equations, 15
Circular-valued observations
Koeberg wind direction,
251–257
Classification, 93
Conditional distributions of HMM,
82–83
Conditional independence
contemporaneous, 140, 144,
225, 231, 233, 259
longitudinal, 139, 140, 144, 259
Confidence intervals, 56
percentile method, 59
Contemporaneous conditional
independence, *see*
Conditional independence,
contemporaneous
Continuous-valued state process,
155–163, 262, 322

Covariates, 145–148, *see also* Time trend; Seasonality
  animal behaviour, 306
  Edendale births, 276–282
  in state-dependent distributions, 146, 254–257
  in transition probabilities, 147, 253–254
Cox–Snell residuals, 104
Cross-correlations
  multivariate HMM, 141, 144

d-separation, 121, 341
Decoding, *see* Global decoding; Local decoding
Directed graph, 15, 30, 33, 267, 300, 341
Directional mean, 228, 229, 242, 243
Dirichlet distribution, 111, 114, 116, 120, 138
Discrete likelihood, *see* Likelihood, discrete
Drosophila melanogaster, 232–233
Dwell-time distribution, 165–185

Earthquakes, 3
  ACF, 29, 55, 62, 100
  conditional distributions, 83, 84
  forecast distributions, 85
  Gibbs sampler applied to, 116
  global decoding, 91, 183, 184
  hidden semi-Markov model, 182–183
  HMMs fitted by depmixS4, 124
  HMMs fitted by direct maximization, 54–55
  HMMs fitted by EM, 74–75
  HMMs fitted by HiddenMarkov, 125
  HMMs fitted by msm, 127
  HMMs fitted by R2OpenBUGS, 128
  HMMs with extra dependency, 151, 153
  independent mixture models, 12–13
  local decoding, 88, 89

model selection, 98, 99, 152
ordinary pseudo-residuals, 106, 107
overdispersion, 6
posterior distribution of number of states, 117
posterior distributions of parameters, 117–119
prior distribution of number of states, 116, 117
prior distributions of parameters, 116
state probabilities, 87, 88
state-space model, 160–161
Edendale births, 275–285
EM algorithm, 70
  applied to HMMs, 70–74
  compared to direct maximization, 77–78
  examples of applications, 74–77
  for HMMs with stationary Markov chain, 73, 79
  Poisson– and normal–HMMs, 72
Epileptic seizures, 201–206

Foetal movement counts, 76–77
Forecast distributions
  bivariate, 95
  earthquakes, 85
  for HMMs, 83–85
  Old Faithful dichotomized durations, 219
Forecast pseudo-residuals, 104, 105
  epileptic seizures, 204, 205
  homicides, 294, 295
Forward probabilities, 38, 65–69, 86, 341, 343

Gamma distribution, rate and shape parameters, 111, 112
Gibbs sampler, 111, 128
Global decoding, 85, 88, 124, 126, 128, 303

Hadamard product of matrices, 151
Harmonic mean estimator, 115, 117, 121

Hessian, 56
  from nlm, 56
  natural parameters, 57, 61
  recursive computation, 58
  working parameters, 57, 61
Hidden Markov process, 30
Hidden semi-Markov models,
    HSMMs, 165–185, 262
  application to daily rainfall
    occurrence, 207–211
  Bernoulli–HSMM, 207–211
  likelihood, 181
  parameter estimation, 181–184
  stationary HSMMs, 182
Higher-order HMMs, 148, 152, 153,
    218
Higher-order Markov chains, 20–22,
    148, 152, 153
  Koeberg wind direction, 250
  Old Faithful geyser, 214–215,
    218
HMMs for compositional data, 138
HMMs with extra dependencies, 150,
    282
Homicides and suicides, 287–296
Hybrid HMM/HSMM, 183

Independent mixture models, 6–14
  likelihood, 9, 44
  moments, 9
  normal components, 11, 13
  parameter estimation, 9
  Poisson components, 12
  reparametrization, 10
  unbounded likelihood, 11
Initial distribution, see Markov
    chains, initial distribution
Integrated likelihood, see Likelihood,
    integrated
Interval censoring, 41, 46, 225

Jordan canonical form, 27

Kalman filter, 156
Koeberg wind direction, 245–257

Label switching, 112, 119

Latent Markov model, 30
Levenberg–Marquardt algorithm, 78
Likelihood
  algorithm, 49
  animal-behaviour model,
    300–302
  constrained maximization, 61
  continuous, 11, 220, 221, 225,
    252
  discrete, 11, 54, 62, 220–224,
    252
  function discrete, 135
  ignorable, 40
  integrated, 114–116, 121
  interval-censored observations,
    62
  multiple maxima, 53
  of general HMM, 36, 43
  of independent mixture model,
    9, 44
  of two-state Bernoulli–HMM, 35
  reparametrization, 50
  scaling, 48–50, 60
  starting values, 53
  unbounded, 11, 53
  unconstrained maximization,
    50–52
Local decoding, 85, 86, 124, 126,
    128, 303
Logit transformation
  caterpillar feeding, 299
  Edendale births, 277
  homicides and suicides, 288
Longitudinal conditional
    independence, see
    Conditional independence,
    longitudinal
Longitudinal data, 187–196
  animal behaviour, 304–305, 313

Marginal distributions of HMM, 32
Mark–recapture–recovery, 317–329
  capture history, 317, 318
  Cormack–Jolly–Seber model,
    318
  covariate process, 321, 325, 328

individual-specific time-varying covariates, 318, 321–324, 328

state-space formulation, 319, 322

survival rate, 318, 321, 324, 327

Markov chains, 14–22

ACF, 18, 214

aperiodicity, 17, 85

conditional likelihood, 19, 215

conditional maximum likelihood estimation, 19, 215, 249

continuous-time, 52, 127

daily rainfall occurrence, 207

eigenvalues and eigenvectors of t.p.m., 16, 18

homogeneity, 15, 17

initial distribution, 16, 37, 65, 71, 82, 148

irreducibility, 17, 18, 85

Koeberg wind direction, 248–250

non-homogeneous, 147, 148

Old Faithful geyser, 214

stationary distribution, 17, 25, 26

transition counts, 19

transition probability matrix, 16

unconditional maximum likelihood estimation, 20, 215

unconditional probabilities, 16

Markov mixture model, 30

Markov property, 14, 39

Markov regression models

Edendale births, 276–285

Markov-dependent mixture, 30

Markov-switching autoregression, 150, 151

Markov-switching model, 30, 150

Missing data, 40, 41, 331

Mixed HMMs, 187–196, 238–241

Mixture models, see Independent mixture models

Model selection

Bayesian, 114–120

by AIC and BIC, 97–100

Moments

general univariate HMM, 33, 42

independent mixture models, 9

multivariate HMM, 144

Poisson–HMM, 34, 42

Motivational state, 297, 298, 314

MTD models, 22, 153, 250, see also Raftery models

Multinomial–HMMs, 136–137

homicides and suicides, 295–296

Multivariate HMM

Old Faithful geyser, 223–225

share returns, 259–262

Natural parameters, 51, 332

Negative binomial–HMM, 133

Nutrient level, 298

Observation-driven models, 275, 277

Old Faithful geyser, 213–225

Bernoulli–HMMs, 216

bivariate normal–HMM, 223–225

Markov chain models, 214

univariate normal–HMMs, 220

Ordinary pseudo-residuals, 104

earthquakes, 106, 107

epileptic seizures, 204, 205

serial dependence, 108, 110

Outliers

epileptic seizures, 204–206

homicides, 294

Overdispersion, 3, 6, 23, 29

Parallel sampling, 116–118, 120

Parameter-driven models, 275

Parameter process, 7, 29

Parametric bootstrap, 58–59, 119

earthquakes, 59–60

Old Faithful geyser, 217

stochastic volatility models, 270

Pegram models, 22, 149, 153

Old Faithful geyser, 218

Poisson–HMM, 41, 42, 45

ACF, 202

Edendale births, 282

epileptic seizures, 201–206
Hessian, 61
homicides, 289
with trend, 290
Posterior distribution
number of states, 114–120
parameters, 111–114
Posterior odds, 114
Prior distributions
number of states, 116
parameters, 111, 116
Prior feedback, 76
Prior odds, 114
Pseudo-residuals, 101–108, *see also*
Forecast pseudo-residuals;
Ordinary pseudo-residuals;
Quantile residuals
normal, 102
segments, 103
uniform, 101

Quantile residuals, 101, 104
conditional, 106

R functions
acf, 43
Arg, 228
atan2, 228
constrOptim, 61, 77, 282
contour, 27
eigen, 62
glm, 278, 283, 284
integrate, 305
loess, 234
nlm, 10, 12, 24, 77, 302
optim, 24, 61, 77, 78, 302
persp, 27
R packages, 123–130
depmixS4, 123–124
expm, 45
flexmix, 10, 13
HiddenMarkov, 124–126
hsmm, 184
MASS, 213, 220
moveHMM, 232
msm, 53, 126–128
mvtnorm, 225

R2OpenBUGS, 128–129
Raftery models, 22, 149, 153, 251
Koeberg wind direction, 250
Old Faithful geyser, 218
Random effects, 187–196, 238–241,
305, 313, 314
Random walk, 230
Regime contribution, 112, 114
Regression
logistic, 147
Poisson, 147
Reparametrization, 10, 61
Reversible jump MCMC, 120
Runlengths
Bernoulli–HMM, 142–144
caterpillar feeding, 311–313

Scaling
likelihood, 48–50, 60
Schur product of matrices, 151
Seasonality, 146, 148
daily rainfall occurrence, 208
Edendale births, 276–285
Koeberg wind direction, 247
Semi-Markov state processes,
165–185
Share returns, multivariate
normal–HMM, 259–262
Soap sales, 24, 62, 79, 85, 94, 110
Soay sheep, 317–329
Sojourn-time distribution, 165–185
Standard errors, 56
earthquakes, 59–60
from the Hessian, 56
of working parameters, 56
Old Faithful geyser, 217
via parametric bootstrap, 58
Starting values for likelihood
maximization, 53
State prediction
for HMMs, 92
State-dependent distribution, 31
State probabilities for HMMs, 86
State-space model, SSM, 155–163,
262
Stochastic EM, 76
Stochastic volatility models, 262–273

    non-standard, 270, 271
    non-standard, application, 272
    with leverage, 265
    with leverage, application, 268
    without leverage, 263
    without leverage, application,
        265
Supervised learning, 94, 199

Time trend, 146, 147
    Edendale births, 275–285
    homicides, 287–296
    piecewise constant, 288–296
Transition graph, 168, 170–172
Transition intensity matrix, 52, 127
Transition probabilities
    estimates on the boundary, 55,
        57, 61
    reparametrization, 60

Underflow, 48
Unsupervised learning, 94

Viterbi algorithm, 89–92
    accuracy, 91–92, 95
von Mises distribution, 136, 228, 251

Wind direction at Koeberg, 245–257
Woodpecker movement
    AIC and BIC, 240
    parameter estimates, 241
Woodpeckers, 238–241
Working parameters, 51, 332

Zero-inflated distributions, 23, 139,
    142, 239

Printed in the United States
by Baker & Taylor Publisher Services